DATE			

NOTHING IS TOO WONDERFUL TO BE TRUE

Masters of Modern Physics

Published Volumes

The Road from Los Alamos by Hans A. Bethe
The Charm of Physics by Sheldon L. Glashow
Citizen Scientist by Frank von Hippel
Visit to a Small Universe by Virginia Trimble
Nuclear Reactions: Science and Trans-Science by Alvin M. Weinberg
In the Shadow of the Bomb: Physics and Arms Control
    by Sydney D. Drell
The Eye of Heaven: Ptolemy, Copernicus, and Kepler
    by Owen Gingerich
Particles and Policy by Wolfgang K.H. Panofsky
At Home in the Universe by John A. Wheeler
Cosmic Enigmas by Joseph Silk
Nothing Is Too Wonderful to Be True by Philip Morrison

NOTHING IS TOO WONDERFUL TO BE TRUE

PHILIP MORRISON

The American Institute of Physics

AIP Press
American Institute of Physics
500 Sunnyside Boulevard
Woodbury, NY 11797-2999

Library of Congress Cataloging-in-Publication Data
Morrison, Philip.
    Nothing is too wonderful to be true / Philip Morrison.
        p. cm.—(Masters of Modern Physics; v. 11)
    Includes bibliographical references and index.
    ISBN 1-56396-363-9
    1. Physical sciences–Miscellanea. 2. Physics–Miscellanea.
    3. Science–Miscellanea. 4. Physicists. I. Title. II. Series.
Q173.M878 1994   94-17906
500.2--dc20   CIP

This book is volume eleven of the Masters of Modern Physics series.

10 9 8 7 6 5 4 3 2

Contents

ASTRONOMY SHINES OUT

SEARCHING FOR INTERSTELLAR COMMUNICATIONS

Part II

ON LEARNING AND TEACHING

About the Series

Masters of Modern Physics introduces the work and thought of some of the most celebrated physicists of our day. These collected essays offer a panoramic tour of the way science works, how it affects our lives, and what it means to those who practice it. Authors report from the horizons of modern research, provide engaging sketches of friends and colleagues, and reflect on the social, economic, and political consequences of the scientific and technical enterprise.

Authors have been selected for their contributions to science and for their keen ability to communicate to the general reader—often with wit, frequently in fine literary style. All have been honored by their peers and most have been prominent in shaping debates in science, technology, and public policy. Some have achieved distinction in social and cultural spheres outside the laboratory.

Many essays are drawn from popular and scientific magazines, newspapers, and journals. Still others—written for the series or drawn from notes for other occasions—appear for the first time. Authors have provided introductions and, where appropriate, annotations. Once selected for inclusion, the essays are carefully edited and updated so that each volume emerges as a finely shaped work.

Masters of Modern Physics is edited by Robert N. Ubell and overseen by an advisory panel of distinguished physicists. Sponsored by the American Institute of Physics, a consortium of major physics societies, the series serves as an authoritative survey of the people and ideas that have shaped twentieth-century science and society.

Preface

Aproper foreword hardly suits this mosaic of a book. How could I outline its purpose, or its audiences, or even acknowledge those who helped form, collect and issue these pieces that flow out of the experiences of four or five decades?

What I can do is expand on what belongs to this book and to this book alone, its title, a saying of that endearing and magnificent experimenter of early Victorian years, Michael Faraday.

I do not know his phrase from any page of print, but from words carved plain in stone on a lintel of the physics building at the beautiful Westwood campus of the University of California, Los Angeles. (My eye has not caught that remark anywhere in the copious writings by or about Michael Faraday, but I accept the attribution to him, as wholly fitting.)

Faraday began his next-to-the-last Christmas Lecture series, *On the Various Forces of Nature,* by reminding the young audience "how wonderfully we stand upon this world." It seems an utter commonplace to stand here, yet once upon a time it baffled the metaphysicians who argued forcefully how absurd it was to suppose that in our antipodes people on a globe would go about standing on their heads. Ergo, a flat earth: the real state of affairs is far more wonderful than that.

Faraday was cogent. Nothing is too wonderful to be true: not an earth ball that pulls centrally on all alike, not the many invisible moons of Jupiter, not the single selfsame form of nerve impulses, whether the internal message they carry is that of sight or sound, not the count of sun-like stars in the visible cosmos, large enough to allot about ten billion suns to each human being alive, not even the slow enduring drift of India out of the southern ice to collide with Asia, there to raise up the mighty Himalaya. Not even the proposal that utterly elusive neutrinos in space supply most of the gravitational pull that slows down the universal expansion of space is too wonderful to be true, though admittedly it is so far only an arguable scheme yet to be thoroughly tested.

There are plenty of wonders that are not true: they are for the most part vulgar wonders, much treated by the media, much relished by the public, including the young. They include the sinister Bermuda Triangle, the UFO kidnappers, the monster of Loch Ness, ghosts and sacred apparitions in profusion, and many more, new ones by the season. The distinction between the one list and the other does not rest on the intrinsic wonder or the strangeness of its entries; it rests, rather, on the evidence and its coherence. The old saw is valid: truth remains stranger than fiction.

Establishing the deep difference between our uncertain efforts after truth and the talented effort to elaborate popular fictional wonder is the chief current task of education in science. The task is not becoming easier, though perhaps more people—still rather a minority—now grasp the difference. But those colorful screens, large and small, small or large, allow equally valid strong visual and verbal support for the wonderful, whether that wonder is compact in real experience, or merely arose in the ingenious minds of the writers and the adepts of special effects. Grainy, pseudohistorical cinema is already a small, mature industry, and digitally-altered pseudoscenes cannot be far behind. Seeing persuasive video images will soon enough not amount to believing, as still photography has long since lost its once self-probative virtue.

Worse, many philosophers and sociologists of science are at work to elide the distinction between the two lists. They claim that fact and belief are not really distinct. That the physicists "negotiate" their results in concert to achieve some agreed consensus is thought to be enough to reduce the claims of physics to the level, say, of contract law, binding only by cultural agreement. Himalayan Mt. Everest is not there any longer; rather, they say, it was constructed by a coterie of mountaineers eager to gain from climbing it. (Here I have swiped a wonderful piece of irony from Steven Weinberg.)

But Everest is there, and neutrinos are likely. With effort one might find them. Most of the wonders of physics are there, like icy Everest; none are too wonderful to be true. Of course not all our proposals turn out as enduring as high mountains, gravitation, and the moons of Jupiter, nor can we know in advance just which of our novelties will be found wanting.

Physics tirelessly sifts its necessarily self-centered world of instrument and imagination for approximate truths, like the gravitational pull of the invisible or the quantitative conservation of matter and energy. Meanwhile a far more self-centered view is dusting, quite unsifted, over much of contemporary thought. Until a balance is better struck there is danger

for us all, for we humans are obliged to delve and spin in a world that in
fact not of our construction. It is one whose structure we must recognize
in more and more detail, as more of us are here to act, all of us in need of
sunshine, food and water, and a modicum of tolerant, complex, and fragile
human unity.

That is the purpose I hope my book of old samples may serve. I am grateful to all of those who have helped me and taught me ever since 1915.

PART I

SOMETHING PERSONAL

Afluent writer, I have, all the same, little talent for autobiography. These two short pieces are a good part of all I have produced along such lines.

There seems to be some reason for this slant. At the age of three I entered a long battle with poliomyelitis, a drawn engagement with the virus and its neuromuscular consequences. First I was in bed, then out of bed, but at home. I went outside to school after a few-year delay, a fast self-taught reader put into the second grade. I was functionally mobile, though afterwards I was never able to run at all; lifelong my walking gait has been awkward, my posture impaired. From childhood on I have taken what aid I could from the use of various braces, often with one or even two walking sticks (now even rolling once outside the house.)

Early on I sought to look past my confined self, first through books, then via radio, and among the wonders of the basement workshop bench. The fortunes of life for the seventy years since have granted me a wonderfully enlarged view. I have never been inclined, therefore, to write much about the cramped foreground near at hand, though some anecdotes of critical or amusing personal experiences are unforgettable, and a sampling follows in later sections.

Radio Days

The well-made wooden lid of the heavy little dark box opened very smoothly. It disclosed the black knobs and silvery dial of *Aeriola Jr.,* a crystal radio—early Westinghouse solid state—that brought to imposing headphones the voices and music of KDKA, perhaps the first regular broadcast station in the country, located at the plant a few miles away. My father, always happy with novelties, had been among the few thousands of Pittsburghers who bought the first department-store radios in the weeks before Election Day of 1920. My fifth birthday was still a few days ahead; but I was hooked on radio from that evening on.

I became a textbook case of the enthusiastic kid with a cellar lab of beloved junk: homemade zinc-powered fireworks and a big bar of unusable stearic acid, a buzzing Ford spark coil, a misused soldering iron, and a tabletop boneyard of radios. Wireless wonder had not faded a dozen years later. With top high school grades, the zeal of a radio ham and control-room hanger-on, I expected my freshman year at Carnegie Tech to lead me into the life of a genuine radio engineer.

Although it never quite worked out it was a wonderfully fortunate start. The physicists won me away from engineering. They all seemed like my ideal: curious, skeptical, open, imaginative, much more interested in how the world worked than in the painstaking care for detail and firm closure that was for me the mark of the engineering students and faculty I encountered. Since the Great Depression offered very little to the college graduate anyhow, I determined to follow my own lead to graduate school and someday to the obscure but intense satisfactions of a career in quantum physics.

Perhaps by chance I had found two admirable paths to the choice of a career: an early but enduring interest, its direction left open to a little shift in recognition of chance and naivete, and a sense of community in mind and tastes with a newfound kindred, those who staff any vocation in which a career is to be spent. For partners and rivals, appraisal and opportunity, are likely to arise for decades to come among people very like the ones you first meet at the entry door. These are both more or less internal criteria. It would be a bold claim indeed that an aspirant physicist of 1934, before fission, before world war, before the fifteen-fold growth in the size of the American community of physics, had any worthwhile insight into his objective future.

Engineers in Kindergarten?

Once upon a time, during the first FDR term, I was an engineering student. Freshman year was tough, physically and emotionally. We all took the same prescribed courses, whatever our aim might be. Of course we had math and English and physics, but they were what we had bargained far, sometimes hard, sometimes delicious. What we had not quite foreseen was a curious collection of half-a-dozen "shops," at the lathe, in the foundry, welding, even laying masonry, during an afternoon or two each week when we stumbled at all those skills under the stern eye of some elderly master craftsman. The idea was that an engineer ought to know something of the problems of industrial fabrication from the foundation up.

I never became an engineer, though both in war and peace I have had a rather engineering-like career even as a theorist of nuclei and of intergalactic space. But I certainly recall the heavy mask through which I squinted at the hot dazzling arc of my never-straight welds, and the foundry molding sand that so stubbornly slumped down just where it shouldn't. It seems old-fashioned today, and I suppose it is. Those techniques are rare today outside the myriad small repair shops of the land. But there was method in the madness. The goal was a good one: you can't be passive at the bench, you can't work from the book alone, you have to think in real time.

The limitation is also clear: you can learn such trades by rote of muscle and mind. That is not good enough for the innovative and goal-seeking work of the design engineer, nor for the alert and adaptive insights of the engineers who mind and operate modern factories, airlines, TV studios, nuclear reactors, or sewage plants.

What the shops brought us was a little inefficient, perhaps, in using up our fledgling days, but it might have been golden—had we needed it! I don't think we needed it much. For most of my classmates came out of a dozen years of basement tinkering, building, planning, and modelling. Most of us knew how to cobble up, trouble-shoot, and make work at least some items out of a long list of faintly raffish contrivances, from charges of home-made gunpowder, to long snapping sparks from an old Ford ignition coil, little planes of balsa and bamboo that flew, unreliable but high-performance radio receivers, or patient grinding of glass or of mineral. We had already learned, maybe inarticulately, most of the indispensable general lessons from that hand-eye-mind concentration such work immerses you in. That it was now drawn from industrial techniques was a small virtue, mostly vitiated by the gap between existing practice and our routine small shops.

As usual the schools, like the generals—we all do it—fight again the war that was won. At MIT where I have been for 25 years, the engineering freshmen students don't weld or turn or mold, nor have they a history of mixing chemicals or wiring little circuits. (Well, thankfully, some still do, but they recognize themselves to be exceptional.) But the crowd of students do study the books, and these days even more attentively the computer screens. They know a lot about computing, about codes and commands, about logic and order, about lists and sorting and memory access fast and slow. Moreover, they know it rather practically, fingers to keyboard, hand on rolling mouse.

So how do we teach them? Just in the same way that I was taught in shop what I already knew in another and deeper way. We demand of them more and better logic, more problem-solving, more familiarity with symbols and formulas in all the courses of the first year or two. And we get it. But year after year, we have reason to worry. The students, especially those with least adequate background, have gained symbol and lost substance. That might not be so bad, you think; the engineer is expected to be more than a craftsman. But the rub comes in this: because the symbols float more and more airily in the abstract world, their learning has become too passive. Correct form supplants meaning. If the form is precise, they get the right answers. But let them slip just once in a command, a decimal, a change of left for right in some diagram, and the form they follow has let them down.

Mistakes are not the issue; every way of learning encounters error. But these students then have no way out. The symbol is all, and once it fails them, they see no appeal. Common sense is a poor guide to invisible currents; intuition has not been built; the tidiness of keyboard and computer screen is simply too restricted a model for the real four-dimensional world full of noise and impurity and error.

A few of us are trying another way, a tentative return to hands-on experience for all our students in the basic sciences they learn now. The mental structures they build for themselves are neat, apt, but increasingly brittle. Resilience comes from approximation, intuition, redundancy, the view from a shifted stance, and it is not often the result of the one best way, the most compact expression, the quick precision of the calculator. The lab and the shop were once there to provide resilient learning. But as symbols grew more powerful, the time-consuming labs seemed less efficient. Their challenge was designed out; reduced to instruction-following, they lost, first their meaning, and then their support.

It was always the openness, the incompleteness, the chance to improve from your own little mistakes that counted for the student; those gone, why bother? It is hard to steer between setting a daunting task that is beyond the grasp of novices, and giving them a well-wrought scheme that leads you straight to good data. Yet that is where essential learning lies, in an engineering career as in the kindergarten.

Let it all come out. The schools in general have the same struggle. Symbols abound, flood, deluge our world, swamped with paper and sixty channels of screened images 24 hours a day. Yet the schools—who must, I agree, teach reading and writing and ciphering—emphasize the symbols their students already have in plenty, and minimize the hands-on real world in its variety and richness. What the students lack they don't get; what they have colorfully and dramatically from print and screen they are given again a little shabbily, for what school can afford to produce what the network can share among millions?

Maybe what the engineering students need is what the kids need: the vividness of the real, uncertain world that cannot come on screen or page. Maybe they need the chance to make something happen for themselves, in their own hands, not quite as the director arranged it, nor even by choosing the right key among a few alternatives at the computer.

This is nothing new, but it is even more apt today than when farm kids were most of the students. They knew first-hand the physical and the living world; what they needed was the organized symbol, in book and speech and number. They could get those in school, and schools worked better, or so we think. School may work better whenever it brings the students what they need. Guinea pigs and circuits and microscopes and knitting patterns and flowing water and cooking flames and moon-watching and mathematical puzzles, too, are all hard to manage and to grade in the schoolroom. But they are what needs to be there, just as a somewhat more specialized mix needs to be right at hand for every young man or woman who sets out to become an engineer.

Is there another word for kindergarten?

THEORIES LARGE AND SMALL

When I came to graduate school I had behind me considerable amateur and student experience in the laboratory. But at the end of college study I was engaged by fields and quanta, and the high excitement they brought to the thirties. I brought that predilection to Berkeley, where it became final under the teaching and the ambience of Robert Oppenheimer's theorists' group. Since then I have been a theorist of a sort myself, with a soft spot for experimental efforts and results, and a rather catholic taste in the natural sciences. I have spent a good deal of effort on geology, biology and microbiology, and especially on astronomy, all from the standpoint of a theorist of nuclear and radiative processes.

I can see now that I was really only a model builder, the explicator and sometime proposer of novel experience, rather than a constructor of new methods, new tools, or new results for the benefit of other theorists. Generality was something I rarely sought or gained, eager for connections more than for theorems.

Partly as a consequence of such tastes I have always felt the deep pleasure and wide importance of trying to explain to others outside the profession the wonderful things physicists know and don't know, and just how we got there. These dozen essays sample fairly widely what I have done, except that none of them, of course, are genuine technical papers from the archival journals or even from textbooks.

Searching for Our Ancestors

T he day was calm. The waves that lapped the shore were small, almost lazy; they were the waves of shallow waters, not of the open ocean. A ruddy sun shone in the hazy sky. The slow stream that came down to the foreshore rippled a little in the light wind, and the pebbles tumbled here and there without much energy. Drama was not wholly absent, for along the skyline all but lost in the distance were two or three volcanic cones. They were quiet just now, but a walk along the beach would soon bring a traveler to a stretch of all but impassable lava, where once, not so long back, the molten rock had oozed and hissed into the waters from an inland fissure. It would happen again, but no one could foresee just where and when the encounter would take place.

The day was calm and the scene was lonely. The beach was devoid of shells. No flies buzzed; nothing at all hopped or crawled along the water's edge. No birds flew; no fish swam in the sea; no clawed creatures scuttled below the tidal waters. The rocky lands inward from the sea were utterly barren of life. Neither lizards nor mice could be found, and neither a tree nor a blade of grass spread green blades to the sunshine. Yet life was present, even abundant, in the scene. It grew everywhere that the shallow waters brimmed out to dry land: dense knobs and sheets of algae and bacteria covered all the shallows, out into the bay and up the stream toward the higher lands. That life was never out of touch with water; it never survives higher than a matter of inches from moisture. Inland, here and there, a few dry old knobs could be found, quite whitened, rocklike—a growing mat of the only life in this quiet land, stranded forever by some shift in the watercourse.

Just such a scene—we can infer the details rather well from the complex fabric of the rock samples—would present itself at the spot, now the western coast of Australia, where the oldest trace of life in all the Earth is found. The time is long ago indeed, a time we can estimate to within a few percent from a secure, mutually confirming set of radioactive decay measurements. The signs of copious algal life, which bears a remarkable resemblance to the same forms found throughout the record of the rocks up to the present day, occur almost as early as the first dated rocks. One must emphasize that this teeming life, single-celled, though colonial in nature, was about all that lived on Earth, not only for the first pages of the record but for four-fifths of our whole past. Not until a time only 0.7 billion years (b.y.) back can we surely see any relic of life more mobile than the algal and bacterial mats. Indeed, they themselves become more complex in microstructure and more powerful in their chemistry over the 3 b.y. of their evolution and change. No life is mobile (beyond the drift of plankton) until about that time, 0.6 or 0.8 b.y. ago. And it requires another couple of hundred million years before anything alive, either plant or animal, can break close contact with the waters, to stand well above the coast, the marsh, or the damp soil. All the forms of life we see in the familiar fossil record, everything so graphically drawn by the paleontological artists, all those feathery sea lilies, bulge-eyed trilobites, all sharks and dinosaurs, all ancient birds and beasts, all dawn palms or big kelp, all that crawls and flies and swims and stalks, all that branches or flutters in the wind, belong to the last 10% or 15% of life's long history on Earth.

The direct rock record supports one major plausible inference: All life we know has evolved from single, small-celled beginnings, forms like those still to be found as copious and vigorous participants in the great geochemical cycles, the blue-green algae and their kin, the bacteria. These had filled the shallow-water world in full vigor even by the time of our earliest evidence. Theirs, of course, is the triumph of lowly biochemistry—not motion, not sensory response, not even structure on the scale of naked-eye visibility. It is rather the microstructure, the complexities at the level of molecular helixes, sheets, tubes and rods, and the complex biochemical pathways they enable, which evolved over the entire first half or so of life's biography. Thus, it is biochemistry we must search out back to the time of the most ancient rocks known, and before, if we can. The winds and the waters, the volcanic fires and the slumping sands, the lava pours and the rolling stones—those were not much different from today. But in fact we do not even know whether human beings—time travelers—could have breathed that otherwise commonplace breeze. Was it oxygenated? Could the old algal mats already use oxygen and sunlight to build their substance in the open air? Or had they not yet made that invention, so that they subsisted in a very different atmosphere from ours, perhaps not yet even making good use of solar photons? We cannot be sure. We know that many bacteria today are fit for vigorous life oxygen-free.

We know, moreover, that all the oxygen of today's atmosphere is turned over very rapidly by the green-plant world. But just when that ability first arose has not yet been fixed. It is in fact the molecular facts we still seek, with much difficulty, there in the most ancient rocks. For it is on that level that the mechanisms of life must have begun, about 4 b.y. back. But the rocks are steadily reworked, buried, heated, and reheated in the Earth's fiery mantle. We have to look hard indeed to find rocks older than Isua in Greenland, which may—or well may not—bear the crucial evidence we seek. Certainly, that search will go on, and with sharper analyses we will find more clues in the ancient, disturbed, and enigmatic samples.

The Depth of Space

Perhaps, we can work our way forward in time to life, forward from the date when the Sun and its planets were somehow condensed jointly out of the interstellar gases. We know that date rather well. The surface of the Moon, a large collection of meteorites that have been trapped from orbit by their chance encounters with Earth, and the strongly supported inferences from the dating of long series of Earth rocks, all lead to the conclusion that our planet was assembled by a complex set of processes just about 4.5 (±0.1) b.y. ago. In the time between 4.5 and 3.8 b.y., we know Earth became the round sphere it is, the rocks of the crust grew solid and were well sorted, and the processes of ordinary geology approached the familiar. This stormy period, with an epidemic of heavy impacts on planetary surfaces, is under detailed study which includes the hints given by the other planets and satellites whose compositions we come more and more to know. The meteors, too, offer samples stored in the refrigerator of orbit by which we try to conclude what material made up the Earth. We are pretty sure that there was plenty of time, for the orbital processes are speedy compared to most geological ones. Free fall is faster than the drift of continents and the weathering of mountains to fill the seas with silt. Even such grand geological dramas take a fifth of a billion years at most. So there is ample time; the physical processes were more or less over and became familiar within a couple of hundred million years (m.y.) or so. But what that earliest stable Earth was like is a question still beyond the grasp of our models; the biologists look to the planetary sciences for the initial conditions of sea and air, and the physical studies still seek a constraining hint from what the biochemists will tell them of the initial atmosphere!

There has been one unexpected finding in space within the last decade especially. It is the universal prevalence of carbon-containing molecules. The linked atoms of carbon were once thought to be solely the work of life. Now we find them overall, in plenty. A substance like ethyl alcohol exists in vast quantities, though gaseous and dilute to the point of vacuum, spread in the huge, thin clouds of interstellar gas. Substances of even more complex and lifelike kind are found in the meteorites that fall now and again to Earth, some of which resemble a blackened cheese, soft and carbon-rich in substance. These are certainly products of the evolution of the meteorites, perhaps somewhat associated with the debris of minor planets or at least with the conditions of the solar nebula where the planets grew. There is a chance that these carbon compounds may have contributed such useful organic resources directly to the early Earth, which was certainly heavily bombarded with gifts from orbit. Most workers believe that the surface of the Earth itself was fully suited for the production of organic carbon compounds from the commonplace atoms, like carbon and hydrogen, which were already in place. It is the density which must be of importance, for complex molecules arise out of the close proximity of several atoms leading to their eventual union, an event none too common within the dilution of space. Exactly that condition (the temperature, too, must be permissive) gives our Earth its liquid water, intimately, indispensably part of life, ancient and modern. Perhaps no other place we have yet found in the cosmos allows the presence of liquid water. Mars long ago had great rivers, we think. Water and carbon-containing molecules fit for living forms go tightly together. The conclusion which everyone accepts is that small carbon-containing molecules form spontaneously, under the right ambient conditions, whether or not life is directly involved. If not the true precursors of life, these molecules are at least patterns for the loom of life. They can be made by a variety of synthetic processes; they differ indeed only in detail from nonorganic molecules, copious but biologically insignificant in the atmospheres of the cooler stars. The molecule is plainly implicit in the atomic world.

We will press the rock record back to its first pages. We will improve the planetary and meteorite studies forward to the Earth in birth. But can we look squarely into that half-billion year gap between the birth of the planet and the oldest relics of life?

The Essential Memories

The extension of 35 years of molecular biology involving the idea of the genetic code and its work, has shown us much of the inner nature of life, especially of microbial life. The effort to frame a general definition of life, which might seem at the core of the matter, is in fact not so salient in the search for life's origin. For what we seek must be the long chain of events which gave rise to that specific complex web of present-day life on Earth of which we are a part. Even if quite other forms are possible, they would seem less relevant to the quest for what happened here. It is no surprise, then, that the key questions have been sought on theoretical grounds. For 20 years people have looked for conditions which, under the little-known natural circumstances of 4 b.y. ago, could plausibly lead to the rise of proteins—the working molecules of life which marshal the linked chemical reactions to build living systems and lend them function. Parallel effort has been spent with the nucleic acids, which on Earth alone form those subtle, long molecules (DNA and RNA), as well as the various copying peripherals that embody and reproduce the instructions for the proteins which implement the order, even the act of code reproduction itself. These complex molecules are universal in life forms today; they have long seemed a wise starting point for the search for origins.

Neither a coded message, whose slow elaboration is the only key to the evolutionary path, nor these protein jigs and fixtures, which alone allow the expression of the inert code within the world of living change, can be the whole story. No code, no way to elaborate the chain of life forever; but no jigs, no action on the world. A biological contract between these molecules, or more strictly, between these two functions in some molecular form, seems the center of the issue. We do not see it clearly yet. The clues from present life are useful, but today the mechanism has become so precise, so well-functioning, so long-elaborated that the steps that led to it from a simpler, nonliving world are hard to guess. It is possible that some simplified intermediate structures will be found; it is also possible that some key contrivance, now entirely superseded, must be found. Some think that the inorganic substructure of some mineral might have offered a crystalline framework for the first systems of molecular self-reproduction. Other essentials besides the message and the action are postulated; or perhaps some early form of the message may have been at least weakly self-acting. The whole topic is complex, central, challenging.

One conclusion seems stronger than ever, touching on the eons of evolutionary elaboration. It seems likely that the first cells of the microbial mat were the unusually small and structurally more or less simple ones which still distinguish the bacteria and the blue-green algae as a group from all other higher forms of life. Sometime in the first 2 b.y., new cell types arose. More or less algal, they were big, hundreds of times the volume of their forebears, like most cells of the animals and plants of today. Moreover, they held the specialized organelles of cells, those which today carry out efficient photosynthesis, hold and take out the message molecules safely and precisely, arrange for sequential reactions of energy storage and liberation, etc., in all higher forms. The smaller cells perform the same functions (indeed, as biochemists they are even more versatile), but they lack many advantages of rate and of diversity in reproduction. There is good reason to believe that here we see another later social contract—the organelles of the larger cells are perhaps the offspring of once-independent organisms, which long ago contracted to dwell within some host cell and share its world. A few such symbiotic arrangements may have made the complex and protean cells of today, the eukaryotes. Their cooperation to form organelles, then whole creatures, has given rise to the multicellular forms of life that are now large enough to make an imprint in the world one by one, and enterprising enough to add swift mobility and varied macrostructure to the chemical virtuosity of that ancient algal mat.

Probably, these several contractual unions are the heart of our problem; so far only the rise of the organelles seems to find some support in the world of life as we study it today.

The Tensions of Research

There is a certain tension in scientific research. No doubt all researchers would like to solve important problems, problems with impact on the mind, or problems whose solutions bear on human needs in a material way. But those problems are hard, often refractory. At any state of knowledge, the researcher is therefore led to seek out, not the most important problems, but the soluble ones. Galileo watched the lamp swing with uniform beat; he could get somewhere with that. But his early effort to explain the tides was not the source of much further work. In the study of the origin of life some balance must be struck. The problem is clearly important; perhaps no other problem speaks more to the common concern of all reflective human beings to learn our place in the world. But it is evidently no simple problem, more so because it cannot be sought within a single discipline. Here, molecular biology meets astronomy, both encounter geology, and each of these major disciplines draws upon the chief results of chemistry and physics over a very wide range. Such a specialty is institutionally fragile; its devotees generally must find their professional niches in quite different, better-focused enterprises—those with degrees, standard texts, students, clear applications. We believe that the maturity and importance of the problem begin to demand explicit support and recognition. The question lies squarely across the major currents of microbiological and of planetary research today. Those sciences could reach no more important outcome than to illuminate the origins of life on Earth.

We might remark also that a focus on the widespread but rather quiet, cryptic life of the microbial mats of the great watery flats is strangely relevant to today's world in which the geochemical and atmospheric cycles themselves begin to show the effects of human intervention for good and bad. The great role of that lowly life in the deep past is to a degree continued today; we know all too little of its complex and rich operations within the changing balance of contemporary nature. As we look back into the past, we will both employ and illuminate the present. But above all, what we seek is not only practical advantage, though that will come. Our chief goal is a kind of self-knowledge, as deep as our oldest myth: how it came about on this Earth that the quick were first parted from the dead.

The Wonder of Time

Philosophy is the product of wonder. . . . And, at the end, when philosophic thought has done its best, the wonder remains.

The essay, "Nature and Life," by Alfred North Whitehead, lively and cogent, repels the dust which gathers now on most analyses of that day, those delicate restatements of a narrower physics which Whitehead bluntly discards as "muddle-headed positivism." The epigraph that I took from his paper, where it is alpha and omega (almost), makes plain why. He was a man of deep thought, irrepressible in curiosity and serious in mental quest, so he knew there was something real going on inside the quiet head, as inside the silent machine, the still flower, and the hard atom. However stable, the world portions are forever in change. Yet it is extraordinary that nowhere in this not very brief piece on nature and life did he set down the word *evolution,* a single word, not very beautiful, which to a physicist a generation later seems to carry within itself some resolution for most of Whitehead's sharply put paradoxes of our world scheme. (Which is not of course to say that the wonder has dwindled. Rather has it grown!)

The present appraisal of perception offers one of the largest divergences from the views which could be put forward by a great man in 1933. We still hold that sense data do not alone provide the data for their own interpretation, that "irrefutable basis for all philosophic thought" we owe to Hume. But evolution has given us the answer plain. Why should the eyes and the ears of a naked ape allow interpretation at any level beyond the sight of the lion and the sound of the waterfall? Our senses have evolved superbly, to infer the likely event from the mean properties of certain large collections of atoms, in a way well-suited to survival. More than that, exactly the same arguments apply to the best instruments of the scientist, the clever galvanometer mirrors and dials of Whitehead's day or the elaborate printouts of our own. No, given the knowable response of eye or meter to color patch or current pulse, we trudge a long uncertain path to find out what the orange mark probably was, or what was the likely circuit response. Not even mathematics—after Gödel we find doubt there too—can give full surety.

More and more of what humans perceive turns out to follow a program "hard-wired" into the brain, like the unconscious processes of comparing image positions from one eye to another which gives us binocular vision. Hardly any perceptual experience is more striking than a long steady look at those dot patterns which very slowly uncover to the inward computer of the mind a meaningful depth within the superficial randomness. Two worlds remain, the world of the shimmering atoms, and the world of human senses, telling us something about averages. But the finest instruments cannot do better; there, too, immediacy and directness are but illusions of familiarity. The perceptions of bat or bee, of spectroscope or amplifier, do not at once agree with the tale of our own five. Again, why should they? In each and every case, perception offers to correlating experience and to careful reason only plausible inferences, whether those of a hunting predator or those of an excited scientist. We are coming to learn how our own inferences develop, as with the work of a Piaget on the epistemology of children. Nothing seems more natural now to evolving organisms, themselves compact of atoms in an atomic world, than this: a very subtle chain must join the richness of primary atomic causes to the limited set of secondary conclusions open to any evolved mind. It is alike for our thought and for the powerful endowed awareness of the animals. Any eyes, say, work only with a flood of photons. The set of possible messages that tide carries is dazzlingly manifold, but they are to be grasped only by some deciphering apparatus evolved and already in place, which yields only a few meanings at a time, through varied statistical schemes, invariably approximate, strongly integrative, and hence always simplifying. Meanings are not rare but copious. The senses do not originate the texts of the world, rather they edit them strongly, using a stylebook.

The solar system, where Newton's rather simple world model dwells pure, nicely accommodates a snapshot, taken at a distance, of all around us. But such passive observation will not survive dinner time! Stronger interactions prevail on earth; being and becoming are incessant, everyday imperatives. It is true that the stability of this world requires a physics beyond Newton. Our solar system follows Newton through two centuries of almanacs of precision because it is so isolated in the depths of space. Collisions with other stars would destroy orbits utterly. That fine gold endures, that ginger is hot i' the mouth, discloses a lasting micropattern in nature, a pattern whose origins in quantum theory were coming to be understood for the first time during the very years of Whitehead's piece. (The heros of quantum mechanics, Heisenberg, Schrödinger, and Dirac, won their Nobel Prizes in 1932 and 1933.) It is the quantum consequence, a pattern that can last, which rules the world of matter close at hand, whether as inert as its paving stones or as lively in structure as a flame, stable even through unceasing change. That stability, of course not total, rests in the end upon the modular stability of the building blocks themselves, the "fundamental" particles, of whose true nature we still have little wisdom, though much knowledge.

How the world pattern is put together from the modular bricks of electrons and their quarky partners into the several hierarchies Whitehead lists is of course the great aim of physics to unravel. But since 1930 a deep historicity has swept across physical science, a search for origins in time, from that of the continents to that of the elements. Sovereign time enters all our studies now: the actuarial knotting and unknotting of a tangle of world lines.

The metaphysician seeks meaning, and we must follow. Right now it appears that meaning itself, like color or taste, is a rich secondary quality, which evolved with mind. With human thought meaning appears fully; it is one response of the world to the gaze of thought, as color is a response to inquiry by retina and visual cortex. Meaning enters shyly with the first minds, those of many animal forebears, for whom at least some patterns in life are real; it came full stage with the dawn of the conscious. Some conjecture—but not the present writer—that we are the first minds among the stars of all the galaxies so profligately flung through space who have emerged to full meaning. There are those who claim even that the physical laws that seem so permissively structured to allow our entry on the stage after ten billion years were designed to that end. We shall see; we may even find we are not alone. It would more modestly appear that our society, whose swift changes are counted by millennia or even by centuries, is governed by the flow not of biology but of culture. Human history could only dimly be foreshadowed by the glacial slowness of anthill and chimp band. Ours is only one among the myriad of possibilities still hidden in the unplumbed potentialities of the laws of nature. Those laws themselves may have evolved long ago in the profound springs of time, to fill the present universe with patterns and processes so fit for wonder.

The Fabric of the Atom

When the undergraduate Darwin was making ready for his voyage aboard the *Beagle,* he called upon the great botanist Robert Brown, a virtuoso of the simple microscope. Brown showed Darwin many sights, among them, we expect, some curious dancing specks under the lens. Those motes were the newly found first direct sign of the unceasing internal motion of inanimate matter. In 1831 the motion was a mystery. But Maxwell, Boltzmann, Gibbs and Einstein, by the end of the century had made plain how revealing was this inner motion. Their work was in the end to do more to amend the Newtonian system of the world than even the fusion of space and time which the young Einstein in the same years brought to completion. For those physicists placed the laws of probability, the rule of chance, squarely in the center of the physical world, from which the strict determinism of Newton's laws throughout their two centuries of elaboration had expelled them. The early workers believed, perhaps, that their statistical form of mechanics expressed less the real world then than the impatience of the mind. In principle, they might have felt, the laws of chance could be replaced by precise step-by-step calculation, as always before. We are less sure of that today.

For a generation it has been clear that probability lies far deeper within physics than even the Brownian motion. The very stuff of light and of matter is in fact ruled by chance. Yet that ministry of chance is under the reign of unvarying law; a kind of fusion between chance and cause, even stranger than that of time and space, lies at the heart of the theory of quantum mechanics, our best and most complete understanding of the world. To make this clear will be one of our tasks.

Too much has been made of the quantum renunciation of classical cause-and-effect. Uncertainty, ignorance, chance, seem to dominate the tone of discussions of the quantum physics. This is a caricature of the truth. Just as relativity, welding time into space, renouncing simultaneity, has given us the first understanding of swift motion, so quantum mechanics, renouncing the old style of causation, has given us our first potent understanding of the fabric of the material world. Quantum mechanics has been, not the physicist's surrender, but his triumph.

The path to that triumph will be sketched. Some of the rich background in experiment, and some of the close and unfamiliar but, compelling reasoning which experiment has demanded from us, will be exposed. We will use words, apparatus, and a little mathematics. Like all modern physics, quantum mechanics rests upon those three pillars of wisdom. Not one can be sacrificed if the structure is to stand.

On the Stability of Matter

Newtonian mechanics gave us a system of the world. Its laws predict with clock-like precision the planetary motions, so many clock hands. But the atoms, which since the Ionian insight we believe to be that stuff of matter, behave very unlike the systems according to Newton. For the solar system has no inherent prejudice; started in any of a wide class of orbits, the planets would have carried them out happily, very slowly running down like the man-made satellites. But the atoms are identical and remain so. History does not affect them; their careers are filled with knocks and bangs, yet the gold of Rome and the gold of a newly minted medal remain indistinguishable. The Greeks knew this and gave to the atom its name— the uncut—to suggest the extraordinary stability, the incorruptibility, of the atoms themselves. That which changes in matter, the rusting blade, the hard coal turned smoke, is but a rearrangement of the same tiny particles. The chemist can recover the iron from rust, or the carbon from the hot gas, and the elements recovered differ in no whit from their counterparts in a sheltered sample. Atoms do not change!

But that cannot be true. We know for fifty years now that atoms are not unyielding balls, refractory, structureless, everlasting. They have a structure. They can be taken apart; the loose structure of a heavy atom, say of gold, can give up hundreds of sub-components, electrons, protons, neutrons, if split by the many means now available. We even know, since Rutherford, that, like the solar system itself, the atom of matter is mostly space, space within which the electrons move, held by forces at least analogous to the force of gravity. No small solar system could behave so;

it would bear the marks of every previous collision. Nor could we expect two solar systems, even granting their planets a wonderful identity, to look the same. The initial orbits either differ, and retain that difference, or must acquire differences reflecting the differing histories of two systems struck by countless billions of neighbours. In this paradox, that atoms have structure and yet remain identical and unchanged, classical physics found its limits.

We have found that in spite of internal motion and spacious structure, an atom somehow remains itself; if wounded, either it heals to its original state at once, or it acquires a distinct new character. It never bears away a mere scar of the encounter, to spoil its resemblance to its crowds of identical fellows, unless it changes to a sharply different system. If the atom were a tiny Newtonian solar system, a sample of gold atoms would be a kind of population, like so many pedigreed poodles, or so many peas in a pod, similar but distributed about a standard type, showing strong resemblance yet distinct differences. Yet real gold atoms remain incredibly alike. It is impossible in Newton's mechanics to have complex structures, differently disturbed, remain nevertheless identical. This is the problem set for every atomic theory. It is no fine point; it is the very basis of the recognizability of the material world. It is the mark of the quantum.

The Graininess of Energy

Matter is grainy. Its grains, those surprising structured and yet stable systems, are the atoms. But the atoms are themselves grainy. They have atoms of their own. Every gold atom is like every other, far more like than peas in the pod. Thus the structures of two gold atoms must be identical. More than that, the atoms of the atom, the electrons outside the binding central nucleus, are also each to each identical. Within the nucleus, too, a grainy structure of identical subunits has been revealed. Electrical charge, too, that measure of electrical response, comes in only one-sized package, one quantum. A particle may bear a single unit of charge, or any integer multiple of that. It may bear the opposite charge—call it negative—in the same amounts. It may have zero charge, strictly. But fractional quantities of charge have never been found.

Such graininess, wondrous enough, seems in the Greek mode. Mass and charge come in lumps, or particles. What is much less of a commonplace is that the measures of motion themselves, quantities like energy, are also found to occur only in the grainy mode. In Newtonian mechanics, energy is smooth, continuous. You can push an obstacle gingerly, moderately, a lot. Energy, after all only an abstract quality possessing many forms, would not seem to be describable as chunky. Yet so it is. On the atomic scale if you strike an atom gently it moves away elastically, showing no internal sign of the collision. When the energy transfer reaches a certain level, then and only then can it absorb energy to change its structure. Any surplus over the minimum amount again it rejects, until suddenly it can modify itself to a third new state, and so on. This is true for particles smashing into atoms; it is no less true for light being absorbed by matter. Light is reflected unchanged unless it has adequate energy to modify the atomic receiver. Every act of absorption of light takes out of the beam of light a particular quantum of energy, depending on the colour, no more and no less. Momentum, too, simple *mv,* can be transferred only in lump sums in any finite system. If we accept this, the stability of matter is accounted for. The billions of collisions among gas molecules, say, leave them unscathed. Only if the odd collision is sufficiently strong can the molecular partners change. Thus the collisions of molecules do not heat them, so that the energy disappears as heat inside the molecular structure. Were the molecules anything like matter on our scale this would be the necessary result. Thus the atomists were forced to impute to the atoms a structureless nature, belied by J. J. Thompson and the rest at the turn of the century. However unyielding that structure, in classical physics it would sooner or later respond to the myriad collisions. Water wears away the unyielding stone. But atoms are not mere rock-hard; they are quantum-rigid; easily changed by a sufficiently large energy packet, totally unmoved by one only a little smaller.

In the currency of energy, too, there is a smallest coin, a farthing. Just as no bank account can change by less than a farthing, so no energy store can change by less than its smallest quantum. The permanence of our world, its specificity, rests on this grand fact.

The Quantum of Action

If a poor man had a bank account, he might notice the graininess of sterling. A farthing more or less in the sum would be noticeable; each would he count. But the Bank of England, whatever it may claim, does not know how many pounds it has to the farthing at any given time. Thus with Newtonian mechanics; the apparent continuity of motion turns out to be a consequence of affluence; the energy currency is so small a coinage that it is not individually detectable in a planet's motion, a locomotive's, a bee's, or even a bacterium's.

Just as for the theory of relativity there is a natural measure of speed—the speed of light—by which we can decide whether or no the subtle effects of relativity will modify a prediction made from the old mechanics, so there is a measure of motion which parts the old mechanics from the new. It is called the quantum of action. If a motion possesses sufficiently many such quanta, the old and the new mechanics must agree; if few, only the quantum mechanics can be expected to apply (looking apart from one or two accidentally simple cases).

The term "action" ought to be as familiar as the term "energy." Both are abstract quantities long used and very powerful in the exploitation of Newton's mechanics, though both belong not to Newton himself but to his great eighteenth-century followers. We can understand action by considering a simple case, the motion of a ball on a smooth table bouncing back and forth between two walls. It is clear that the ball moves at uniform speed between bounces. If the space traversed is L, the time taken is T, and the speed $v = L/T$. It was Maupertuis who first showed that this actual motion makes minimum the quantity $mv^2 T$ which he called the *action*. We have to require that the energy remains constant. Under this condition, the least action, the least $mv^2 T$ is obviously the actual straight line motion at unvarying speed. It is plain that another expression for the action is given by $mv(L/T)T = mvL$. Now, the quantum of action, named after Planck, is a bit of this same quantity. It is so small that if for my bouncing ball I were to use a single pollen grain, for the space a stretch invisible to the eye, and allow the dust to drift at a snail's crawl, the system would possess more units of action than there are pennies in the annual income of all the world. It is no wonder that the quantum of action is unrecognized in the world of everyday sizes, or in the solar system.

But on the atomic level, the quantum of action, Planck's constant h, is the sovereign measure. The motion of an electron in an atom, with the velocities appropriate to the motion under electrical forces, has just one or two or three such quanta. For that motion, not Newton, but Planck, Bohr, and Heisenberg have given the laws. Let the action of the motion grow, its energy-time product grow, and the system will obey as near as anyone could want, the time-honoured legislation of Newton and his peers. Once again the world is conceptually split by a fundamental constant. As c (the speed of light) parted old space and time from the new relativistic fusion, so h parts Newtonian forces and trajectories from the fused cause and chance of the new mechanics.

The Fusion of Cause and Chance

Is there a more profound opposition in everyday thought or in classical physics than that between cause and chance? The almanac is certain on the time of the eclipse. For this we owe a debt to Newton. But the old book can be thumbed in vain for the position of the clouds. The old physics thought it was enough to bring the clouds, too, under the same causal law; difficult, but worth hoping for. The new physics has built on a new foundation; the atomic world of small action is ruled by a fusion of cause and chance.

The most commonplace source of events of small action, ruled by the quantum, turns out to be light. The eye, the photographic plate, the iconoscope surface, all display an atomic form of recording. Not a smooth response, but a grainy, individual reaction is the mode of all of these light detectors. Nor has this been fully accounted for by the undoubted atomic nature of the matter of which they are all made. There is clear demonstration that the light itself can transfer its energy and momentum only in packets, called photons. These packets, whose energy content is colour-dependent, given by $E = hf$, cannot be split. Every effort to split them—by half-silvered mirror, by polaroid filter, by chopping with swift-rotating blades, by allowing space to dilute out the light—none of these have ever worked. The number of photons can be made small, when the light is weak, or great, when the light is strong, but the integrity, the stability of the individual photon, is never altered. Light has atoms, too.

The picture is now apparently simple. It was old Newton's. Light is merely a corpuscular beam; crowds of independent photons simply flood the lenses and the air, going by chance as they have been aimed. But what of the nineteenth century? What of the optical patterns of Arago and Maxwell? Are they merely mistaken, to have doubted Newton? Not at all. For we can split a beam of light between two slits, photon by photon. And yet where the light goes will depend precisely on the existence of an open slit through which the photon bullet might, or might not, have gone. These remarkable "bullets" can cancel, if you wish, as no Newtonian projectile ever could. Still stranger, they can cancel even coming one at a time! In fact, the marvellous apparatus of the classical theory of wave fields turns out to control with precision what light does in the long run. The patterns predicted by Maxwell, strictly causal, always appear, painted out by individual photons, no matter how long the wait or how fierce the torrent. But just where the next photon will strike, just when it will be counted, are matters which obey in all detail the mathematical laws of chance. For every photon, too, we must not forget, the energy and the momentum strictly satisfy the overriding and precise laws of conservation, which govern particle and field mechanics alike, and extend their undoubted sovereignty to the individual events of quantum mechanics.

Matter Waves

That light behaves somewhat like waves in ripple tanks was the nine-teenth-century's verdict. It stands. But we know that the model is strik-ingly incomplete; we could equally say that light behaves somewhat like the bullet-atoms of Newton or the Greek atomists. Of course, light is nei-ther; the wave is far from tangible, determining a mere probability by its strength; the particles are far from isolated, but deposit their invariable stores of energy in accordance with the probability pattern defined with stringency by the wave-like field. This is a kind of duality, wonderful but not paradoxical, once the completeness of a mechanical model has been surrendered.

What of electrons? Charged bullets in the television beam, surely they are the paradigm of Newton's corpuscles. Ask the question not in thought but of nature. As J. J. Thompson showed the electron beam charged parti-cles, responsive to magnet and battery exactly as so many tiny charged projectiles, so his son, G. P. Thompson, showed conclusively that the electron beam displayed the interference patterns dear to the wave theo-rists of light. But the slits could not be man made to test this property; they had to be of atomic size. The wavelength of the field pattern was tiny; it is given by $\lambda = h/mv$. Note that if h were zero the photon energy would approach nil, and the particle wavelength as well would go to zero. If the photons had much smaller energy, they could never be seen as indi-viduals; if the electron waves were arbitrarily short, we would never de-tect the field patterns. Compare, as a deep-going analogue, the foot-long waves of sound with the microscopic waves of light. A man singing as he walks past the door is easily heard around the corner, but never seen so. Only in wave-sized slits does light, too, bend corners.

Not electrons only, but neutrons, protons, whole undisturbed atoms of helium, all obey the relation $\lambda = h/mv$; all form wave-field patterns when the experiment is tried. We believe today that all the particles, from neu-trino to uranium, all the wave fields, from sound itself to light, gravita-tion, and the rest, share this new nature, of quantized wave field, to give its formal name. Mass, energy, momentum, charge, and some other such quantities are always causally conserved, never missing, never new-born. But when and where these grains of quantity are to be found is ruled by a wave pattern. The wave-field pattern is always to be found in space and time, laid down without fuzziness or error. But how it permits the transfer of one or more of the basic grains is up to chance. This is the fusion of cause and chance, seen over the whole of physical phenomena, provided only the mechanical action is not much greater than the touchstone h.

The Principle of Uncertainty

Now we approach the scene of renunciation, usually billed as the key of the tragedy of modern physics. Uncertainty! But we play today not tragedy but glorious chronicle, not Lear but Henry. The word *uncertainty* has become too strong. To be sure there is knowledge we may not have, but that is nothing new. It is more a surrender of what we never had than it is an admission of defeat, like the ineffably beautiful color of the concept *17*.

The issue was touched in the previous pages. If we need dynamical knowledge, the energy and momentum of mechanics, then we will seek in vain, for the position is space and time. The converse is true as well. Here the analogy to ordinary waves becomes irresistible. Where is the wave in a piano string? Why, everywhere along it. When does it pass by? Any time. Only if we distort the string, not with its usual note, but with a sharp pulse, can we hope to localize a packet of energy in space or in time. We can see that the harmonics of the violinist, very short, high, but pure tones, have a rapid vibration as they sound. To build a localized, restricted pulse of energy on the string, some such very short wavelengths must be used. But they cannot do the job alone; some longer ones as well must be present. It is not difficult to show from the simple picture of waves on a string that to build a sharp pulse either in space or in time, a single frequency—or vibration rate—will not do. Nor will a single wavelength. Instead you must superimpose on the string a whole smear of frequencies, with a range extending over a band from some minimum, say f_{min}, to some maximum, f_{max}. With such a smear, it is possible to contrast a pulse which passes us by between its arrival time, say t_a, and its completion time, t_c. A remarkable but simple inverse relation holds:

$(t_c - t_a) \times (f_{max} - f_{min})$ approximates I.

We usually write this more compactly, calling $(t_c - t_a)$, Δt and $(f_{max} - f_{min})$, Δf to get:

$$\Delta t \Delta f \sim I,$$

where $\sim$ means "about equal to."

The same argument gives for the length of a pulse, Δx, in terms of the wavelength spread we must employ to build it, $\Delta \lambda$, about an average or mean value λ_m,

$$\Delta x \frac{\Delta \lambda}{\lambda_m} \approx \lambda_m$$

These two say that a sharp pulse, in space or in time, must be made of a wide smear of frequencies or of wavelengths.

The product of the time duration of a passing pulse by its frequency spread cannot be made to vanish; indeed, it remains always greater than unity. The translation of these simple enough wave results into quantum mechanics is made by the ruling quantity, h,

$E = hf$, so $\Delta E \Delta t \sim h$,
and
$p = h/\lambda$, so $\Delta p \Delta x \sim h$.

Now these relations are potent. For the first one says $\Delta t \sim h/\Delta E$, that if any system has a well-defined energy, with ΔE small, the spread of time Δt during which this energy is known to be present must be large. Thus a sharp energy level must persist for a long time; a transient system, on the other hand, cannot have a single unique energy, but rather a widespread ΔE. Here we are touching the secret of the stability of matter. Its identity and persistence are quantum marks; the energy is fixed, and the state must be permanent. We cannot hope to follow the 'motion' of an electron in the orbit of a normal atom. If we could, we mark time in its passage. But then the energy would spread, and the atom be in no well-defined energy level. On this theme we can play the variations of the whole of the atomic world. Uncertainty?

The Physics of Identity

It is an old theme of metaphysics, East and West, that the distinction and relation between the single individual and the class was a rich one. I believe that the most striking result of quantum mechanics, insufficiently emphasized in textbook accounts, is to have found this very point not merely metaphysical, but deep in physics. The fabric of the world could not be woven without it. We have more than once stressed that the grainy nature of quantum systems permits them to be in such states that all representatives, say all electrons spinning parallel to a given axis, or all stable hydrogen atoms, are truly identical, not merely similar, as are peas in a pod. Of course, no direct measurement can establish identity; it can certify merely as to the smallness of differences.

The closeness of all quantum mechanical theory to experiments, to possible ways of knowing, is well brought out in discussing identity. How could we tell each from the other the peas from one pod?

First, we could look carefully, say with a magnifier.

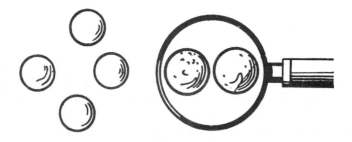

Small differences always appear. But they cannot appear for the atomic or sub-atomic structures we are describing. Failing that, we could force a difference, say by marking little numbers on the surface.

Obviously, we cannot label an atom save by changing it drastically; there is no surface to mark lightly, no sub-atomic ink or tags. Next, we might simply take note of two peas lying at different points in space, and follow them through all their history, never allowing ourselves to mix them up.

Finally, we could box them, and label the locked boxes.

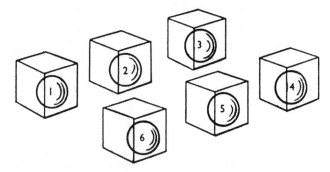

Only the last two remain for the atomic world. But we know from the uncertainty principle that we cannot follow with precision the tracks of two particles which are going to overlap in space.

There will arise a time when we must give up following, or disturb the motion. If we cage two atoms, we can keep them distinct all right, but there is never any context in which our success makes any difference.

The theory agrees that if two particles are never in danger of confusion, they might as well be different. But within a single atom, or a metal grain, say, electrons must mix and our distinction soon be lost. Now the theory has a remarkable result. For we know that what is measurable is only the square of a certain number $(a)^2$. If a consists of the amplitude summed for two electrons, $\alpha(\text{Dum}) + \beta(\text{Dee})$, and if no measurement will ever tell us the difference between Tweedledum and Tweedledee, or—wonderful to say—it *could* change from positive to negative, for $(-a)^2$ is the same as $(a)^2$. Electrons, protons, neutrons, all the stable constituents of matter follow the latter rule always; photons, phonons, π-mesons, the first.

With the first rule, if two electrons were placed into one and the same state, so that $\alpha = \beta$, it follows that the total amplitude $a = \alpha + \alpha = -\alpha - \alpha$. But that means $-a = +a = 0$ and the probability vanishes! No two electrons can ever have exactly the same α. It is this property which keeps the

atoms and nuclei from collapsing into a kind of undifferentiated sticky mass. The periodic table and the nuclear charts derive their marvellous numerical and physical structure from this condition which flows, not out of mere fiat, but out of the general theory plus the consequences of true identity. Particles which possess this property are the generalizations of the old idea of particle—impenetrability; other particles, like photons, which follow the other rule, turn out to love to be in one and the same state. These gregarious particles become in crowds simply the generalization of the old idea of waves, whose energy content can be as great as the source is able to make it.

The Quantum Fabric of Matter

The whole of the material world, with radiation, falls under the domain of quantum mechanics, from the scale of the nucleus within the atom up to the structures of living beings, wherever nuclear physics or atomic chemistry plays an essential role. In the large—from engineering to the solar system—quantum mechanics is not newly involved, except in the properties of materials. The whole domain is ruled by forces familiar to Maxwell—the electromagnetic—plus the newly found strong but minutely ranging forces specific to nuclei. Every sort of structure, each nucleus, atom, molecule, crystal, has its proper energy levels, calculable by the laws of quantum mechanics from the nature of the elementary building blocks and their familiar forces of interaction. The calculation for stable structures is in concept simple: it is necessary only to ask which wave patterns will yield the minimum total energy for the total system, taking into account the potential energy of interaction, the kinetic energy of the localization of the particles, expressed by the uncertainty relation,

$$\Delta p \approx \frac{h}{\Delta_x}, \quad \text{and} \quad KE \approx \overline{p^2}/2m$$

all as modified by the requirements of identity, so that no two electron states, for instance, can be identical. The rest is an exercise in energy economy. The onion-like shells of electrons in the atoms represent one such valuable approximate description, giving an understanding of the periodic table. These are, of course, not orbits; the electron does not move in any track. Were it to do so, the atom could not remain unchanged. But here the detail of quantum knowledge is breath-taking. The hydrogen atom can be understood from the general laws, and the properties of electron and proton alone, to an accuracy in the energy levels of a part in 10^{10}; the frequency of its radiations can be predicted to what amounts to a millisecond in the year.

For more complex systems, there are heavy difficulties. To carry out the computations is a task too heavy even for the computers of tomorrow's dreams. Instead, the chemists learn to appreciate regularities which enable them to make adequate approximations, simplifying their problems. For example, they recognize that the electron patterns tend to form regular bonds of similar sizes and shapes in analogous molecules. They learn that electrons can leak from atom to nearby atom, binding them even into the great arrays called crystals. Even waves, like waves of sound, obey the rules of the quantum. At normal temperatures, however, soundwaves in crystals have an energy spacing so small that the Brownian motion from the impact of the surrounding molecules hides the levels. Thus quantum levels are obvious either at atomic scale and smaller (recall $\Delta p \sim h/\Delta x$) or at extremely low temperatures, where the external noise is reduced. Here the superconductor shows on a large scale exactly how the electrons unceasingly move in an atom, without any loss of energy; there is a permanent current. And superfluid helium shows eddies of millions of identical atoms, like quantized smoke rings.

All of matter obeys the structural dictates of the quantum.

Creation and Destruction

So far we have emphasized stability and form. But change is equally part of the material world, and equally well treated by the theory. Atoms, for example, emit photons whose energies reveal the nature of the atom, their spectra. In the old physics, all charges in non-steady motion emit radiation. The same consequence of charge holds in the new physics, whenever particles with charge change their states, that is, adopt new wave patterns. Plainly, the photons are not, as the electrons are, present in the atom before emission. Rather they are created at the moment of change. In absorbing light, the reverse occurs; the photons are suddenly absorbed, destroyed. Their energy and momentum, of course, are always accounted for. The atom can never be 'emptied' of photons; given energy anew, it will always emit one.

The nature of the description is what by now we might all expect. A wave pattern which corresponds to the photon slowly grows from none, pumped by the radiating charge. But when and where a photon actually appears are governed by chance only, given the expected probability from the wave amplitude. The atoms decay at random, even if they are identical. But their *expectancy* of life is fully calculable.

The presence of a photon induces emission of others just alike, for identical photons are particles of the gregarious type, related to the classical field. But even if no photon is present, a system which has the right charge pattern, and energy to lose, will emit one in time. The wave pattern of photons is not zero even when no photons are present. Empty space— "the vacuum"—contains a noisy, random, fluctuating field; but a real photon, with finite energy, of course cannot appear unless some other system will foot the energy bill. But when very many photons are present, their crowd will paint out with apparent smoothness the same electric and magnetic field old Maxwell would have called for. Electron and photon alike are fully described in the quantum theory.

Other particles, too, can be created and destroyed under certain restrictions. Electrons, for instance, are born or die always twinned with positrons. A photon can pass its energy and momentum on as a legacy to an electron-positron pair, itself disappearing at once. But it cannot do so in free space, for the energy and momentum cannot there be balanced. Some external force must be present. If this were not so, photons could not come to us from the edges of space. The typical seat of this process is the region near a nucleus, where the strong electric field contains transient photons, called "virtual," which are mere transient disturbances in the electric field, with no energy to pay for a separate career. But they can add momentum to the photon-pair creation, to realize it. The momentum in the end kicks the heavy nucleus a little.

That electric center which is a nucleus, and the very electric field it possesses, should in the full theory all be treated as wave patterns. This is normally done only when stable structures and bound states are not involved. At the highest energies, then, used for the probing of atoms and nuclei, a complete quantum view of all phenomenon is taken; but for the quieter world of the chemist or the biologist, the electrons and nuclei are never created or destroyed. They are regarded as given, and their wave patterns refer only to the motions they undergo.

Elementary Particles: The Elusive Ultimate

The theory relies upon the existence of these truly identical entities, electrons, protons, neutrons, and a score more of unstable "particles." Their nature has been illumined, but still only dimly, by the theory we are sketching.

In some sense we could think of an electron without its charge. Switch on the charge, and the electric field appears. The "bare" electron is now "dressed" (in the contemporary jargon), but its cloud of virtual photons appearing and disappearing with enormous frequency. The energy bill is not paid, for the uncertainty relation assures us that $\Delta E \sim h/\Delta t$, and if we do not ask for the energy loan for a greater time that Δt, we can overdraw the energy account by an amount $h/\Delta t$. For an electron, this time of imbalance is some 10^{-21} seconds, based on the total energy available to the particle, its mass $m_0 c^2 \sim 10^{-6}$ ergs. Some of the photons make electron-positron pairs in the same transient and irresponsible way, so that around the electron is a flickering double layer of electric charge. In the theory all these effects, while really predicted, cannot be reliably calculated, but a consistent way is known to introduce the experimental results. The charge cloud of electron-positron pairs around a proton which is so heavy has been directly demonstrated by its influence on scattering other charges. We can see also that light may interact with light via the transient charges.

We know a whole clan of particles, of various families. Each of them dresses itself with a cloud-cortège of all those others with which it maintains a strong energy of interaction. Here we see the rise of structure, even if the bare particle is assumed to be a mere structureless but interacting point. But here, too, there is by no means finality; there is only the promise of order. No really crisp results exist about the particles themselves;

their compounds alone are well understood. Some particles, of course, may yet turn out compound. The theory is not at rock bottom; but no sign of the failure of quantum mechanics has yet appeared, only a frustration of consistent application to the various interactions.

Strange enough, this story. The vacuum, no longer empty, but, as for old Leibnitz, a plenum of potentialities, crowded with virtual particles, is our space. In it particles of every sort are created and destroyed, each event ruled by chance, yet overall obeying the high laws of conservation, and fulfilling in statistical detail, the space-time predictions of the theory of wave fields. One generation has lived with quantum mechanics, the most potent, as it is the most subtle of all the theories of the physicist. If change lies ahead, and it always has, it seems to me that nearly all we have said will yet retain its value. We shall see in the brightening glass what lies in the regions beyond. That we will discard our picture of fused cause and chance, of conservation and virtuality, I have no present reason to expect. The quantum will always remain.

Why Man Explores

I f you ask, Why do human beings explore? I would answer, as I think the Greeks would answer, "Because it is our nature." Now I am anxious not to make the mistake of thinking that the term "human nature" is explanatory, that it covers every activity of our species, the most diverse ethnographies, the artifacts that grace the museums, and the publications that crowd the newsstands of Los Angeles. "Human nature" is an impoverished description of all that diversity; but there is one feature—for me it is perhaps the only feature—which does define human nature, which parts our species (and a few vanished species of our family related to us) and has parted us from other creatures for surely tens of thousands of years, maybe for a few hundred thousand years. We are beings who construct for ourselves, each separately and singly, and as well together in our collectivities, internal models of all that happens, of all we see, find, feel, guess, and conjecture about our experience in the world.

A clear context in which this was put for me is a beautiful ethnographic work by a woman called Elizabeth Marshall Thomas, who lived for many seasons among a small group of the wandering peoples of the Kalahari whom we call Bushmen, people whose inventory of physical goods is very small indeed. They own nothing that sits still. They carry all that they have, all that they make, in a pouch of hide which they bear on their shoulders. They wander forever through life, stopping now here, now there, to sleep in a kind of nest, to try the fruit of this tree, to scratch up that waterhole, to meet for a ritual encounter with their wandering friends, and so on. These people, whose minds are full, though absent writing, absent crowds—in fact they are few—live in small bands of extended families. Each band tends to stay within a region about like that of Los Angeles County, an area of a thousand square miles or two, in quite desert country. From their point of view they are by no means poor; they manage to make an excellent living, as the time-and-motion study people have demonstrated to us, while working rather less hard than the Harvard anthropologists who watched them. Their skill is so great, their understanding and their wants are so well controlled in the environment, they are so beautifully adapted to their situation, that they need not work harder.

The one need they constantly discuss as they wander through the cool mornings, the cool evenings, and as they rest in the heat of the day, is to know exactly where they are. They discuss it always. They note every tree, they describe every rock. They recognize every feature of the ground. They ask how it has changed, or how far it has been constant. What story do you know about this place? They recall what grandfather once said about it. They conjecture, and they elaborate; their minds are filled; their speech elaborates exactly where they are. You see they have built an intensely detailed, brilliant, forever reinvigorated internal model of the shifting natural world in which they find their being. What that simplified case suggests I dare to extrapolate to all human beings everywhere. I see in it, I think, my own behavior; I hope it will be so for others. It is fair to say that our language, our myth and ritual, our tools, our science, indeed our art, are all expressions translated in one way or another by the symbols of our communication or otherwise of certain features of this grand internal model. The presence of that internal model and its steady need for completion, the obviously adaptive need of its leading edges to have continuity, not to fade off into the nothing or the nowhere: this is the essential feature of human exploration, its root cause deep in our minds and in our cultures.

For me exploration is filling in the blank margins of that inner model, that no human can escape making. Of course, we can rest content within the margins; then we live with a shadow of uncertainty at the edge of the map. Indeed a culture is free to do that, as many cultures have done it—I should say a little more about that later. I want to make quite plain that an internal model is not the only way in which complex accomplishments can be produced. I suspect that we are not the only creatures to show this quality, although we show it in quantitatively distinct form; but we need not fear comparison with other creatures. There is another way to construct even complex architecture without ever having an internal model; were we built that way, we might yet in the course of sufficiently long time evolve all the complexities we have, even if we would not explore. It's conceivable, save only that the universe might not last that long. It is the speed, which is our way to change, that eventually marks us.

When I was a schoolboy, I learned (from a very bad book, I am now sure) that one of the distinctions of truly high civilizations is the ability to construct the true arch, that curved arch with the keystone that holds everything together—not the lintel beam which the Mayans had—but rather those things which Greeks and Romans and other proper countries had which made them high culture and restricted the others to the first chapter of the book. I soon grew away from this kind of provincialism, which was more common a hundred years ago when the man who wrote the book was trained.

I was most forcefully struck by the work recently reported by some French entomologists who have studied in South Africa the work of certain species of large termites. Termites, of course, are social animals of considerable power and prowess. The structures these particular forms build are great things. They are 15 and 20 feet high on some occasions; they dot the landscape like so many termite skyscrapers. They are large and enduring architecture. Layer upon layer hidden within this termitary which rises out of the ground, are true arches, curved arches which support the next floor, and then more arches for the next, and so on, exactly like the crypts of a building somewhere in Italy. You have to ask yourself the question, Are termites then such thinkers and philosophers as we? That would be the most fallacious view; the reason is not that we can dismiss their accomplishments. As with the qualities of human beings, you cannot judge only by what they have done. You have to judge them in the sense of potential, because what they have not yet done, what is contained in the internal model, is the key.

The termites, of course, always do the same thing. They have done their thing now for twenty million years without changing very much. Mind you, they build the true arch—in the dark. Blind animals building arches in the dark! There is no architect, there is no building-code inspector, there is no critic. All there is is a little hollow in the ground and a thousand termites milling around in the dark making pellets. There is a built-in instruction: "Make pellets out of the discarded leaf matter, the fecal matter, which lies around on the floor." They form lots of pellets. Each one by himself makes pellets. If it should so happen that the density of pellet construction in some region is greater than that in the neighboring region—of course, it must happen that way sooner or later by the laws of chance—then the instruction is: "Leave your pellets which are few and go to where there are more fragrant pellets, a few inches over." Pretty soon they divide themselves into little groups of pellet builders, all making piles of pellets. In between they have stopped making them; those termites gather around the larger piles. Now the piles grow to columns; they stick them together. The next instruction says: "If, as your pillar gets pretty high, you detect another pillar higher still, stop yours and go to work on one that has crossed a certain limit." (We reconstruct these rules by watching their behavior.)

Pretty soon you have many half-finished stumps of pillars, but you have also a few rather high pillars sitting on the floor. The next instruction is: "If two high pillars chance to be reasonably close together, get on top and build each toward the other." That's exactly what they do. So, of course, in each layer the number, size, and placement of arches is different. No great architect has seen where they will be, no one has counted them, no one has decided on them; but the work overall is adaptive, improves the termitary, its strength and its ventilation. So they go on building arches; they will do so for tens of millions of years on end. There is no internal model within any termite, or even in the collectivity, for how those arches should be built. There are in the DNA, in the chromosomes, some kind of simple rules that tell them how to make arches in a broad general way—not the making of the arch itself but the giving of rules of the kind described. There is never an arch present until one appears by chance; whereas when we build arches, or anything else, the arch is in some sense present before it ever exists. That is what I mean by an internal model. Now the need to complete that internal model—to extend and fill in its fringes—is, I think, what we mean by exploration.

I recognize that this deep need to complete the internal models is certainly expressed differently in different cultures. Sometimes it lies very quietly. The pioneer Alpinists who came in the early 19th century to Switzerland found villagers who had lived there all their lives and never had searched their peaks. But once the visitors raised the idea that it might be worthwhile, it turned out that among the villagers there were a few young men who had quietly ventured into the peaks even before the English gentlemen came to hire them. They became the first guides. Climbing wasn't celebrated, it didn't butter any parsnips or feed any goats, but it was needed somehow to complete a model. I believe those cultures which manage to show some public concern for filling in the edges of that model, for extending the margin of the map, are those in which we now live, and those in which we shall live for most of the time of human history. Democritus said, "I would rather find one cause than be emperor of Persia." That is a statement which a physicist can beautifully adhere to; were we to lose that feeling, it would indeed be a heavy loss.

There is one problem which Viking, the prototype of what I am describing, does not solve, that is, access for a wider number of persons to this scheme of filling in the edge of the incomplete internal map. We have founded such great social structures to pyramid our exploration upon that those at the base often do not get to see the stars shine above the apex. This problem we will step by step come to solve. Finally, for me, human beings explore because in the long run, time after time, when we wish to adapt to the world as our inner nature has evolved, both by genetics and by culture, we can do nothing else.

Two Dials

Inscribed on a cave wall in the south of Spain, not far from what was the edge of a receding ice cap ten or fifteen millennia ago, are series of ochre handprints—many, many prints that overlap to make a kind of frieze. Nearly every book on Ice Age art offers an illustration of this frieze, and as a visual image it is quite moving.

Some of the handprints lack a few digits. Whether these omissions are an artifact of the drawing, reflect some sacrificial act by the artists who made them, or represent some curious medical phenomenon of the time, no one really knows. What these three-, four-, and five-digit hands do suggest is a count, fingers counting very long ago in human history.

It is no accident that the term *digital* derives from our ancient habit of counting from one to ten on the fingers, or digits, of our hands. Back then, such counting was both digital and visual: a set pattern, neither symmetrical nor wholly asymmetrical, an assemblage on a cave wall that captured the eye. In a sense, two different ways of looking at the world, of studying and comprehending all of nature, are capsulized in that single frieze of handprints.

Quite literally, the frieze of hands may be seen as a digital mapping, a series of distinct but discontinuous signs (not drawing so much as tallying) that maps some part of the world. It is quite plain that these signs represent a language of numbers from which the human mind has evolved algebra and all of the other mathematical symbols and numerical structures that are characteristic of our time—the digital world of powerful symbols, a world that grows more and more pervasive each day.

But these hands represent more than the world of the digital. As powerful visual art, they also represent the world of the image, the symbolic gesture—the *analogue*. From the digital to the analogue—this spans all that we think and all that we do in the sciences, and in the arts as well.

Both are modes of the mind's expression, and it is quite plain that these two poles in some sense merge. We can map abstractly all that is geometric and all that has spatio-temporal extent by the digits alone—merely the fingers of the hand. For this we have strong evidence, particularly in this country: The television screen, in all that it visually conveys, presents only a complicated collection of digitized pulses (that isn't completely so yet, but it will be in a few years' time). Those digital impulses can spell out everything.

Even printed words are not as they used to be, strings of letters that some artist carved by hand and chiseled into the soft steel from which type was cast and finally inked to form an *o* or *t* on the page. That doesn't happen universally anymore. The *o*'s and *t*'s on the pages of this magazine are nowadays never formed in metal, but are generated in the form of little bright dots flickering on a television screen before a photolithographic plate. This dot pattern defines by a photographic process the offset plate that prints copies of many magazines. Look with a hand lens at the printed letters, and you will see that they are not bound by the straight lines and circular arcs of a properly carved letter, but by a scalloped edge comprised of little, merged dots, small enough that the eye can't see them unaided.

All we write can be put into dots; dots mean pulses, and pulses mean yeses and nos—the full digital scheme is nothing more. Even images and pictures are dots. A half-tone picture, as everyone knows, is nothing but a succession of digital statements about black and white (with perhaps some gray half-steps), and that's all. Color, too, is merely a few sets of dots on the page.

Popular psychological theory is fond of suggesting that there exists a significant logical difference between these categories: digital and analogue. It is often said that these dual attributes of thought and expression are assigned to different hemispheres of the brain, and so to separate levels of logicality. One—the digital—is associated with logic and reason and the rational world (the left hemisphere), and is generally called algebraic. The other—the analogue or geometric—is associated with the imagination, with dreaming, with activity of the inner mind (the right hemisphere).

I don't believe their theory for a moment. If I have learned anything in these matters, I learned a long time ago that it is very difficult to split algebra and geometry into this hand and that. For every logical proposition in geometry, a strictly parallel statement can be framed algebraically, and the reverse is also true. Each can be fully rationalized on the logical plane; each can be used to express imagination. The only difference lies in the form of expression.

To see in the digital the way of reason and in the analogical the way of the dreamer is plainly the claim of a writer. Writers are fond of seeing cogency in the string of words they put out. Such a person is likely to think that art—visual art—is a rather playful, spontaneous activity, preliterate, for the kindergarten. Once out of kindergarten you are a serious person. Writing, be it typing, editing, or plying a quill pen to meet the deadline, is the kind of serious mental labor done by the fully conscious, thoughtful person.

But artists by no means find the business of art trivial. It is writers, by and large, who believe that art rolls off the canvas like water from a duck's back, that it requires little effort, no strain, no determined repetition, no careful screening—none of these things. Because it is writers, not artists, who write, we encounter the frequent claim that language is where logic stands and that art, that alternate world of form, comes through to us in snatches, without much human intervention, from some mysterious source. In that last claim writers are right, but the source of both language and art is the same human mind.

What we are discussing is a profoundly complementary relationship. Both digital and analogue fully tax the mind, both are deep in the culture of all society. I doubt very much that, unless the masters of nucleic acid mix us up quite a lot, we will find ourselves at any time abandoning this profoundly complementary duality that marks the expression of all human thought. We are committed to both the analogue and the digital, the visual and the geometrical, by the nature of our being. Perhaps such nature itself is grounded in the universe from which we evolved and to which our being must in some rough sense conform. As we learn more of the mysteries of the mind, we will learn more about the ways these geometric transformations enter and are maintained by some subtle digital network. The two are there, intertwined, and the people of this epoch are beginning to see that they work together toward a truer perception.

The most familiar of all analogue-digital contrasts is, of course, the watch. In *The New York Times* every day, and especially in its Sunday *Magazine* before Christmas, you can find a museum of watches to study. Dozens of pages display scores of complicated timepieces that, I daresay, people give as gifts to friends and relations. Christmas is the watchful season of the year.

The fundamental analogy embodied in the watch, curiously concealed for some four hundred years, is that the moving hands of its dial are the analogue for that great celestial dial on which the sun moves, marking out each day and night while we wake and sleep. The clocks of the thirteenth century still had one hand that boldly circled the dial every twenty-four hours, just as the sun circles the Earth, or the Earth spins on its axis in the face of the sun, as you like it. Every twenty-four hours, one orbit. Between the twelfth and fifteenth centuries this cycle became the hours of the day and night—two twelve-hour periods. But the watch, alas, does not go around the dial once each day. My wife claims she cannot really enjoy a watch until they bring back the twenty-four-hour cycle, so that she might sense the great dance when she looks at it. It is a marvelous analogue and should have been kept. But it is the way of our species to gild even the lily; the dial is circled twice each day, and the great circle of the sun has been lost to nearly all our clocks, which remember only the second harmonic.

Much more striking still is the novel form of watch peculiar to this age of the silicon chip. Of course, I mean the digital watch, in which the sun is forgotten entirely. Nothing at all goes around even *within* this instrument, save perhaps an abstract vector in some Platonic world to represent the voltage within the oscillators. The sun is far removed; the watch dial might as well tell us the time in words or wish us happy birthday (and some of the more intricate models do). The digital system can, as we have seen, map any relationship.

For these reasons, I have spoken on occasion as an antagonist of the digital watch. How can we celebrate the spin of the Earth and still consult a device wholly disconnected from that phenomenon?

Now I think I have been hasty. True, earthspin has faded from the world of my watch. But the device retains a close analogy to a profound natural phenomenon, perhaps even more intimate than the rhythm of day and night. I have come to see that the prototype of the digital beat of my watch is the very heartbeat of its wearer. Again, a discrete sequence of relaxation oscillations, not circular but periodic in the most digital way, rules the new watch dial as it rules the heart. Even more, the visible beat of the watch—one second (however cleverly generated by counting to-and-fro twitches of a tiny quartz crystal)—is surely derived from the near-second rate of the human heart. These foundations are clear: If they are not so old as earthspin, they are more intimate for that. In nature no one consciously counts heartbeats (apart from the physician). It is nothing new to have the beats, but perhaps it is new to have the circuits that sum them up. Or is it? Is not our lifespan itself the sum of heartbeats taken a billion or two at a time, adding up to that unknown interval which ends at the day of all days for each of us?

The circle is closed; the function has returned to its previous value. The purely digital has given way to an old analogue, and the analogy touches us to the core. I can no longer claim that the watch on my wrist must have a slow-turning hand to recall that great source of life, the sun. The quick-flashing dots recall life as strongly as the heartbeat itself.

The sun's rhythm is shared by humanity with all organisms that bear chromosomes like our own, down to the glowing plankton of the red tide. But the beat of the heart is nearer, largely within our vertebrate kin, a device only a few hundred million years old. Insects have a rather distinct pattern of pumping; perhaps one could be more generous, but our chambered heart with contractile walls seems one vertebrate patent. To find it mimicked, not in flow to be sure, but in respect to time, restores my tolerance for the new-fancied watch. Heartbeat is no less a wonder than earth-spin; even a physicist can accept that.

Thus the digital beat and the analogue hour-cycle both echo the natural world, echo great sonorities indeed. The conclusion we are drawn to is surely right: Neither digital nor analogue form will serve alone, in the sciences as in the arts. Our knowledge is compacted of both; only their dual use, their subtle congruences and contrasts, spans all our understandings.

Science and the Nation

Before the House of Representatives had yet assembled in the still-incomplete Capitol, the President of the United States had publicly and privately sought scientific advice from a scientist of the greatest distinction. Alexander von Humboldt, admirer of our Republic for all his Prussian title, was invited in 1804 by Thomas Jefferson to visit him at Monticello and in the White House for long discussions. Jefferson, himself more knowing in science than any other President has been, wanted especially to learn from Humboldt, then fresh from a year's travel over Mexico, what he knew from that side of the poorly-marked southern regions of our new-bought Louisiana: "What population may be of white, red, or black people? And whether any & what mines are within them?" Humboldt was forthcoming, producing a wealth of maps and journals for Madison and Gallatin to study and copy. But Humboldt was no mere watchful traveller, he stood for science; he, first of all men, had carried magnetometer and barometer to the highest peaks and deepest thickets of the New World, and had caught the wilderness within the enlightening net of growing human knowledge.

I see the role of science in our government still by those two distinct lights: we need to know many specifics, so that we may act upon such knowledge, but we also need to recognize that science is not merely a means to our several ends.

It shapes those ends as it shapes our view of the world and of humanity's place in that world. The scale of this city and of the nation have vastly grown, yet we face the old needs, which we still hope to help meet by our science and technology: food, fuel, health, the control of birth and death, communication, and transportation. We have as well seen intense intimacy arise between modern science and the weapons we wield.

Let me be explicit. Allowing for the roughness of fitting events into the arbitrary pigeonholes of decades, I can summarize the past thirty years in the two domains, that of science as it gains knowledge and power to extend the mind's grasp, and that of science—perhaps it is more closely technology—as it produces means to deal with socially perceived needs. Up to and in the 40s, physics I think was the major science, and its analysis of the inner structure of matter and radiation became wonderfully powerful. With that went the wartime impact of radar and of fission, as applications. In the fifties, the world of science was changed most deeply by the famous discovery of the double helix, the foundation of biology today; in applications, by thermonuclear energy—then and still only for bombs—and the proliferation of solid-state electronics, which made possible modern computers, and with them all the intricate electronics still in such fast growth. In the 60s rocketry and its uses for peace and war dominates applications, but it was marine geology with its evidence of moving continents which was the sharpest edge of discovery. Copper mines on the sea bottom are only the first fruits of the new geology; no applications of the new biology are yet commensurate with its promise. They will come, for good and bad, as we employ it.

It seems clear to me that science and technology demand Federal attention at every level, not only as immediate problems or as quick fixes, but as long term investments of human and material resources. Nor can the economic and military return alone measure our interests; the world of mankind would change with science, even if we did not make the applications it promises. Consider the single fact that we today see our earth and all life as billions of years old; Jefferson still reckoned under ten thousand years since Creation. It follows that science and technology must and do pervade the Federal scheme. How would Transportation or Defense or Health act without research and development for their own ends? Yet science is wider than any such Department. It makes no sense to deprive these executive branches of their own dedicated investigations and tests. Since these will remain the bulk of Federal expenditures it is unwise—impossible—to sequester science and technology in a department of their own. On the other hand, it is plain that the competition of departmental aims for limited resources, and even more important those consequences of discovery which transcend the categories of the *Federal Register,* demand an explicit understanding of science at the highest levels of policy formation. How can we gain that?

For Congress, continued hearings, both systematic ones and improvised to examine new issues, are needed and real. Internal education of elected members and their staffs is a constant and growing need, even if we expect to encounter few Jeffersons in any decade. (The Office of

Technology Assessment represents an important step whose time had come.) The structures of law dominate the training and the minds of most elected officials; it has always been so in our Republic. Since it is under law we are governed, that is most reasonable, but it is equally clear that a world more and more dependent on new technology demands a new flow of information for our law-makers, if they are to govern with an eye to the future.

The courts work case by case; this method has always used a rotation of experts on call. I am not sure what if any changes might be useful here; certainly the notion of the friend of the court, the expert panel to make special reports, can go far to clarify the intent of the Congress. In the Executive Branch, which employs above a third of all American scientists and engineers directly or via contract, there cannot be, if I am right, any Department of Science and Technology: rather, they must be strong within every department. There can hardly be a permanent policy center in so special an agency as the National Science Foundation, whose function is predominantly a granting one, working mainly case by case with peer or panel review. The whole problem seems rather to suit a device of which we already have some experience. Within the Executive Office of the President, but in response to statute, and made in some degree responsive to the interrogation of Congress, there needs to be an Office of Science and Technology, its aim the shaping of policy in the short term, and in the long. Experience suggests a single responsible leader, who might or might not become personally close to the President, but who would always be responsive to the Chief Executive. He cannot work alone, though he can take responsibility. He needs a small panel of diverse and able people, on repeated call, prepared to consider what he sets before them. They should be able to gain experience, and thus to serve for some years. There is no substitute for a small, devoted, expert staff, freed to study any matters within the government, including as a major area the relevant affairs of the Department of Defense, and of the Energy Office.

Science is no general affair. True, it has always about it a certain hopeful skepticism, but that is not enough to define it. There is no simple, general path to scientific, as to any other sort of wisdom. It consists in minute particulars, but seen broadly by people with hard work to their credit, to test the specifics of every claim. For this reason alone, it is essential that a free flow of ideas come to the government, be it the Congress or the Executive—I would hope to both—from the diverse community of science and engineering.

Finally, science does not work by terms of two, four, or six years. Careers last decades, human plans must ripen and will change. As an astronomer, I am in daily touch with matters of prodigious time span; the laws of Nature are enduring. The immediacy of our years sometimes seems petty. But our lives are brief and our issues cannot wait; they are matters of human life.

On the Causes of Wonderful Things

Acentury before the Royal Society, even before Galileo's Academy of the Lynxes, there met in Naples a group of amateurs gathered by the learned and ingenious Giambattista della Porta, among whose inventions was a low-power two-lensed instrument, forerunner of the telescope, probably made while Galileo was yet a child. Those *otiosi* (men of leisure) in Naples made it a condition of membership that "each man must have contributed a new discovery of fact in natural science." Theirs was the first scientific society in modern times, although its main legacy was its founder's publications, books hugely popular over all Europe for a century or two. For della Porta the practice of Natural Magick (so the first English edition of his book of the same title called it in 1658) was nothing at all like "Sorcery; an art which all learned and good men detest; neither is it able to yield any truth of Reason or Nature." Rather, it was "the practical part of natural philosophy, which produceth her effects by the mutual and fit application of one natural thing unto another."

First among twenty rather specialized chapters of *Natural Magick* ("Of Beautifying Women," "Of Distillation," and "Of Strange Glasses" are three of them) is a theoretical introduction, "Of the Causes of Wonderful Things." The structure of the world, it argues, lies in the internal elements and forms of things, their powers drawn and ruled by the stars. The importance of time and place is obvious, and the richness of combinations can open many valuable new paths. *Sympathy* and *antipathy* are general relationships and can be widely applied. Thus: "A Dog is most friendly to a man; and if you lay him to any diseased part of your body, he takes away the disease to himself, as Pliny reported." On the other hand, in one rare but terrible crisis, "A Dog and a Wolfe are at great enmity; and therefore a Wolves skin put upon any one that is bitten of a mad Dog, asswageth the swelling of the humor."

For nearly a century now we have treated rabies worldwide with excellent results. The modern therapy is on the face of it less plausible than della Porta's, though it shares with his scheme certain unexpected qualities. We prepare our magical inoculum, the spinal cord extracted from a sacrificed rabbit, itself dying of rabies. That tissue is treated according to a precise ritual of times and dilutions with a solution of formaldehyde. Samples of the substance so obtained are placed day after day within the bloodstream of the patient. If treatment is in time, during the few weeks between the infecting bite and the appearance of the first symptoms of the disease, nearly every patient will recover. There is profound antipathy at work here, all right; but we know it as the generation of antibodies by the immune system of the patient.

That repeated inoculation with tissue from a rabid animal treated with a stinking poison like formaldehyde can indeed stave off death from rabies is surely wonderful. Della Porta could have hailed Pasteur's technique as a development out of his own crude ideas. In the bizarre nature of its procedures our practically applied science does not differ from magick. "Nothing is too wonderful to be true": that remark, ascribed to the meticulous experimenter Michael Faraday, appears carved on the physics building at UCLA. He was surely right. The "mutual and fit application of one natural thing unto another," often strangely specific, can work wonders, effects really unmatched within the ordinary noisy experiences that inform what we call common sense.

Della Porta places his therapy within a generally causal theory of matter, as do we. Della Porta relied upon earlier reports of experience, as do we. Della Porta employed only the objective properties of "natural things"—not, for example, their names in Italian, or their relationships to the life of some saint, or other properties that we could reject out of hand as failing common-sense tests of invariance or causality. It is not easy to point to any broad methodological distinction between his natural magick and modern applied science. In fact, if someone could show, say, that certain stable antigens in most wolf pelts might be involved in some subtle induction of the immune reaction, the Porta "remedy" could even be assimilated to our theories. Indeed, a placebo effect of the magickal therapy might be invoked, so that the patient internally turns on the healing flow of antibodies.

No, the weakness of wolfskin therapy, like our success with a carefully inactivated virus, does not rest in generalities of method. It certainly cannot lie in the unexpectedness and incongruity of the procedures; the preparation of rabies vaccine seems closer to eye of newt and toe of frog than does plausible and cozy wolfskin. It does not lie either in the power of results achieved by quiet means. We often speak of rapid and internal cures as magical, "working like a charm," whenever they seem to go beyond common-sense experiences, yet the replacement of a single loose wire or the use of the right antibiotic or hormone can induce such results, quite causally, whether in man or in machine, by "mutual and fit application." The truth is instead deep within the hard-worn detail. How many cases did old Pliny see? Had della Porta any better sample, or was he generalizing on an old citation without skeptical questioning? (Even for us, the bite of a rabid dog does not always inject live virus, and so assessment of the value of treatment cannot flow from a single happy case. A big sample itself is not probative under such circumstances, while the demand for a series of untreated control cases that logical inference requires is too stoically heroic to be met, except by chance. Here is one reason, of course, why tragic conditions tend to induce bad theories.)

Next is the question of texture, of the robustness of a theory under diverse tests of its inferences, tests that vary experience under the guidance of theory. Does it matter if the wolf is male or female? Might a cat's skin or a coyote's be used if necessary? Of course, the old scholars made no such tests, nor could we in fact expect them to. But the Pasteur method is by now robust. Nowadays one can use virus from a sheep, or horse, or even from a tissue culture alive in a glass vessel. The formalin inactivation can be replaced by ultraviolet light. The virus particles can be seen in the neurons under the electron microscope, and the antibodies titrated. Practical success leads to refinement of techniques; we expect the statistics to become themselves impeccable one day, unmarred by even a single inexplicable failure in a long series of cases. Step by step down a long causal chain we chase the virus and the antibody; even if here and there logical gaps still remain, we have found a detailed procedure, specific but not capricious, robust to theoretically admissible changes of detail, improving in practice sometimes even under statistical control. By now rabies vaccine is embedded within a much wider view of disease and host, a structure so wonderful as to include the role of the silent thymus gland, the ancient evolutionary sequence of the higher organisms, and a strong molecular dose of protein structure.

The sixteenth-century pioneers fixed one beacon we still steer by. Della Porta sought to explicate, to reveal the secrets, to be open about the use of wolfskin, even to tell why. His style is in strong contrast to that of today's miracle men, such as the psychic surgeons of the Manila countryside, who plainly carry on a successful practice which elevates secrecy and sleight of hand to a major role. When a psychic surgeon takes out the offending tissue through the uncut skin, in a gesture which mocks so reassuringly the deftness of the real knife surgeon's frightening incision, the bloody chicken parts he throws into the fire are calculated to have an effect on patient and spectators. That effect would change if he were to show beforehand how the stuff is palmed or held in some magician's gimmick. If any fraction of such cures is placebo-related, their hidden procedure may well be of practical value. The very same attitude, if less deceptively applied, is still found in medicine, either as a relict of the old ways, like the Latin of the prescription, or as behavior held to fit the social role of a healer with respect to a patient, or as a conscious decision of the doctor to meet his view of the patient's personal need. Trade secrets, military-industrial secrecy, and craft processes of many sorts are more or less distantly related to the healer's covert actions. Here the aim is rather a competitive advantage; some grasp of the natural world is being guarded as a piece of property. Often the hidden procedure is itself not understood; one particular source of a well-known ingredient functions better than any other, say, though just why is not known. The famous discovery that gelatin derived from cows who had grazed on fields golden with mustard made more sensitive photographic emulsion could have been a trade secret—for all I know, it was—before it became clear that its sulfhydryl impurities formed many electron traps in the silver halide crystal grains.

There is a whole spectrum of such occult matters in every complex fabrication. Often the expert himself is not even conscious of his hidden maneuver, for it lies in the black boxes of a system he has not fully understood. That phenomenon is familiar in the homeliest of contexts, too; there is often just the right way to start an old car or to turn a worn-out switch. Empirical results are not in themselves a road to understanding, only to success.

It follows therefore that we must ask more of any procedure than just that it works, if we are to learn *how* it works. High fees, glowing testimonials, accounts of visitors and friends are not firm demonstrations of why something works—whether it is the topic of a TV commercial or an operation by a psychic surgeon. Either there is step-by-step openness, subject to change and deductive questions at each point, or there must enter the machinery of rigorous induction, with careful statistical control and well-designed experiments and double-blind clinical trials. For error in the conclusion can be not only that of the patient who is being denied the truth of his operation by the deliberate stage skills of the psychic surgeon, but even that of an unwitting, sincere experimenter or clinician.

That much is evident; nothing new. There is more. It must be admitted that there are circumstances in which deception, either of others or of the self, can in fact improve the claimed results. The logic is clear; it is even conventional offered in support of the withholding of the diagnosis of a terminal disease in many cases. The argument is very old; the doctors of the Skidi Pawnee invariably preceded their big seasonal ceremonies with displays of sleight of hand, taking a quartz crystal out of the nose of this or that awed spectator, turning a mud duck into a living one by dropping it into a vessel of water, or swallowing a deer's head, antlers and all. They freely employed the techniques of the carnival and the stage, with accomplices planted in the crowd, prearranged mistending of the fire to darken the scene at the crucial moment, and so on. All this was unabashedly known to the inner circle. But these men maintained that their little show was exactly the right way to prepare their audiences for the serious religious ceremonies of the group. They had done so for a millennium or more. In a parallel if trivial context, the classroom experiments of Craig the magician in "To Believe or Not to Believe" touch the same facet of human nature. The students preferred their mystery; Craig as an amateur sleight-of-hand artist was an example of rational, practiced, and dextrous craftsmanship, not especially remarkable. But Craig as a partaker of hidden and mysterious powers, powers which might—who knows?—descend magically upon anyone, was quite another kind of wonder. The practical outcome here gained from the deception was that the occult thrilled more than those undisclosed but all-too-human skills of planned misdirection and sleight of the hand that are such hard work to perfect. Once the conjuror pleased his audience by faultless and apparently effortless mastery over cards or coins; now he is more likely to claim the need for the warm sympathy of the audience for a trick that will fail once or twice—until the "vibes are right." Most of the time he in fact does exactly the same thing, but he attends closely to the social and psychological climate of the day, a climate which varies from year to year, place to place, and group to group within our complex society.

For science, for natural magick even, there can be no more admirable style than one of openness, bright lights, sleeves rolled up, all really fair. Every departure from it must be seen as reducing the strength of evidence. When the departure is taken in the service of some end other than understanding, as may indeed be necessary in many human contexts, the strength of the evidence declines. This remains true even when the ends are better served by some degree of concealment. What in fact happens overall when one cannot follow an open path from start to completion must be judged finally by that statistical skepticism of elaborate controls.

We cannot expect ever to understand procedure that remains intrinsically less than open. We might eventually prove that it works or fails to work, though even that proof is by no means cheap and easy.

These arguments of course are in the sunny spirit of the extended Enlightenment, an ambiance very congenial to science. But we do not live in the day of Diderot. One wants to recognize that fact and to admit widespread interest even in what old della Porta did not care for: sorcery. They are, above all, *hidden* powers that sorcery invokes. No arguments are likely to shake theorists who impute such occult layers to the world. Even as a matter of logic it is not easy to demonstrate the error of those who explained away the fact that the stainless-steel pins that held some healed fracture within remained visible in x-rays, even after the "surgeon" had assertedly removed them, spilling only chicken blood. The defenders of the procedure observed that the painful "essence" of the implants had indeed been removed, leaving only their physical shadow. A radiologist can hardly dispute such a theory. And if one wishes to argue against such postulates (except by the razor of old Ockham) he had better be able to show that the relief rates do not support extraordinary claims. Enter controls and statistics again, that slow and costly logic of skepticism.

The urgency of healing has always made it a source of bad theory. This is all the more true because it is a safe bet that most healing procedures, whether in the most scientific practice or in a frankly magical one, are unneeded. For what patients present are usually self-limiting conditions. Most medicine has other purposes than a cure, which in time comes naturally. Perhaps the most important is just the assurance that the condition is not something worse, with a poor prognosis. But it is to be kept in mind that a patient is a very complex system indeed, which science by no means understands in detail. When the manifold therapeutic effects of cortisone became known a generation ago, some scientific meetings took on the air of revivals. Witnesses testified all over the hall as to the rapid and disparate cures the drug effected; sudden freedom came to the halt and the lame. We learned then that endocrine systems can work powerful changes in many organs and tissues. The autonomic nervous system as well is capable of much control over normally involuntary bodily functions; the brash positivist denial that an adept can pass into a self-induced passive state of minimal metabolism has been refuted right on BBC-TV. That over-simple model of the complex human body had not taken sufficiently into account the powers of our inbuilt electronics. But there is no mystery here, only wonder. Plenty of hard-wired circuits cross from brain to brainstem to transmit command to the heart and to the breathing reflexes. So little do we know yet about the details of the relationships between mental and bodily states that we can afford to be tolerant of many claims involving them, secure in the hope that in the end they will be explained by mechanisms as yet unknown to us. But that is not necessarily to admit their existence without powerful proofs. These are even more strongly required, since the system is complicated and its initial state little controlled. We need not expect any forces to be at work save those that flow along the bewildering network of electrochemical pathways, conforming to, but surely extending year after year beyond, what we already know.

The human organism possesses a set of very sensitive amplifying systems. Some of these are chemical, capable of mediating gross physiological changes under stimuli as small as the entry of an invisibly small microscopic sample of antigen. Even a mere symbolic signal, properly impressed by light or sound on eye or ear, can work—provided that earlier learning has allowed the form to take on a powerful internally stored meaning. It is of course this fact that forces an ambiguous interpretation on many unorthodox procedures of healing by no means absent from often-used techniques in what we think of as scientific medicine. Placebo is powerful; the will to benefit offers much relief.

Many of the paranormal topics are not so much applied science (medicine is of course an important example of that domain) as they are purported accounts of the world as it is, what one might call science without regard to applications. The question of UFOs is a clear case in point. Here the key figure is not a patient, but a close relative of that variable human being: a witness. Witnesses are also remarkably complex systems. What they say is connected to the sensory input that gives rise to their testimony by a very complex and unpredictable set of links, links that certainly can join today's statement to events even as long ago and far away as the infancy of the witness. For language is the usual form of testimony (allowing a few gestures and drawings), and its categories have long been held in every witness's mind. But a lawyer could tell us much more about witnesses and their value than any physicist can.

The first issue is one of competence. Granting veracity, could the witness have learned enough from the experience to draw convincingly the conclusions he offers in testimony? The air crew sees the fireball as a mile or two away, even though it proves to be two hundred miles distant. They were not dishonest, nor even confused, but simply had no way to judge the distance of a light in the sky at night, unless they assume something of its physical size, speed, brightness, or the like. When a fireball is viewed as though it were an aircraft ablaze, its distance cannot be rightly gauged. In such cases perceptual physics is invaluable.

In more social matters, where the degree of disinterest of the witness, the requirements of corroboration by others, and more subtle questions come into play, no natural scientist is very fit by experience to judge. Here the more worldly are at home. One magician as knowing as James Randi is worth a dozen eager nuclear theorists. Once, a London group of academics was testing out the claim that a certain wonder worker could cause a Geiger counter to fire by occult means. The task was accomplished, to everyone's amazement, and the chart recorder that had duly marked the mysterious counts was examined later with care. The paper chart showed strong counts; indeed, the pen had been so strongly deflected that the experts could be sure that the amplifier used did not deliver enough power to move the pen so fast and so far. Energy had probably flowed in from another source. Choose your own occult pathway! The sober view is that the deflection was made by hand, by a spectator friend of the adept who firmly pushed the pen over at the right time. Nobody was steadily watching the recorder; it had been (naively) assumed that all the action was at the counter end. Such an attitude is totally that of the physicist; no magician and few lawyers or detectives would take such a trusting stance.

Much testimony is written and published. A model for the internal and external examination of printed sources was offered us by Larry Kusche in his account of the "media hoax" called the Bermuda Triangle. A book is of course simply a limited form of witness, whose testimony is fixed in time. Cross-examination cannot be used, but the evidence is open to detailed study. It becomes even easier to ask how the witness came to know what is placed on the page.

In either case, whether it is evidence offered by witnesses who were there, or by those who have mainly documents for their own sources, one method of testing is generally applicable. A step-by-step survey of how the statement came to be made on the basis of the experience recounted is the deepest form of analysis. Questions of motive and judgment remain secondary. The whole topic is a rich one, extending so far that it includes the methods of science themselves. The scientific paper is the written testimony of experimenter and theorist. We know how much we question every novel conclusion, and with what good reason—for most of them turn out somehow to be wrong. New territory is usually opened to science step by confused step. It is ingenuous to demand less from new claimants of grand domains of experience, long sought for and long and unconvincingly claimed, than we ask of authors who offer only a single new proposition within a well-explored discipline.

Certainly fraud, hoax, and cupidity are real phenomena often found in human affairs, especially when substantial gain (of fame or fortune) is foreseen. That gain is enhanced in our times by the very existence of widespread means of communication and publication. Popular books can sell tens of millions of copies, offering genuine wealth to the few really successful authors. It is certainly relevant to question motives, and to draw inferences about their influence on truth telling and completeness. But they ought not to be held decisive. A true statement can be made by a generally deceptive witness or compiler, and a false one by a person of indubitable integrity. The touchstone for truth is not the honesty of the speaker, but the evidence presented. Sometimes it can take the form of independent repetition, the replication dear to the experimental scientist. But this is not always a possible form of verification. What is essential is a sufficiency of objective evidence, although the criteria for that sufficiency are not simple or clear.

It is certain that unsupported statements by people with an interest in their outcome are not enough. Even in ordinary science, the person who puts forward a new result is by no means disinterested. The tale of Alfred Russel Wallace and his spirits is a perfect example, though drawn from the paranormal. The dry remark made about his stance applies more widely: "Personal conviction is a very fallible guide." The creator of any idea or result is almost surely a partisan of it. Truth is hard to win, and the sense of success is rewarding for any researcher. How can he or she be indifferent to the future of this infant brain child?

In normal science that bias is assumed; another worker must be free to examine the evidence critically and in new ways. Sooner or later the matter will either be dropped as error or become widely accepted. But only a long history of testing and success offers really firm support, while certainty is hard to gain outside of mathematics. We hold every idea with some degree of firmness; the more carefully tested, fruitful, and richly textured the idea, the more firm.

This spectrum of conviction is perhaps the chief difference between the parascientific enthusiasts and the scientists. The latter are willing to doubt, if only a little bit, even the conservation of energy. But in parascience the criteria are far too lax. Every claim of success is seized upon. The burden of proof is freely shirked by those who would put the evidence of a few witnesses, or of a compiler with a livelihood at stake, ahead of manifest contradiction of precisely verified regularities from dozens of contexts. The whole structure is thus weak.

Consider the marvels of the pyramid form. Assertion or even less—a mere suggestion—is the main basis for large claims. The matter is governed rather by a will to believe than a will to *doubt*. It is the fragility of evidence, not the fact of self-interest, which lies behind my pervasive doubt of most of the conclusions in most of the books about it.

Purity of motive is not a necessity for progress in science. Openness of method, skeptical reception, full publication, and much-tested data are more nearly requisites. An overall judgment can be made only when the overall view is clear. Nor is it right always to expect large discoveries; truth enters more modestly, a little at a time. By and by the fuller meaning becomes clear. In the account of the astral journeys of one subject (An Addendum on Double Standards) we learn that the unique adept moved away, so the experimenter was forced to give up on the tests. It would have been quite reasonable to expect a different behavior, more partisan and less detached, from the experimenter. If one had found a person whose perceptions could leave the body, as we are led to believe, for a distance of several feet (to read a paper lying on a shelf above the sleeper) it is unreasonable not to have put him to more tests. To do that an engaged researcher would certainly follow the subject "to Zanzibar or beyond." It is hard here to turn aside the suspicion that the experimenter himself was not passionately convinced; suspecting some defect, he let the discovery rest at the modest test he describes. His result, if true, was enough to shake all of physics—but not enough to demand of him a difficult rear-rangement of his time.

I feel a similar sense of guardedness in the millions of readers of all those elaborated books of unlikely marvels. The believers do not truly believe; they entertain the possibility as a diversion from the hard facts of daily life. Somehow the paranormal has a large audience that remains suspended between conviction and tolerance. Science knows such a feeling very well at the frontiers; but the history of its growth has been to watch the seep of certainty across the plains of doubt. Atoms were first speculation, then hypothesis, then convenience, and now hardware. So far, the paranormal has had no such growth. The oldest books we know are just as good a digest of the occult as the latest volumes of the specialists; indeed, skeptical critics in the past argued about as well as we do today on many of these topics.

In the end, belief is social. One who holds unique data cannot convince others in science; the data on the I.Q. of separated identical twins by Sir Cyril Burt were all his own, and most proved to be fictional. But the belief system that still supports hereditary intelligence does not much attend to any empirical data; its strength arises quite elsewhere. So it seems to be with most of parascience. It is not the self-interest of the authors that most vitiates their theories; it is the overall weakness of the inferences, and the ready acceptance by a public whose purposes in reading are broader than the test for truth, topic by topic. In science, too, the experimenter's natural self-bias is of no great consequence; the community will test claims to the point of satisfaction, over the long run at least, or simply neglect the claims until some process of confirmation can go forward.

Purity of motive is more a question of biography than of science. The rise of belief should not follow from mere admiration, as the depth of doubt ought not to come only from the judgment of ulterior motive. Larger stakes are on the table.

It is no surprise, though it is a kind of tribute to science, that the "miracles" of our day are for the most part a sort of modish natural magic. This is surely the recognition of the richness of modern science and the power of its partner, technology. Sorcery, with its appeal to personified powers, survives, to be sure. But most of the widely known claims closely follow the model of science. They shadow the subtle interactions and the invisible structures—even the statistical tests—that science and long experience have teased out of this world. Quantum mechanics, with its subtle and surprising limits on the analysis of atomic events, is made to serve as rationalization for poorly demonstrated claims of telepathy and power over remote matter. The implied mathematization of this fuzzy world is seen as no barrier, but a way to prestige. The wonders of the EEG and EKG are made the basis for a doubtful theory of biorhythms. A firm belief in the veridical nature of one scholar's version of very old and often obscure texts is made the vehicle for a wild extrapolation of the possibilities latent in Newtonian celestial mechanics. The original treatise of Velikovsky was so little concerned with the details of science that the molecular combinations of hydrocarbons and carbohydrates were conflated, even though therein lies the enormous common-sense difference between butter and bread, or gasoline and wood. For such scholars in the beginning was the Word, not the category it named. The careful grids and trowels of the archeologists are condescendingly subsumed, and all is given a fantastic and hyperbolic gloss, tending to suggest, perhaps without saying so, that the old folk, those lesser peoples far from blond Northern Europe, could not lay masonry or carve monumental statues without celestial aid. On the other side, the same painstakingly detailed science is loaded with a charge of unexamined wonders, like the Atlantean columns off Bimini, which appear to be old cement barrels when looked at closely without rosy diving goggles.

That real science and its real wonders acquire such an airy comet's-tail of lesser matter is not without its warnings for science. We have let the wonder become too deeply buried in the mechanisms and concealed in an economical style of publication, too austere for its own good. A serious and responsible popularization is needed that sets a style closer to the best human experiences of science than to the letters in *Nature*.

Perhaps Frank Drake's wonderful prospects for finding thinking and erring beings elsewhere, not by magic saucer but by real radio dishes, offer a glimpse of science and exploration to come. The giant squid, perfectly real in the depths, a few specimens modeled in incredible full size in our museums; the basking shark, which in death and decay looks like the sea serpent; the hint of a truly giant octopus not yet fully known: all these offer more in the way of real monsters than the dubious Yeti or Nessie or Sasquatch or sea serpent. And delvers in the Rift have found hard bony evidence of the true "wild men" who were our forebears a million years ago, not monsters but our antique selves. The cetaceans intrigue us with their faint hint of kinship. All these offer evidence that reality can still combat fiction as well as it ever did.

But it takes more than dry briefs to do the full job. Science and technology require their artful exhibitors, too. As travelers leave Orlando Airport, a cautionary pair of road signs confronts them: one points to the fantasy theater of Disneyworld; the other offers the Space Center at Cape Kennedy. "This way," says the latter, "to reality." And you can see Saturn stages lying there, huge cylinders near the vast Vehicle Assembly Building, a giant's country all real.

Of course, our science has its deficiencies and its defects. First is the troubled state of modern world. The problem is imputed to technology, though surely the responsibility is shared by the structures of social power. The faults of character in science are evident: arrogance, pomposity, insensitivity, the natural faults of preoccupation and highly specialized intellectuality. Complacency is not easy to maintain in the current precarious fiscal state of science, and it must not rule our judgment any more than our hopes. We can be sure, for example, that some valuable part of the fabric of science we now accept will be held naive and erroneous by those scientists who come after us.

Just which part we are not given to know. To me it does not seem likely that it will come in one of the "paranormal" fields. But it is no betrayal of science to admit the certainty that somewhere we are wrong. The challenge is to show where, not to turn the wish for immortality or omniscience into a pseudoscience. If I am asked to make one guess, it is that we will come more to appreciate the steps toward science made long ago, perhaps even by our cave-dwelling forebears. They might turn out to have known a great deal more than we now credit them with, perhaps something like writing, calculation, and certainly naked-eye astronomy. But they will have learned it all, as we do, without extraterrestrials as guides, by hard thinking and sharp seeing. Human beings painted the caves, no one else.

Those cave paintings are wonderful, but like everything we know they are not too wonderful to be true. It is their reality that gives them wonder, and while there will never come a time when some of us will not wish for more than we can have, the happiest of us will wait confidently for other tangible finds. We treasure the cave at Altamira where a century ago a little girl first saw the great painted bison. New caves will be found, year after year, in lab or clinic or sky or ocean depth, or even in ancient markings. That is the promise of real science, which cannot allow wish to rule mind, but nonetheless finds unendingly wonderful things.

The Simulation of Intelligence

Mankind once distinguished life from non-life by purposeful motion, until seemingly magical machine simulations voided that criterion. Here and now, in an age of computers, how shall we understand the uniqueness of life? And what will we have to say of our own minds if machines come to think?

The eighteenth-century simulacra of life established once and for all that it is not by intricate, purposeful motion that living matter is unique. We see now that nor is life unique in having many internal states, for our electronic machines have that. The uniqueness of life is not in the irreversible separation of parts of the organism either, though 40 years ago, this was a life/non-life distinction: in any machine take a part out, and put it back again, and the machine works. Take a part out of a man, and put it back in again; he does not work. By now this seems a very naive criterion, wrong on both sides, but 40 years ago it was a true distinction.

Introspective self-examination? We do not have much of that in our machines. We do not now have machines that reflect even in a small way upon their inner states. They ought to, and certainly if machine design is done in a philosophical spirit, that is the one property that ought not to be left out. Even now, the mere existence of memory-rich machines makes it extremely hard in reality to ignore inner states and accept any stimulus-response theory, any input-output view in which the processing takes place in a "black box" about which nothing need be known. This is not likely to be adequate to represent a machine, a mammal, or a human mind. So much depends on the memories on a computer's discs that the black-box approach of the behaviorist never occurs to a machine repairman. The machine is a function of its history. The repairman would be quite lost even in the presence of a simple machine without some sense of what I would like to call the introspective concerns of that machine—

No, that is not quite true yet; I have jumped the metaphor one level more than I am entitled to for the present. But the gap seems only an economic one. In the efforts ahead in artificial intelligence, I would demand, as would an old psychology test, that some trace of self-examination be provided. A machine lacking that goes far toward overlooking the principal philosophical differences between behavior, and behavior with consciousness.

Machines and Mind

We must now ask the question: "Can a machine be comparable to the human mind?"

I fully believe that we will find no barrier to success in any aspect of machine simulation, in any feature of full machine reproduction of any canon whatever of human mental states. I think that what looks to be true will probably turn out to be true; namely, that the human mind can be described as a slow-clockrate modified-digital machine, with multiple distinguishable parallel processing, all working in salt water. Yet I will say that all those subtleties philosophers and artists talk about, that make life unique and distinguish it from non-life, are true after all, foolish as they may sound when you are young and enthusiastic, and have a naively positive view of science. I offer a position of tension, a determined yes, machines will simulate life, but . . .

Mind as Questioner

We begin with a statement from a considerable scientist of our own generation, Sir Fred Hoyle, who very stoutly said to me once that the only important thing in science is to ask a good question. An answer will appear; it is eternally implicit in the process. The person who asks the right question is the truly great scientist. That is exactly what is written on Cantor's gravestone in Latin, so I can cite even a better authority than Hoyle, namely Cantor himself: the question is the essence of science.

Then the notion that thought can be judged by the answers to an interrogation process is much too low a level of test for respectable mental behavior. I will not be satisfied with the machine that claims to simulate human mental life until it asks important questions, raises new problems. That is what I regard as fully human behavior (or at least part of it): not merely answering, however cleverly, however subtly, however neatly picking up literary allusions to Shakespearean sonnets. Merely answering questions is not going to do the job. I can almost conceive of a question-answering machine, but a question-asking machine? That is more of a challenge. On the other hand, such a machine, because of imperfections, because art is long and life is short, will lack some of those complex essential features which together made a whole enduring social being out of ourselves after five billion years of Earth's history.

But wait: there are strongly limiting constraints on this oracular delivery!

First, a machine simulating the human mind can have no simple optimization game it wants to play, no single function to maximize in its decision making, because one urge to optimize counts for little until it is surrounded by many conditions. A whole set of vectors must be optimized at once. And under some circumstances, they will conflict, and the machine that simulates life will have the whole problem of the conflicting motive, which we know well in ourselves and in all our literature.

Second, probably less essential, the machine will likely require a multisensory kind of input and output in dealing with the world. It is not utterly essential, because we know a few heroic people—say, Helen Keller—who managed with a very modest cross-sensory connection nevertheless to depict the world in some fashion. It was very difficult, for it is the cross-linking of different senses which counts. Even in astronomy, if something is "seen" by radio and by optics, one begins to know what it is. If you do not "see" it in more than one way, you are not very clear what it in fact is.

Third, people have to be active. I do not think a merely passive machine, which simply reads the program it is given, or hears the input, or receives a memory file, can possibly be enough to simulate the human mind. It must try experiments like those we constantly try in childhood—unthinkingly, but instructed by built-in mechanisms. It must try to arrange the world in different fashions.

Fourth, I do not think it can be individual. It must be social in nature. It must accumulate the work—the languages, if you will—of other machines with wide experience. While human beings might be regarded collectively as general-purpose devices, individually they do not impress me much that way at all. Every day I meet people who know things I could not possibly know and can do things I could not possibly do, not because we are from differing species, not because we have different machine natures, but because we have been programmed differently by a variety of experiences as well as by individual genetic legacies. I strongly suspect that this phenomenon will reappear in machines that specialize, and then share experiences with one another. A mathematical theorem of Turing tells us that there is an equivalence in that one machine's talents can be transformed mathematically to another's. This gives us a kind of guarantee of unity in the world, but there is a wide difference between that unity, and a choice among possible domains of activity. I suspect that machines will have that choice, too. The absence of a general-purpose mind in humans reflects the importance of history and of development. Machines, if they are to simulate this behavior—or as I prefer to say, share it—must grow inwardly diversified, and outwardly sociable.

Fifth, it must have a history as a species, an evolution. It cannot be born like Athena, from the head full-blown. It will have an archaeological and probably a sequential development from its ancestors. This appears possible. Here is one of computer science's slogans, influenced by the early rise of molecular microbiology: A tape, a machine whose instructions are encoded on the tape, and a copying machine. The three describe together a self-reproducing structure. This is a liberating slogan; it was meant to solve a problem in logic, and I think it did, for all but the professional logicians. The problem is one of the infinite regress which looms when a machine becomes competent enough to reproduce itself. Must it then be more complicated than itself? Nonsense soon follows. A very long instruction tape and a complex but finite machine that works on those instructions is the solution to the logical problem.

One cannot say that the slogan does justice to the extraordinarily complex structure of the "DNA dogma," in which we have the DNA "tape" and the mechanisms whose instructions are encoded upon the DNA That microbiology is positive science. The rest is slogans, but the slogans are powerful ones. They establish in the language of computer science at least the logical possibility we see realized in life: the self-reproducing system. And they do so by a means that makes evolution possible.

By the time I have described all these attributes of a machine to simulate life, I suggest that I have described something which is more like the human mind than it is like our image of the machine. Yet life is not mocked. Mechanical this simulation of life will nevertheless be—as clockwork motions are mechanical.

Nobody believes today that the slightest insight is offered into the nature of organic life by the fact that a clockwork mouse can run across the floor. But there was a time when the wisest people thought that the existence of a simulation of purposeful motion had to be demonstrated. Now it is only a proposition in the transfer of energy and forces on the floor, and we know too much to accept that as a special property of life.

In the same way, I think, we will see man creating a simulation of other aspects of life, yet without creating a *doppelgänger* of the human mind.

Aspects of the Simulation

In one respect, the human mind works in a way that hints at the methods of a computer. Abstraction of perceptual information comes very early in the human mind's processing of sensory data; it begins at the level of the peripheral organs. Images are not transmitted from the eyes to form a picture in the brain so that an homunculus in the cortex can look at the picture to decide what it is. Descartes in his day found that idea problematic. Since Hubel and others, we know that aspects of the image are coded right away—abstract items like diagonal lines. An object running out of the field of view might induce a signal that appears at some point—I don't know where—in some junction box behind the lateral geniculate, or some other complicated anatomical region. But quite early the inputs are transformed, not into point-for-point spatiotemporal representations, but into a more abstract language, one suitable to the machine.

There are, though, chasms between computer design at present and the apparent workings of the mind. Contextual information concerning the meaning of symbols seems indispensable for any economical machine program that has a chance of being "intelligent." The notion that it can all be done by logical manipulations of these symbols, without any reference to their contexts, is inadequate, for the presentation of many distinct contexts to the central processing device is characteristic of all living beings, characteristic especially of human beings in a social environment of extraordinary richness. But it is absolutely uncharacteristic of the kind of machines we now design and build. Filling that gap will perhaps be one of the greatest steps.

New Genesis

When machines acquire a diverse, self-knowing, active behavior—which is question-asking—and can evolve—which is based on a society of mechanisms out of which some sort of language grows—no one will need to ask if our salt-water machine works in just the same way. We do not argue about the fact that a little mechanical mouse runs along the floor more easily on wheels, not on legs. It is locomotion, all right, differently realized.

We can have creative and personal machines, structures which will act and reflect, and they will share to a degree the attributes of the kind of persons we ourselves are—attributes which we gained by evolution. If the machine does not share those properties, it probably will fail to attain the special high functions of the mind which I have described. My argument arises only out of a sense of the deep unity of the world, a unity which does not demand similar structure for similar functions, but does demand the kind of coherence to which I offer homage.

I do not myself expect to see, but the world may well witness, four kinds of "life." First, there is our own kind, life continuous by descent over some four billion years, with a heritage of certain antique biochemical ferments. There will be a second kind, born in a glass test-tube. It will have a new and discontinuous genesis, on quite a different path of information transfer. Perhaps the amino acids will be the mirror images of our own! There will be life of still another sort on another planet. This may be only the Martian fossil plants of the cold pole, or it might even be a fully-conscious life evolved around another star than ours, made known to us by a marvelous microwave link. Finally, there will be a synthetic device, far from biochemistry. It will not have been designed from the beginning by some human programmer, but begun at a higher logical level by humans, to evolve its subsequent internal hierarchies out of its own structure and experience. Once complete, it will behave in the ways I have outlined, in a manner akin to our own nature. It will not be the same as us. But will it be wholly different? By the strength of analogy and faith in the plenitude of the world do I foretell these beings; not by any surer insight.

The father of those that know, Aristotle, wrote: "Mankind is the measure of all things. The hand is the instrument of instruments, the mind is the form of forms." He was right, not because man is separate from and above nature, but because human beings are part of nature and have been engendered by nature over several billion years. We can expect then to take from our own behavior lessons which may one day lead us to the synthesis, the wondering synthesis, of machine beings, somehow alike and yet very different from that consequence of a cunning, age-old, half-unerring, yet half-random, chain of evolution.

The Actuary of Our Species

I t is the job of the actuary to look with a cold eye upon life, upon death, and try to strike fair odds. It is perfectly plain that we are in a special position here, because the genuine actuary derives this credibility from plentiful statistics, from the fact that while no human life is predictable, human lives, as a class, satisfy the weak law of large numbers. They can somehow be predicted; people make money on mortality in the life-insurance trade. If we are, as we now think, the only species of our self-aware kind, then actuarial possibilities do not even come into question. With that slight apology for my statistical metaphor, I want first to outline the central point: science has been able to make a broadly reliable judgment of the place of human beings in space and time, a grand assessment of just where and when we are.

Homo sapiens in Space and Time

It follows from that assessment that we can offer roughly persuasive predictions and assertions about events we have never seen, never experienced, at least in comparable degree. First of all, consider our place in space. Barring a small fringe of true believers, that is nowadays a subject met with considerable objectivity in the public view. Nobody cares much about it. And what we have to say is commonplace. But it was not always so. We are told that the philosopher Anaxagoras was exiled from Athens because he taught that the sun was a fiery sphere and as big as the Pelopennesus. Since the sun was ascribed immortal properties, his claim was regarded as impious. Perhaps he had political enemies as well, so out he went. We now know his views to be sound, if rather understated, and nobody now seems to be excited about that very much. By the time of Newton it was clear what our position was in space. We then understood what I still think to be one of the priceless jewels of scientific knowledge: namely, the clear fact, which we can demonstrate, that the sun is a star and all the stars are suns. That class is one class which nevertheless contains both the hot disk that we see every day, the source of human life, of all life on earth, and also that collection of bright points, twinkling in the night sky, their utility at best for prognostication or navigation, but not for anything else. The fact that these two are the same, and that they differ solely because of our relative position of view, is a remarkable truth. It is the kind of truth which is inescapably present in many (but not all) results of contemporary science, not arguable, but permanent, hardly subject to the general critique of the ebb and flow of thought. The historians and philosophers do not often talk about that. They prefer to discuss more theoretical matters, abstract, even mathematical issues, so complex as to partake of more subjectivity. But I doubt very much that the time will come when reasoning people will think other than this. Huygens was one of the first to publish these arguments, but the view was widespread in the seventeenth century. It was not unknown before then; there were hints and predecessors ever since Copernicus, perhaps since Nicholas of Cusa in the fifteenth century, but it does not much matter. By the time of the speculative philosophy of the eighteenth century, based firmly on the huge success of Newton, Halley, Euler—say, Kant—it was already conjectural that the little patches of light that the few astronomers had seen in their telescopes which were not stars, elliptical in form in the feeble telescopes of the days before photography, were galaxies like the Milky Way galaxy. Therefore we lived in an island universe, where Milky Way after Milky Way was scattered through space, each an archipelago of suns. This was still conjectural. It was hotly debated over 200 years, to be settled only in the first half of this century. Now we know it to be so; almost no one doubts that the extent of space we have to deal with is measured at least in the billions of light-years. That is a truism of our day, which every child knows from television.

So much is only a static backdrop. It says nothing about change. One point does remain, a very important point. But it really requires some physical theory to appreciate it before I draw the next conclusion. Judging from space alone you would not think that time was involved. But we know that through the velocity of light, an inescapable and absolute limit of velocity in which we have high confidence, if not absolute certainty, any extension in space implies extension in time, through the connection made by light. If we can see things that we argue are billions of light years away, then (unless an extraordinary special creation was made with the light set halfway in space, which opens a way of altogether getting around any philosophical arguments about the past) you could not admit time could be shorter than the transit time of light across space. These two views connect to give rise to a truly profound gulf in space which has been occupied by human thought only quite recently. Newton did not know that light had any special role in velocity. He conceived of possible infinite velocities; all scientists really did until the turn of this century and Einstein. Newton was quite prepared to accept, and indeed worked systematically and quite well, on building a historical chronology of his own, to patch up the Bible by using the documents of other Middle Eastern cultures and of antiquity to give us a firmer chronology. He published a formidable book, in which he established a chronology, tolerably in agreement with other theologians of the day, the famous 4004 B.C. Day of Creation. (That date is not to be taken as a dogmatic final result, but just as one result by some particular scholars with a lot of influence on King James's Bible committee, and so included in the Book.) It was much debated, just as geological dates are debated still. Newton himself believed approximately the same thing: our earth was created some 10,000 years ago, give or take a few thousand years. This time scale did not change during the Enlightenment. Even the Enlightenment scholars first saw no way of getting at it and mostly they let it go. Of course, as science grew there came a succession of thinkers on geological time. A conspicuous and early one was Halley himself, Newton's friend and collaborator, who computed from the salting of the sea that geologic time had to be measured in hundreds of millions of years. It could by no means be 10,000 years. Generally, unbelievers and the Enlightenment understood something was very wrong, but they were unable to present any secure physical age to substitute for the learned historical one.

    With the growth of systematic geology at the outset of the nineteenth century, the whole situation began to change. Still, time was not marked well in geology. As Mark Twain made deliciously clear, the geological calculations, especially in those days, had to proceed by naive linear extrapolation. You found that the sea bottom near Dover, say, would go up a fraction of an inch in a century or whatever it was, and then you said, "Well, since it's up a thousand feet and you can multiply linearly, anybody can exhibit wonderful depths of time." But of course this is an extremely imprecise argument; it gave rise to much ridicule. Twain's is one of the best. He points out that the Mississippi River can be shown to shorten its course by washing out oxbows, cutting across them at the rate of a mile and a half a year. You can show that a certain time ago it must have been sticking out above the Gulf of Mexico 50,000 miles! It is clear that linear extrapolations of this kind are not very happy. One can be more serious than that, too. For when Darwin came on the scene in the middle of the nineteenth century it was the great time, the Victorian equivalent of the Enlightenment, when the geologists had become the implicit champions of the secular. Darwin took on the same role when he said there was no special creation even for living things. (But the geologers had already had their opprobrium and their triumphs, for example, at Cornell University. It was founded as a nonsectarian school, much criticized by its neighbors for its godlessness, lack of compulsory chapel, and the rest. The library of the geology department at Cornell is the only decorated chamber within the frugal original foundation, the only lovingly funded portion, where the architect was allowed a little license to spend a few thousand dollars to decorate, to make something handsome and grand. Around the geology library are cast-iron reliefs of the then new paleontological reconstructions; I look at them to realize that here is the spirit we see in the proud Roman design of the Massachusetts Institute of Technology, or the elegant design of many nuclear laboratories of the 1950s. A certain air of importance is imputed to the scientist by his surroundings.) But the Victorians did not have a good time estimate. They knew it had to be pretty long. All the biologists kept saying was that geological time is infinite. That was the best Darwin could do. (He was poor in arithmetic and either did not quite grasp the formal meaning of the word *infinite* or at least did not care to be precise about it.)

A struggle began between the physicists and the evolutionists of the day over the question of time scale. Lord Kelvin, a prodigious worthy, a grandee of science, the inventor of the laws of thermodynamics, more or less, a man with no small sense of the pronunciamento, insisted from good thermal calculations that the sun and hence the earth could not be above a few tens of millions of years old. And indeed within his lights he was right. It was only at the turn of this century that physicists had the scope to understand that indeed the geologists and Darwin were right all the time: geologic time was a hundred times longer than Kelvin said, which was all they meant by "infinite." They wanted some billions of years, and that is just what they got.

Lord Rutherford has a superb reminiscence over the point. As a young and enterprising physicist from the colonies, teaching at McGill University, he was invited to London to give a distinguished lecture on his new work on radioactivity. He was very proud that he could carry about in his pocket a rock, in those days one of half a dozen rocks on the face of the earth about which one could say: we know the age of this stone. He knew that the old Kelvin was in the audience. He was very worried about what he would say because he had shown that Kelvin was manifestly wrong. Kelvin had calculated the age of the sun from the maximum energy release possible under the use of gravitational forces. Of course, no chemical fuel will do; if you tried to burn a mass of coal or gasoline the size of the sun—imagine the oxygen is magically supplied—it will only last, at the sun's rate, for a few hundred thousand years. The nature of the sun is not that it puts out so much energy, it is not that it is so bright. It is bright only because it is so big. In fact, the metabolism of the sun is a good deal slower than your metabolism or mine. Per unit mass the sun develops only one part in a few thousand as much heat as the metabolism of a human being; the sun is not much more active metabolically than a piece of cold granite. The trick about the sun is that its output endures over a very long time without an external supply of fuel: not much power per unit mass, but sustained for a long, long time, the indispensable condition for our slow evolutionary rise. Kelvin before 1900 had no idea of that at all. Ernest Rutherford had to explain that Kelvin had been ignorant of radioactivity. He feared flatly to contradict the old boy, who would sit up front, perhaps to pound his stick on the floor when a misstatement was made. But Rutherford happily hit on just what to say. Nothing is more important, he said, than the great work of Lord Kelvin, who had given an absolute limit to the lifetime of the sun under the assumption that no new energy source was available. Now we had found radioactivity, just as this remarkable prediction had implied. The old man nodded cheerfully off to sleep.

In fact geological time is limited, the sun's age too is limited; now we know even more remarkably that the galaxy's age is limited as well. All these points show us that there was some kind of cosmic birthdate. That need not, to be sure, mark the birthdate of all that is, which is what theorists now like to think. They are probably wrong now, as they were before. And I will not join them in that. But the existence of the starry universe in which we now live is definitely finite. The galaxies all had a birth; we know that date crudely; I do not think we will ever change our view. We might somewhat modify the estimate. The age of all the discrete objects the astronomers see, all the stars and galaxies to the limits of the telescope, is in the order of 10 billion years, a couple of times the age of sun, moon, the meteorites, and the earth's fabric itself.

The Tree of Life

The geologic record is plain: there has been life on earth, in an unbroken genealogy, one biochemical kinship, for more than 3.5 billion years. Most of that time—say about 2.5 billion years or more—all life was a proliferating, complex low mat of colonial blue-green and green algae, and the associated microorganisms. Life never left the extreme lowlands; life was a phenomenon of the shallow waters, whether salt, brackish, or fresh. It spread over the tidal flats and the stream banks, bogs, and marshes of the forming continents. Never did it become actively mobile; but steadily its biochemical powers grew. Very likely it began to change the entire chemical environment, modifying the atmosphere itself to enrich it in time with oxygen to near-modern value. It hoarded iron, calcium, and phosphorous, which once were mere dilute traces in the weathered rocks and the volcanic gases. But life endured everywhere in the shallow wet world.

Of course there were ecological catastrophes in plenty: floods, volcanoes, tidal waves, even prodigious meteorite impacts. Yet life spread. Finally it pushed beyond the shallow waters, first for the shelves and then the depths; it left the water margin itself in time, beginning some half billion years ago to carry its essential internal fluids onto dry land. A couple of hundred million years before that land invasion, out in the shallow seas and the coastal waters, mobile life had begun, the first animals with structures strong enough to allow easy locomotion. Plants too gained the strong polymer framework which would allow tall stems, sun seeking beyond the edge of the lapping waters. Life now covers the earth's surface from near-pole to near-pole, and from the darkness of the depths to bright Alpine snows. In all this time, though species have come and gone in their billions, the fabric of life has remained continuous and enduring, by virtue of its diversity and steady reproduction, not by a mere rock-like toughness. Rocks, all rocks, erode.

What is striking is that the system of life has plainly withstood every environmental change during a couple of dozen orbits around the galaxy. Not the cosmic environment, not internal changes in our sun, not even the more intimate complex unfolding of earth, both of gradual and of catastrophic change, have ever brought an end to life. The extinctions which are so tantalizing a feature of the record of the past—What happened to the dinosaurs?—have never been general. The population as a whole, the overall biomass, may well have changed drastically; any given species, and many whole orders, even whole phyla of living forms, have come to an end. But the living fabric, which has been raveled, worn, even ripped, has never been wholly severed. The tree flourishes though the leaves fall.

Spectacular changes, earthquakes, tidal waves, huge meteorite collisions must have been more or less local in time and place, at least limited in their effect. The more widespread changes, changes in the solar photon, in the ambient temperature, or in the chemistry of air and sea, have mostly been so slow and so much within the range of adaptability of life forms that they have not brought life to an end over all that time. There is plainly a powerful feedback loop at work. When the air lacked oxygen, anaerobic cells flourished. As toxic oxygen grew in abundance, the cells found what had been a dire poison to be the very stuff of opportunity of increased vigor. That is the context in which human life, the life of a few species— but most novel ones—is to be seen. Our distribution is certainly wide, from pole to pole, by land and sea; our response grows rapidly, once we see danger ahead. It seems probable that we have, or will soon have, cut our ties with natural selection, with slow-driving forces of organic evolution. For better or worse, we have taken our long-range future under some sort of control.

So much everyone believes. We need to look at the endings which remain possible, in the domain of the environment, and in the biology of our species itself. Above all, we need to ask what the powerful culture which has given us some freedom from the old laws of change ironically entails as a source of novel danger to the longevity of the species.

Mother Earth and Father Sun

The long past has evidently witnessed cataclysms on a scale beyond our historical or even human experience. The modern eruption of Krakatoa is just about the largest volcanic explosion we have ever read about. In late August of 1883 that island volcano in the straits between Java and Sumatra blew up fearfully. It flung out overnight about twenty cubic kilometers of rock as dust and larger chunks, leaving a caldera three or four miles across. Its tidal waves inundated the nearby coastal towns with a wave 100 feet high, drowning 35,000 people. Its high-altitude dust spread around the earth, yielding blue moons and red sunsets for a couple of years, during which the world temperature declined. But the long-dead volcano whose cold caldera is now the wide green pasture called Valle Grande near Los Alamos, New Mexico, spewed out in its time ten times more ash and pumice, leaving a collapsed opening much larger than that of Krakatoa. We cannot say that it happened all at once, though that is quite probable; the event took place about 1.5 million years ago.

The point is clear, all the same: catastrophic eruptions, earthquakes, hurricanes, even tidal waves are all localized in effect, even allowing for the great variation possible over geologic time. There is no sign of internal energy sources which could be mustered to modify the earth as a whole. A city, a nation, even a whole oceanic culture could be ruined by such disasters. But the species can hardly be at risk. A tidal wave might damage the coasts of a whole ocean, with London, New York, Cape Town, and Rio all heavily inundated. But Madras would hardly know of the event without its tide gauges and barographs. The people of the high plateaus in Colorado or Kenya would remain largely unmoved. Life, and the life of humankind, is too widespread to fear such an end.

Consider the Universal Deluge, the Noachian Flood of Scripture. It is not likely that the waters were in fact available for a worldwide flood. The Book is building myth out of limited experience in the ancient lands between the rivers. That story is compelling in imaginative force, but the geologists have found hard evidence of real events almost as striking, almost as far beyond mere recorded experience. (This story is in fact yet not quite sure, but it is based on strong new evidence.) It appears that the whole Mediterranean Sea itself, Mare Nostrum of Rome, was at no great time in the past not a sea at all but a low desert, a huge arid salt-covered basin like the Mojave Desert, though much larger. This was only six or eight million years back, a time before our hominid kind evolved, but certainly witnessed by our forerunners, the apes. After a long persistence of this desolate barrier between Europe and Africa, one fine day a high barrier at Gibraltar finally gave way. Over that huge lip, over high cliffs, the green Atlantic Ocean poured into the dry low Mediterranean basin, as the Colorado River poured into the Salton Sea in California on a merely local scale before World War I. The flow was prodigious; the falls of Gibraltar were scores of miles wide and a mile high. In a hundred thousand years the sea was filled. A couple of hundred million years farther back, the Atlantic itself was only a narrow rift, like the Gulf of California.

The earth indeed changes, but the changes, dramatic to the point of hyperbole, are nevertheless coherently within the domain of our science. We need not adopt a narrow methodology, like the uniformitarianism which could characterize a less mature geology. Flood and fire are not to be excluded; all we need is evidence for them, the linked circumstances of a coherent view, and the familiar laws of matter in motion yield extraordinary novelties. But life is more persistent than mountain or sea.

Take the kimberlite pipes, where diamonds are mined. They appear to be a sort of cold volcano, suddenly arising without warning, anywhere, even far from the rings of fire which include most volcanoes and earthquake epicenters. But they do not threaten our species. It seems possible that one fine day a square mile or two of rock could erupt to fly high into the air, the fierce jet from the deep hot mantle of earth spalling the rocks above, to fall back down over a ravaged countryside. It has happened a thousand times or so before, over all the time we know. A city might go with it, as cities have fallen within the last decade from mere earth trembling, but the event remains local. The diamonds are a sort of rainbow residue of the disaster.

Of course, even larger changes are to be expected, and even read from the record. They are generally slower, perhaps taking millennia. The most evident is the worldwide climate change which brings on polar glaciation. Our species is a creature of the glacial age: we hunted woolly elephants across the tundra of France not so long ago. That was skillful indeed for a bunch of cunning primates who went bipedal in the hot African savanna not so long before that, as the earth counts years.

In spite of their many claims, I hold that the climatologists do not yet know what is the real cause of such big changes in climate. In any case, even if London and Moscow and Peking lie as ruins under a mile-thick layer of ice in a thousand years or so, that in itself can hardly end the species. The coasts will widen as the water leaves the seas; our posterity might raft freely down the Baltimore Canyon across the plateau of the old continental shelf. Even if the change is lightning-swift (some hold that the ice cap might form in a century, though most expect a slower change) we would surely survive it in part. The population might decline, but hardly to zero. There are hopeful engineers who imagine we could defer or modify the climate change, by damming the Bering Straits to confine the coldest waters, or by dusting over the new ice fields with soot to melt them in the summer sun, or in some other way. These are possible delaying tactics; it is not so likely that we can always manipulate clever triggers against such big events, and real macroengineering has shown no signs of viability. We humans still build small perforce; any single well-known mountain dwarfs the whole history of human rock and earth piling, any big crater our hole digging. We are still geologically small, and so are all our works. Rather it is our diversity which seems to mean safety here. Biology itself works in no other way, and we humans share it. Our spread from pole to pole, from alp to atoll, seems to assure the future at least for a good share of our whole genetic and cultural treasury.

One can imagine even chemical changes of worldwide moment. Volcanoes might exude carbon monoxide or some other poison. But the air would quench it pretty well. More subtle changes, like some new substance in small quantities, which would cause the ozone layer to dwindle—as it is feared too much freon might do—could affect earth habitability in the large. Against this sort of mysterious effect it is clear that the long geological record of continuous life is our best argument. It has never happened, or at least not beyond the resources of the fabric of life to maintain itself.

Ozone is one small-quantity feature of the world which seems all but necessary. We depend—our crops as well—upon that thin stratospheric layer of ozone to screen out the damaging harder photons of solar ultraviolet. Less ozone would indeed modify life below, though again probably not so far as extinction. After all, we spend only a small fraction of our hours of life outdoors. And farmers could learn to do most of their work in twilight, or even by night. No, not even ozone loss would do us in. The limit would be set by the domestic plants; but they surely vary in response. Some new crops would grow. Just the same, ozone is perhaps the rarest of the known substances on which we heavily depend. The whole earth possesses about one billion tons of this active substance.

I have largely avoided the quantitative till now. But the figure of a billion tons of ozone provides an exemplary case. Human effort produces cereal grains, wood, petroleum, coal, iron ores, stone, sand, gravel, and cement, all major bulk products, in the rough amount of a billion tons each per year, some rather more, some rather less. That—with their waste products—measures perhaps what we can add to the environment, bar some important unusual trigger, like radioactive emissions, which are potent against life in small quantities. But this threshold for human effect on gross features of the environment seems a good guide. There is besides ozone no other rare atmospheric ingredient of known importance present only at the billion-ton level. (The oxides of nitrogen might be included, though their effects are unclear.)

The next rarest substance of world geochemical importance is perhaps carbon dioxide. We exhale it, as do all our fires and furnaces. Plant life depends upon it. But the natural air contains it at a level a few hundred times higher than ozone. We only now begin to modify the carbon dioxide by amounts which compare the natural variations of the span of earth history; that may become climatologically important, but it also does not seem to spell an end. The system can adjust to likely changes over a modest time. It presages only a transient crisis. The oceans provide a reservoir of the gas in solution, and a big if slow source were the atmosphere deficient. Climate change might mean worldwide economic change, but not the end. Human actions begin to approach the scale of volcanic output in some substances, and especially in the heat we release through fire. Perhaps that is the clearest sign of approaching limits on human population. Surely they are not sharply defined by so vague a match; no one has put up a model of catastrophe from such a cause. It can well be argued that we are approaching a limit: ten or so billions of industrious (and industrial) people may mark the prudent end of our growth, whose many signs we now read.

Intrusions from Without

The vacuum of space isolates us on our big planetary spaceship—obviously not completely. We remain creatures of a majestic potentate out there a hundred million miles in space: the great nuclear reactor of the sun. We control it not; yet it controls our life. An irregular cycle of poorly known nature and origin modifies its output in details; we call it the sunspot cycle. There is good evidence that those spots are correlated at least weakly with broad changes in earthly weather. The most successful—but still uncertain—theory of the ice ages sees the glaciers as triggered by perpetual small changes in the earth's orbit and inclination caused by the big planets. These might enhance winters or summers. Some have found causes in even more remote features of the cosmos, in the great cosmic dust clouds among the stars, into which our sun might drift as over two or three hundred million years it orbits the Milky Way. All these proposals are plausible to a degree; none is proved. What is clear is that humanity was a creature of an epoch of glaciation; whether we arose because of the ice or in spite of it is not clear. But it seems pretty sure that neither the sun's inconstancy—if it exists except in detail—nor the tiny orbital dance has ever threatened to rip the fabric of life, whatever they might have meant for this or that species. In short, we cannot well change these phenomena; but perhaps we can adapt to them, growing our crops in the tropics instead of Montana, on the plateaus instead of the coast, blackening the snow fields, or whatever the realities demand. No ultimatum there.

Nor is there much sign that the slow orbit of the sun among the stars of the galaxy—over the galactic year of some 300 million solar years—sees much outside influence. True, we may have come from time to time within the sphere of influence of a great star explosion, a supernova event. There is suggestive evidence that the very origin of the sun was in part mediated by a nearby supernova, whose nuclear debris found entrance into the more normal gas mixture which became the early sun. Since then, the event has not been repeated, or at least not with any strong sign. The most sensitive systems on earth to such distant radiative events are probably life and the chemistry of the atmosphere, especially the trace chemistry, like that of ozone. The handful of supernova neighbors implied by statistics during all earth history might have had some effect traceable in life.

The proposal has been made many times, but no direct support has ever been offered for such scenarios, to my knowledge.

In the same way, galactic dust clouds may fog up the solar system from time to time. Here again the proposal is plausible, the details not very persuasive, and the evidence lacking. It does not seem that outer space has so far meant anything central to our life, apart from the indispensable endowment of atoms at the sun's birth.

Within the solar system, space is not so dominant over matter. It is rather crowded around here. Worlds in collision are real enough, though that is not to offer the slightest support to wild notions of recent planetary encounters. Every moon crater was made by an intruder. Some of them are big indeed. Meteors strike earth, and in the past there were certainly truly big ones, leaving craters of subcontinental size. But that was in the earth's childhood, before ever life began. Just the same, the errant chunks of the solar system—the most important and erratic are comets—can and do hit earth. A body of mountainous size will produce local catastrophe, and some effects worldwide, from tidal waves to persistent dust in the atmosphere, like Krakatoa, or a super-K. By now we have all heard that such a comet might have ended the rule of the dinosaurs. Certainly there is evidence that the marine life changed everywhere suddenly, so that silt came down instead of chalk on the sea bottom at the end of the Cretaceous era. How ingeniously this has been connected to the great dying of that period is not our story; the case is good, if not yet certain. But it seems to leave the story more or less unchanged. Even a worldwide darkening by dust which lasted for years, so that green plants were mostly killed, would not end human life, it would appear, provided the overall continuity of life was maintained. We would find ways—at least some of us—to live on what was left, the stored seeds and stems, the plants and creatures which could survive the dimming. The actuary also sees that orbital wanderers are no longer so many nor so big; the big erratics in orbit were long ago swept up. All the comets and asteroids which remain do not add up to the mass of a small planet like the earth. Of course, only a tiny fraction can ever collide with us, or even pass nearby. It looks as though there is no major planet wreck to be foreseen, though the effects of past collisions with minor bodies may have been decisive for some forms of past life. Thus the sustaining sun, through its radiation, and some members of its retinue, by direct collision, can affect life; more distant objects probably cannot, and no intruder can annihilate.

By Our Own Hand

Today's forebodings of catastrophic nuclear war are not to be ignored. Indeed, unless international relations take a substantial turn from past patterns, it is hard to avoid the prediction of a large-scale nuclear war some day, whether in decades or in centuries. The coolest argument for this is based on the observation that the energy yield of the weapons grows year by year, without ever declining. But the area of the earth and the volume of our atmosphere do not increase at all. That most typical of all growth industries, the world manufacture of weapons, must saturate if there is to be a long-range future; so much seems tightly entailed by the events. That arms reduction could take so many forms, but it must arrive somehow. The most pessimistic would say the stockpile will go down because the warheads are consumed in use, the population going down along with the weapons stock. The most optimistic see a simple desuetude, or even a final rolling-back of unused and unwanted arsenals. History will decide. Of course, there might be a whole series of calamities, recoveries, escapes. Here the question is simply the chance of survival.

In the last years, grim new attention has been drawn to this possibility of the coroner's verdict of species suicide. In summer 1982 an informal study was published (in the authoritative Swedish environmental journal *Ambio*) by two climatologists, P. J. Crutzen and J. W. Birks. They realized that it is not only some rare constituent of the global atmosphere that can change, but that a trace of some novel but potent stuff might be added. The substance of their concern was familiar enough, merely soot and black smoke from fires burning the organic matter of the forests, fields, cities, and stores of petroleum. A small portion of dark soot spread throughout the upper air can dim the sunlight of life until the soot has been diluted and fallen or washed out of the air. That may take months. In the meantime, the land masses will suffer an unexpected winter. Should the dark clouds arrive in summer, crops will die; so may the animals and plants quite generally. If the smoke-induced winter is protracted enough, even the tropics would freeze, and the effect could spread between hemispheres. Only one cause of fires so widespread is thinkable: general nuclear war. Since the first, several studies have tentatively borne out the early conjecture; nuclear winter is indeed a terrible possibility, even after use of only a modest fraction of the present nuclear weapons stocks. A thousand big explosions over Northern cities could ignite fires that would in days burn an area greater than that which burns in the world's forests during a whole year. The natural processes of dilution and recovery could be overwhelmed. The uncertainties are large; our ability to forecast the weather long-range is not yet a firm one. But a prima facie case has been made that nuclear war puts at real risk not only the targeted peoples, but everyone. Perhaps the most important point is that this hazard, surely not in itself so strange, had not been addressed by the experts during decades of attention. What else has been overlooked? We might learn only too late what our model had neglected.

Yet nuclear weapons, the most present of all dangers, I believe, do not seem likely literally to end our species. Should the worst happen, I would expect a severalfold decimation of the two or three big warring nations. They would sink, their cultural properties as well. All that would remain in the United States, say, would be a few portions of certain less-populated regions, amidst vast unvisitable wastelands. Barren, dying fields and forests might be found worldwide, amidst general hunger. But it still seems to me that some of our far-flung species would survive, its population, wealth, and hope sadly dwindled, but not finally ended. The Southern Hemisphere is actuarially the better off—airborne radioactivity and the dark clouds cross the equator slowly—though it would not remain unthreatened. The best guesses are still that a factor of ten, though not much more, stands against the death of the whole species by nuclear war. Certainly the average danger does not strike every individual; and distances and environmental differences are great. Once again, diversity is the key to survival. The false winter or two have come and gone; the ozone layer is at risk; epidemics of tumors and birth defects persist for centuries; the best croplands are unusable for a long time. The very air holds chemical toxins. The tale is genuinely tragic. Yet there will be people left, plenty of them miserably surviving here and there by their wits on the unused storehouses and the residual life.

Other weapons can be imagined. Plagues and contagions do not appear general enough, though one cannot be sure. There are powerful natural immune mechanisms which have been a long time evolving. Here the nations have been more sensible; biological warfare has not been well developed, and is now proscribed. We can hope that example of good sense is permanent. We could poison some important living forms within the world ecological cycle; the death of the marine plankton by new toxins added to the sea was one scientist's cautionary invention. But the seas are big and poorly mixed; I do not think we could empty them even if we tried; it cannot happen by accident. Our industrial and weapons systems are not big enough for such effects. Once again, we are looking for trigger mechanisms: radioactivity, ozone reactants, and the like.

What of the will to survive? Physical attack by weaponry is only one means of self-destruction. It is not necessary to invoke the macabre vision of a universal Jonestown to imagine an end to the species by quiet consensus. All that is required, of course, is a gradual end to new births. That seems to have happened in one or another small culture once placed into an intolerable bondage by a powerful invader or colonizer. Now that the imperative instinctual arrangements of biology have been cut away from procreation by consciousness and culture, this possibility is opened more widely. Surely a physicist is here on poorly understood ground. Once more the diversity of the psyche can be counted upon to avoid the danger, all the more since the feedback loop, awareness of the predicament, seems easy to activate: local information is enough for alarm. It is true that the growth of the unity of mankind, plainly evident over the last millennia, suggests the possibility of a psychological unity, mediated by travel and communication, which was not at hand when the races of mankind differentiated under strict geographical isolation. The slow return of one world—we had it in the Rift Valley when the genus was young—holds the sort of risk under discussion. But the turmoil of the present world of resurgent nationalism and regionalism seems to presage just what sort of limits arise to the dreams of a universal brotherhood and sisterhood. Given genetic diversity and experiential variety, we will not become all of a mind. If some become sheeplike, others will become more solitary or even enterprising. Fluctuation is not merely a thermodynamic notion; it has mechanisms deep in culture.

We are predators worse than wolves, man to man, but predation does not often end species, except for the power of human culture, and the old human rivalry for a niche. Even the latter is perhaps more an end by incorporation than by hostility, with one example perhaps in the relationships of Neanderthal and early *sapiens* in Europe.

We risk terrible damage to ourselves; it is in every tax return and news story. But self-destruction, down to the very end, does not seem indicated by our powers or by our nature.

When the Sun Dies

Two paths of endurance are plain. The one is that which has saved the horseshoe crab and the works of Aristotle: multiple copies widely dispersed. The other is the path of the sun, or of the Blue Ridge Mountains: construction of such durable material on such a scale that the tooth of time must gnaw for a long while. But of course in the end the mountains will erode to basins of silt, and the sun will die. The death of the sun is likely to be a fiery one, we believe, not a mere cooling, but an expansion and a feverish glow. The earth will be ended by that, let alone the life upon it. Even the outer orbits where spaceships might dwell will have a hard time adjusting to the erratic and gusty course of sunlight and solar wind.

Will we emigrate from this home system out among the other stars? Science fiction, of course, sees that as a commonplace. After all, the sky is filled with stars in our galaxy, among them a couple of hundred million old enough, stable enough, single stars, with temperatures like our modest sun. They lie far away, indeed, but not beyond any hope of reaching them by human transport over generations. So one admits the possibility that our species will become wanderers in space seeking homesteading rights. But that will almost surely be a sporelike venture: the whole of our number will not go, but only a few representatives, able perhaps to maintain the memory of earth.

Even more plausible than this plot, which implies forecast over billions of years until the sun does undergo serious change, is quite another. Are we alone as self-conscious appraisers of our future? Or among the suns are there many like us in insight, however different in biology? It seems probable that we will know that within a modest time, perhaps by catching radio signals from Others far away. Perhaps we are The People, solitary on our planet-island, someday soon to be surprised by other beings from far away, with radio beams if not white sails, easily our masters in many domains of life. It has happened before: it may happen again. We do not know, but perhaps we will find out by an active if vicarious radio or laser search. To be sure, those Others will not be of our species, but they will by hypothesis share the niche we hold: conscious beings, surviving and spreading by a new form of evolution, changing more by hand and eye than by the gamble of inheritance.

We have never seen anything like that before: a new species which shares our functions though not at all our image, our descent, even our biochemistry. If our culture and our history pass into the keeping of a wider society than earth's, the issue of species survival becomes somewhat moot. Here we touch questions of philosophy, of the very definition of the end. What does it mean? Would we be held to survive if our new partnership with natural selection led to a filiation into something rich and strange? The older speculators, say George Bernard Shaw, saw the deep future as the rise of an etiolated but powerful race of philosophers, perhaps even of pure thought: the more silicon-bound speculators of our day foresee the inheritance of our culture by subtle machines we set on a path of machine evolution.

I shall not worry this point unduly; here imagination is hardly constrained by experience. It would appear that such changes will not in fact preserve us; that they might keep our echo, the kind of echo we hear faintly of the Olmecs or the Etruscans, without any lineal descendants, is plain. We will have been transformed away, reborn if you like, through a channel as compact as a seed. The likelihood seems high that we will share a wider interstellar culture one day, unless indeed ours is first of all the hundred million suns to engender the carriers of thought. That case is logically possible; but it seems a poor bet.

All that I have argued has been built on the science of our times. Surely the past teaches us that much of what we hold true is in error, and more incomplete. How can this be a guide to actuarial forecast? Here is the most troubling of all the estimates I have made. In spite of the clear precedents, it is hard to see where new knowledge will change these remarks to a serious degree. What will we learn before science itself comes to an end? A closer approach to the ultimate structure of matter, to the ends of space and time—if they exist? The inwardness of the development of living forms? The architecture of the mind? Surely all these and more that I cannot foresee, with a plentiful triggering out from every present scientific branch. We will find cause in the domain of chaos, and no doubt chaos behind the apparently causal, as we have often done.

Yet in a curious way science has told us what we need already. It has defined the space-time volume of our life, the nature of stars and galaxy, the age of earth and sun, the epoch of organic evolution. Just as we can be sure that there is no mountain on earth a kilometer higher above sea level than Everest, I believe we can be sure of the minimal extent of our space-time history, even if we cannot yet see its farthest boundaries, or understand its initial phases or final fate; its infinities are beyond us. But can they matter? I do not believe we will find how to work magic. We fly now with thrust and wing, not on carpets or by wands. We travel through space freely, but at the cost of time, and never through time against its current. Mind indeed masters matter, but not by telepathy or by spell, only by the cunning disposition of structures, circuits, and energy sources. Unless the writers of fiction prove correct—time travel, starlike intelligences, instant mind contacts over arbitrary space, and other timeworn marvels of wishfulness—I do not see new science confounding what I have somewhat cautiously outlined. If we cure all disease, make life freely, travel at near light-speed, all those things will not free us from the constraints and the supports of time and space, from sun energy or a surrogate fuel. We will remain finite beings, with finite powers, bound like the cosmos itself by the laws of thermodynamics. I doubt much that we will turn the galaxy into a park, or visit the ends of the cosmos. It costs too much, in energy, time, and plan.

The best glimpse into the dim future we can hope would come, if it comes at all, from the signals of our peers and superiors out there among the suns. If that happens, the actuary will have some statistics. Until it happens, the best guess seems to be that we will survive, surely for longer than our written history so far, surely beyond the return of the ice, probably long beyond that to a time as long as our evolutionary span so far, the couple of hundred thousand years since *sapiens* came on earth. If we allow for changes, our modified cultural posterity might with good fortune go on for geological epochs, as far as I can see. Maybe our new evolution can persist as long as has the cockroach or the cycad. The odds go down with time, but slowly. One day, the sun dead, the universe itself perhaps collapsing to a fiery rebirth, it will all be over, long before matter itself decays away as likely it will do. But we are now a long-lived, widespread, and marvelous species, all the same, marked by a symbiosis unique on earth, living intimately with a creature of wonderful talent and sinister cunning, our Monkey of the Mind.

Cause, Chance and Creation

W hat you can read here has little to do with the headlines that tell of applied science, like the space rockets, nor even with discoveries like new atomic particles. Here we discuss instead the very roots of physics, the sources of its strength and its fruit. They are at the same time its anchors. With them physics can sometimes take firmer hold on the real nature of man and universe than even the wisest of the seers and the philosophers can do without its aid.

How wide is the scope of physical law? Do life, thought, history fall within its orderly domain? Or does it describe only the inanimate, the remote and the very tiny? It is the claim of contemporary physics that its laws apply to all natural things, to atoms, stars and men. There are not two worlds: the cold, precise mechanical world of physics, and the surprising, disorderly and growing world of living things or of human existence. They are one.

This is no new claim. Since George Washington was a young man, there has been a view of the world based on Isaac Newton's physics. That view was mechanical; the world was a great clock rather like the newly understood motions of the planets. Feats like the prediction of the time of an eclipse of the sun, possible with high accuracy for a thousand years ahead, became the hallmark of a powerful science. Its proponents claimed that one day they could so explain everything. But there is a clear and common-sense retort. Life is not at all like solar eclipses; novelty and surprise, building a tangled complexity of events, are much more the essence of the world than is the serene dance of the planets, however intricately they weave.

So the physicist's view became first less plausible, then hardly more than absurd. Even in the last decades, the mechanical theory of the world is often taken to be the last word of physics. If this were so, I should make no wide claims for the implications of our science. For the world is patently not clockwork, not even clockwork with a few loose screws. In fact, what has happened is that physics has come nearer maturity; it can understand not only the neat and the mechanical but all the tangle of events in the everyday world in one and the same way, a richer way, a subtler way, than it first learned to chart for the motions of planets.

I will try to show how physics gained this new view and what it is. Along the path, we shall have seen as well something of how the isolated, rather technical bits of laboratory lore merge into a world view powerful enough to claim its share of the attention of everybody who thinks about the great questions of man and his place.

Let us begin by taking stock of the old philosophy of the physicists, the view that the world is only a great piece of clockwork. This point of view meant much to the history of our republic, which was founded by men who had been brought up in such a climate of thought. To say only that much is to prove that such a picture is not wholly tainted; the careful checks-and-balance mechanisms of our Constitution owe something to that philosophical view, however much they may owe to real political experience. Yet no one can fail to be repelled a little by the fatalism which this viewpoint implies.

It was Laplace, Napoleon's great official mathematical physicist, who expressed the mechanical philosophy in just the language we want. Said Laplace: "An intelligent being who, for a given instant," came to know all the forces of nature and all the positions of every particle, could predict every event to the end of days in one sure formula. "Nothing would be uncertain for him; the future, as well as the past, would be present to his eyes." Laplace saw in the astronomy of the day, with its predictions of planetary motion, its tables of eclipses from antiquity to the year 3000, a feeble foreshadowing of that state of total knowledge which he could view as the ultimate goal of science.

That sort of world is a hard world to breathe in; for clearly there is no use to make up your mind for good or for evil. It is all written in the orbits. Laplace's "intelligent being," if he is physically possible, mocks our hardest decisions with his complete book of all our future actions. It is a world without real novelty, without real change. Yet such a great Victorian as Thomas Huxley, the defender of Darwin, seemed to see the world so; he wrote once that even in the play of the spray on the waves in a stormy sea, all was predetermined to the last degree. Here is, I think, at least part of the reason why science and the humanities are so separate in our times, far more than they were in the eighteenth century. This cold doctrine became so difficult to hold during the wildly changing and difficult century which followed that educated men withdrew from science. They felt that science and its mechanical model of all things could say nothing plausible about the great choices of the time. If physics, as I believe, can now give indispensable guidance to any who wish to know the nature of man and his place in the world, it must have found a more reasonable way to explain a world bursting with novelty. It has. To trace the growth of this new view, and to try to see where it leads, is the object of these pages.

The physics of the twentieth century bears one signature writ very large: Albert Einstein. But Einstein's relativity, which shook the educated world far beyond the corridors of the laboratory, made no deep crisis for the clockwork picture of the world. Einstein showed—and we have found his every claim fully supported by experiment—that the notions of space and time held for two centuries were naïve. Neither distance in space nor passage of time could retain the simple meaning which we once thought part of "common sense." Space and time were curiously mixed together by his theory: what appeared to one watcher as a difference of a second in time between one flash of light and a succeeding flash might be measured by familiar, sound and sensible procedures perhaps as a half second, or as two seconds, by another observer who is moving rapidly past the first one. What value the moving man found for the *time* interval would depend on the distance *in space* between the two flashing lamps. Two flashes could be simultaneous to one observer, but successive to another. The concept of *simultaneous* is thus merely a relative one.

So much is well known. But nonphysicists seldom appreciate that Einstein's relativity, which plays so fast and loose with space and time, regards other familiar physical concepts as absolutes. What concerns us most here is that relativity insists that one relation must remain unchanged to every observer: every cause must always be separated by a minimum time from its effect. The order of flashes of light from two lamps may be relative, but the order of snapping the switch and seeing the light must be retained. The order of events *not* related as cause and effect *is* relative to the motion of the onlooker, but every cause remains distinct in time from its effects for every possible observer. In this sense the sequence of cause and effect is not relative but absolute.

The eternal web of cause and effect, then, is just as unbroken for Einstein as for Laplace's prophetic "being" and the Newtonian physics upon which Laplace rested his case. Yet, after Einstein's mighty shaking, two cracks appeared in the foundations of the older view. First, Einstein taught once and for all that even familiar and well-tested notions must be rethought and retested in every new domain of experience. The chain of cause and effect familiar from the eclipse predictions and nautical almanacs of Newton, Laplace and their successors was not in fact snapped by Einstein. But since what had once seemed the self-evident truth of the old ideas of space and time proved only approximate, one had learned not to be cocksure of anything. There is here the suggestion, which I think correct, that what we regard as "common sense" and the "self-evident" are only those matters which we have all early learned and which we were all well-taught. The physics student of today finds Einstein common-sensical enough; a generation back, only the few experts grasped what the theory meant.

Second, the idea of an instant of time became a little less simple. Laplace's all-knowing being, his prophetic demon, was to predict the whole of the future once he knew what was happening everywhere at one single instant. But Einstein showed that information could flow to any observer with a speed which at maximum was the speed of light. The imaginary being of Laplace could collect his data at his desk, to make his prediction, only after a great deal of time, enough time to bear the news from the far corners of the universe. Some part of his wonderful predictions become mere eyewitness accounts, for he cannot make them until the predicted event has already occurred. Predicting demon is thus both prophet and historian, for with Einstein the sharp differences between *now, the future* and *the past* have become blurred. He cannot hope to make observation and prediction simultaneous.

A still-greater blow to the universal clock was in the making for a generation before Einstein, and came to fruit in the generation after his great theory. Between the experiments of the early molecular physics, around 1870, and the maturity of atomic physics in the 1930's, we learned beyond doubt that the world is an atomic world. All matter, and even the seemingly smooth flow of light itself, is grainy, all made of tiny units of one sort or another whose myriad combinations and interrelations weave the fabric of events.

At first view, this seems agreeably Newtonian. Laplace spoke of "the positions of the entities composing nature" as just what his marvelous predicting observer had to know. Since indeed nature is built up of separate and distinct "entities"—atoms, electrons, light quanta and so on—this modern advance might appear only to confirm the Newtonians. But it is the very completeness of the atomic victory which makes trouble. The data which must be known and recorded if the world's future is to be predicted are prodigious in amount. Laplace, one fears, never examined just how much would have to be known by that Great Intelligence. The sheer bulk of the data he needs is itself no real criticism; for Laplace, while he had in mind no supernatural being, but one far more skilled than humans but "as finite as ourselves," was willing to allow him to be as well-equipped as necessary. The trouble is, the Observer and his notebooks and his yardsticks and his stop watches and his very brain must all be made out of atomic matter or of grainy radiation. In preparing the data for his omniscient view of the universe, he must consume great amounts of material. If he wishes to economize on material, he can do it only by splurging on radiation, on energy. Try what we will, we cannot imagine him as the subtle, remote observer Laplace had in mind, but rather as the master of a great laboratory.

For knowledge itself must somehow be expressed, and there exists no more refined form of expression than embodiment in the grainy stuff of atoms or of radiation. As long as only human knowledge needs expression, the problem is not noticed. For atoms are so numerous that a human brain, for example, needs only a minute portion of its energy for the sheer material expression of its knowledge. Almost all of the work of the human brain is a consequence of the particular living process by which it works. In principle, the energy or matter which is used for knowledge is an absolute necessity for the brain, but it is unnoticeably small compared to its own needs. But faced with the gigantic task of noting down every atom in the universe and its motion, the Predicting Being of Laplace turns out to become a major, even the dominant, portion of all the universe. With that enormous bulk his credibility disappears. For now it is possible to show that he cannot even know fully himself and his own laboratory, let alone all the rest of the universe. He is beaten by the atomic facts of life, the hard fact that no measurement can be made without the expenditure of some very small parcel of energy, no result recalled without the use of some atomic particle of minimum mass.

The eighteenth century was misled by its experience of science. For them, science was represented by the astronomer—at his telescope on the dark, quiet hill, with his pad and pencil and his tables of logarithms—plotting the paths of the undisturbed distant stars. Today the great humming machines of the nuclear laboratory, and its extravagant electric bill, serve the ends of the physicist who needs to study the heart of the atom by taking it apart. That contrast may make concrete what was described more abstractly in the preceding paragraph. Measurements of the numerous and tiny atoms require a more active intervention than mere distant rough observations of the great and remote.

Such intervention has, moreover, more important effects in the world of the small particles than it would have in the realm of the ponderous planets. A small delay in the first atomic event of a long chain may delay the whole chain, and thus a small cause, a part of the biography of one atom, may grow to have a giant effect. This amplification of causes is one of the characteristic features of the atomic world, and almost by itself would make foreseeing the future of a collection of atoms a calculation of probabilities, within the purview of statistics but not of the large-scale measurements of astronomy.

The world of the atoms is in fact far more predictable for us than for Laplace. He could not really understand the behavior of a single chemical substance, whereas, standing high on the shoulders of all those who have built science over three centuries, we are able to see far and wide in the world of matter. We are able to design new material of many kinds, building them up according to theoretical principles from commonplace atoms to serve many ends.

But the predictions we make are, without exception, for atoms in their crowds. They are not the individual predictions of the almanac, planet by planet, but rather the mass certainties of the actuary, who knows not when any one man will die, but who can set reliable expectations of life among very many. The most familiar example is that of radioactive decay. Here one may be sure that of 10,000,000 identical radioactive atoms, nearly 5,000,000 will disintegrate in the first week, say; 2,500,000 more the second week, and so on. But when only ten atoms remain, a week may see five more decays, or three, or even eight, following the well-known mathematics of random events. When only the last atom remains, though it has an average life expectancy exactly the same as all its fellows, it may by chance live for weeks, or it may expire in the next minute. No one can predict the moment of its demise. But if one takes a great many atoms of this kind, one may be certain that they will on the average live just one week.

This rise of statistical prediction, of probability, is perhaps the most characteristic of all the developments of twentieth-century science. It represents the realization that we cannot claim to know all the causes of things, for those causes are far too numerous. We have not measured enough, we have not recorded enough, we have not interfered in the past as much as we would have found necessary if we had tried to carry out the program of Laplace. But we have not lost our sense of the orderliness of the world; we have gained rather a sense of its complexity. And in that gain physics at one step found new and awesome powers. At last men could see how a world of novelty and surprise could yet be a world of scientific order.

There was one more step to go. That step was taken between the two World Wars, in the development of the modern quantum theory, during what was quite possibly the most fruitful generation of work in all the history of science since the Babylonian priests charted the morning star.

Quantum mechanics is subtle and versatile; by no means have all its difficulties been solved. But I think it is not too hard to try to give some sense of what this last and deepest revolution has meant.

We have already spoken of the world of atoms. A closer look at the atom is needed. Let two facts be recalled:

1. All the atoms of a given species—say all atoms of gold, are identical. This is evident even in common experience. A newly mined nugget from California and a worn Roman coin yield the same pure gold to the assayer. He may discard some dross, but the gold that remains is one and the same. But how different are the series of knocks and bangs the two samples have suffered! Yet even precise measurements do not reveal any differences between individual gold atoms, though they do show us that different mass samples of even pure gold differ slightly, say in density. The near-identity in bulk depends upon the true identity of the atoms.

2. Yet the atom has a structure. If it were, as the Greeks thought, some indivisible—a-tom means *without cutting*—whole, we could understand the identity. Yet out of a gold atom we can pluck hundreds of electrons, protons and neutrons. It is a relatively complex structure made up of moving, interacting parts. If it were like the solar system, for example, we could not expect two atoms, which had suffered different series of knocks and bangs, to remain identical. Perhaps the hard knocks would break them apart, and we would simply no longer count them as gold. That indeed happens. But there must be many softer blows of fate in the lifetime of an atom, each of which ought to modify it a little, yet leave it still whole but distorted, as the solar system would be distorted by a passing distant star. On this view, which is surely the only one consistent with all we have so far said of physics, a sample of gold atoms might form a kind of population, rather similar as all peas are similar, or all pedigreed poodles, similar, and yet differing slightly one from another in more or less every property. The similarity of complex systems is never precise, in the old physics, but only approximate.

Thus do atomic identity and atomic structures conflict. Atoms are complex structures, yet they come in great armies of truly identical individuals. Only the quantum theory has explained this, in a fairly complete and certainly a deep and unexpected way. It is a way which contradicts flatly the old mechanics of Newton, and yet in a strange way bears it out.

In the modern mechanics of the atom, only certain definite patterns of internal motion can be long-lasting. These patterns differ from one another, not simply by gradual, almost imperceptible differences which shade smoothly one into another, as might the positions of the planets in a great number of similar solar systems, but by real gaps in energy, real differences in shape. Such distinct patterns are stable and self-maintaining against small disturbances.

If an outside disturbance—say, the shock of collision with another atom—is sharp enough, it can throw the atom into a differing pattern. But if it does not become that serious, it makes no lasting effect at all on the atom, which remains in its original state after the collision. The pattern either heals completely, so to speak, or it dies and is replaced by another. There is no such thing as a mere wounding, ending in a scarred recovery, a little bent or distorted. Such behavior admirably explains the identity of the atoms.

But what of the atom's structure? How are the electrons moving in the atom? The answer is strange but quite definite. The electron motion is fixed by the whole pattern of the atom, but it is fixed only in a statistical way. Just as one cannot predict the day a given policyholder will die, so one cannot at all predict where the electron is in any given atom in its normal pattern. Nevertheless, if a great many of the identical atoms are inspected, to locate their moving electrons, a perfectly precise and definite pattern will be revealed in the statistics so compiled, just as the life tables give a regular change of mortality with aging or residence or occupation. The electron's every position is contained in the pattern of the atom. But it is contained only potentially: the pattern has the potential of showing up the electron now here, now there. What the very next result will be can be predicted only statistically. Yet statistically the precision is as complete as one could wish: every feature of the pattern is fully and precisely realized as the number of individual identical atoms inspected grows and grows.

The story can be told from the other side. It is possible to "prepare" a normal atom, confining it with electric forces, say, in such a way that one knows exactly where the electrons are. But then that atom has no long-lasting pattern; its electrons will move about in a way which depends upon every detail of the original preparation, as they would in the Newtonian physics. No sign of identity would remain. No two atoms would be sure to evolve in exactly the same way. Whenever atoms are found identical, then their electrons are not to be located except by stating their potential positions; as soon as the electrons are indeed located and tagged, then the development in time becomes specific to the initial set-up, and the remarkable self-healing identity is lost.

There is no use trying to blink away that this dual behavior of atoms is a new state of affairs in physics. It has been studied and questioned in a thousand diverse contexts, and has so far withstood every test. The consequences turn out to apply to crystals or to fluids; to molecules, atoms, and their constituents; to light—to the whole range of phenomena in the universe. And the predictions that follow from it merge into the older physics, with a boundary between the two so smooth and neat that no joint can be seen.

For the possibility of true identity depends on the presence of the little gaps in energy between one persistent pattern and the next. If these gaps are larger than the energy transferred to the system in its encounters with the outside world, the structure will remain unchanged nearly always, and its pattern will heal. But if the gaps, the quantum jumps, are too small, then each external influence can make some lasting change in the system, and the whole complex will behave according to the old physics, showing the individual scars or distortions which it has acquired in its life. Then true identity will be absent.

Now, the size of the gaps depends on the massiveness of the system. Everything of atomic size will possess sizable gaps, and even hard knocks by similar atoms will generally be resisted. But anything so big as a dust grain will easily yield to disturbance and must then behave in the old way. Remarkably enough, even structures the size of gem crystals reveal their quantum gaps, and assume a structureless identity, whenever they can be sheltered from the ordinary atomic collisions. Such a quiet state is achieved at very low temperatures, near the "absolute zero" where molecular motion would, according to the old physics, cease.

The success of this new theory is the final attack upon the fatalism of Laplace. For we now see that the myriad particles which indeed do make up our universe form an unfolding pattern of events whose every *potential* future state may be predicted. But just what events are realized *in fact* can be foretold only in the statistician's sense, with a sureness that grows the greater the larger the piece of matter, or current of energy, with which the prediction is concerned. There is room to breathe in such a world. Yet it is no world of caprice or chaos. Chance and cause have been wonderfully married into a point of view in which precise pattern governs potential events, and yet in which the variety of the potentialities allows the full growth of that novelty which we know to govern the world we live in. Many have written of this richer theory of the physicists as based on the principle of uncertainty. *Uncertainty,* if you like; it seems a poor word to describe the richness of valid prediction which the science now possesses. It is becoming common sense.

True, some great questions still lie beyond the frontiers of our knowledge. One of them is part and parcel of the problem of atomic identity. For the building blocks of the atoms, the electrons and protons and neutrons and their like, are themselves marvelously identical. Yet in electron or proton the kind of inner structure is like nothing we have encountered before, if it is structure at all. For the parts we can take out of the proton, for example, are truly particles with mass and charge, yet can leave the proton unchanged. The proton can be made to yield up particle after particle and still remain the same, like the purse of the happy Fortunatus. Only energy need be supplied. This is no assembly like an atom or like the parts of a watch. Quantum theory alone gives us a start to understanding this, and even it has no full answer here.

By now we have surveyed most of the range of physics in our time; and the Laplacean clock that ticks away our lives by rote has not remained. In its place we have a few great new unexpected laws, which are in fact far more powerful, and yet somehow closer to common sense, than the simple and complacent physics of Napoleon's day. We might recall them in a terse list. The labels of time and space are not absolute, but cause and effect remain connected for all observers. Knowledge of the physical world is no mere passive, ethereal thing, but always is based upon action, measurement, and must be recorded or expressed in the only medium physics has, a grainy, atomic one. Cause and change are caught up together in the world of atomic scale: predictable pattern is found ruling the identical atoms, but their individual changes can be predicted only by the actuary's powerful schemes. Such is the subtle picture of the world which modern physics has drawn.

Now we come to temptation. Newton and Laplace built a picture of the universe which was based on their brilliant successes with clockwork and eclipses. We know now that the universe is not merely clockwork. Modern physics—Einstein, Boltzmann and Bohr will do as names of pathfinders—has a far grander picture, one that will one day beautifully explain atoms or stars, birds or bees, television cameras and antibiotics. But the universe is wider even than that. Every attempt to build upon the results of science a picture of the universe which will sustain the life and the values of men is risky. For science can never be complete, never without its insecure frontiers.

This is the essence of science. Notwithstanding, each generation must try. And I believe that in our times we have come to a scientific picture wide enough and supple enough to form a background for the philosophy of life of every man. Science cannot itself be such a philosophy, but it is the proving ground of every philosophy today. For the ideas of physics are no less invasive than the machines and the power lines which they have made possible. Just as the devices of modern technology press their way into ancient lands, so will the great ideas that lie behind them find their way, year by year and land by land, into the old ways of thought, into the long impregnable fortresses of belief and assumption everywhere. In the end, no view of the world can remain unchallenged by the physicists' findings: the future of every philosophy can be measured by the degree to which it can admit to the world of the mind the physical map of the atomic universe.

The aims and the values of men are not and will not be made by physicists. But if those values are to remain the basis of a way of life, they will need to come to terms with physics. Physics no longer sees its universe as a mere whirling automaton, but rather as a changing, chancy and yet an impersonally ordered and patterned place. In this place action and knowledge are not distinct; full certainty and complete hazard are equally unreal. Such a picture of the world admits the play of moral choice, yet it requires that even moral choice can be studied at least in part with the tools of science. We think often of the novelty of western science and technology as it sweeps violently across the old lands. But we do not see so clearly the ancientness of our own systems of values, standing as they do upon religions ten or twenty centuries old, or upon philosophical systems mainly older than modern science. The modernization of our own ideas may prove to hold as much of difficulty for us as the industrialization of the peasant ways has held and will yet demand of Asia or Africa. But like their hard trials, ours too may bring hope and the promise of a newer and a finer life of heart and mind.

On Broken Symmetries

The Idea of Indiscernibility

I touch on my subject with diffidence because it is so large a subject; I can introduce it only in a very sketchy and incomplete presentation. What I hope to do in pursuance of an aesthetic perspective is to suggest metaphors and connections. These are less than explicit, and probably less than entailed logically, but they seem to me to exist deeply in what we notice in the world. Much of what I have to say will be familiar to students of physics; less familiar to others. I make little claim for the novelty of these remarks. They belong in any insightful survey of our experiences in art or in science, most particularly in consideration of architecture and industrial design because those products must work as well as be—a condition not imposed on other works of art. Function requires that the building at least stand up, and usually it ought to keep out the rain: a requirement very close to physics. I would by no means like to restrict my remarks to architecture. What I have to say will be of relevance to people concerned with the understanding of art in any domain of human action, from history to metallurgy.

To define the idea of symmetry is certainly not simple. I shall not try to make an all-encompassing or precise definition, against which a contradiction can quickly be brought, but rather to put out a kind of statistical approximation and then to enrich it by example. The idea is still alive and growing. We don't know all that the concept implies. I like best the idea of the seventeenth century philosopher Leibniz, who talked a great deal about symmetry not much employing the word, using rather what for him was a more epistemological word, a term at the heart of the matter. For Leibniz, symmetry is related to the indiscernibility of differences. Once you walk into the hall of a Palladian building, you can't quite remember whether you turned left or right. They look just the same inside. The two wings are indiscernible. They become discernible, one from the other, only when you understand that one is the mirror image of the other. But if I place the building up to a mirror, then the one becomes the other; once again the distinction is indiscernible in that sense. You see, indiscernibility stresses the idea of perception, which is why I want to use it as part of the definition. What is symmetrical under one aspect of perception may not be so under another. Symmetry has subjective quality. If I am color blind, I cannot tell the red-marked side of the boat from the green, the port from the starboard. The boat is, for me, perfectly symmetrical. I may say that the symmetry of the boat is fully present. But once I can see color, the left-right symmetry of the boat is broken by the two distinct colors, and therefore—most important—the pilot who approaches by night doesn't hit it. He can turn the correct way to avoid collision. It is no trivial matter: one is required to destroy the symmetry, the actual symmetry of approach. Without the lights you cannot tell whether the boat is approaching or receding. So naturally we mark the two sides to make them discernible. Discernibility depends on the channels of perception you use. Nobody could deny that red paint is inherently different from green paint: it contains a different chemical pigment, you can buy it from a different maker for a different sum of money, etc. Nevertheless, discernment does place some load on the observer.

We perceive the world by means of a complex system that we poorly understand—ourselves, of course: the eye-brain complex. Let me cite an indispensable aesthetic experience. It is the remarkable experiment presented to us *only in the last decade* by the work of Bela Julesz at the Bell Telephone Laboratories who presents to view two random dot patterns, one seen only by the left eye, one by the right eye. These present no figure, no structure. Each one is simply a set of five or ten thousand little black and white squares, like a much enlarged checkerboard. The squares are not colored in a regular pattern, but black and white at random. The right eye sees the dot pattern. The left eye sees a subtly different pattern, which indeed looks quite the same at first glance, a textured field of black and white. Within the left picture, however, a central square portion of the right picture has been inserted, exactly as it is on the right, except shifted a few columns to the left. The rest of the field is merely repeated. What you have then is a random dot field for each eye, except that a certain portion has been shifted *in toto,* maybe one thousand or so of the five thousand dots. When you view these two fields, one with each eye as is usual for inducing the stereoscopic effect, you gradually acquire a new and unexpected perception. The key word is *gradually.* Sometimes it takes seconds, or even tens of seconds, for the experience to form. You simply stare at the pictures without conscious effort of any sort. Something internal and undetected is going on within you, for after the delay you will see the square of dots, which was repeated in shifted form, float in space, a random texture above the random texture of the surrounding dots. Exchange the pictures between eyes, and the central square will float below the field of the others, farther away. It is a genuine and striking presentation of a binocular clue to depth, without any defined edge or familiar form.

The extraordinary thing is a sense of watching yourself because you're not conscious of any internal mechanism slowly coming to recognize the remarkable fact that the apparently random dot fields actually contain strong correlations. The little bit of difference in all these hundreds of pairs of dot distances is just enough to make the figure stand out in depth. Once you have that experience, you will never deny that there is some kind of internal, unsensed action. (It would be fascinating to develop some ability to perceive what's going on.) Something powerful is happening. After a few times the form comes more quickly, for a more complex structure like a figure eight or a bow knot, but it still takes many seconds for the depth judgment to jump out at you.

By extension, the idea of perceptual indiscernibility and therefore of discernibility, which is, of course, the other side, is most important. It has a great deal to do with our fundamental aesthetic perception of symmetries of all kinds and our delight in them. It is probably built into various symmetrical, "indiscernible" features of our mental processing: the logical circuitry, compiler, assembler, and all such tricks, in the metaphorical language of the computers. Of course, we know that constant scanning of the eye axis seeking lines and all kinds of other features is built in. Only slowly do we come to understand in any real way what is going on within. Be that as it may, I think it no miracle that human sensibilities are attracted to symmetrical presentations, for symmetry plays a major role in the world of which we are a part, of whose parts we are made. This is no philosophical mystery. Only if you see man as imbedded into a world of which he is no natural part, do you think that very strange. The relations of mathematics are exemplified as much in ourselves as in the rest of the world.

The Symmetry of Identical Modules

The first kind of symmetry I want to talk about in detail I must still touch delicately. For although it's most important, it has nothing directly to do with perception on the level of works of art because it is at a scale too small for us to perceive directly. Like the unconscious computer which we carry around, it lies so deeply within experience that it determines that experience in many ways. I'll try to adduce one or two of them, although we never directly perceive them at all.

The principal, most important symmetry of the world, in the sense of Leibniz, the chief indiscernibility, is the fact that the world is modular. I mean this: the particles that make up all our world are supplied to us in a few models, but in incredible numbers, all of each type identically the same. You cannot distinguish any single electron from another. Upon enlarging a little bit, you can, because you can measure its axial spin, but then you can only distinguish two classes of electrons: those with one hand of spin and those with the opposite spin. Among all with a given spin, you can't distinguish any one from its fellows. Indeed, you can keep one apart very carefully; for instance, put it in a box, and make sure it stays in the box. If you never allow it any chance of getting mixed up with another one, then, of course, you'll never lose track. Call the one in the box Joe, and you'll keep him straight! But that is a trivial victory because you might as well have called him Bill. It makes no difference. If ever you allow him to get into position where you might exchange two, where you take your eye away for a moment, you won't be able to tell the two electrons apart. There is no tag, no marker, nothing at all that we know to distinguish one electron from another, or one proton from another, or one hydrogen atom from another. Allow the possibility of a spectrum of states in which each of these structures can be found; in the case of the electron-proton it would be two states, in the case of the hydrogen atom maybe an infinite number, but all states with labels. I can say, "Yes, this hydrogen atom is in state number seven," and then I won't confuse it with any other which is in state number five, but all the *fives* and all the *sevens* are enormous populations of identical objects within which I can never distinguish.

That is the most profound symmetry in the entire canon of symmetries. It lies at the heart of our world of matter and radiation. It is this, and only this, which guarantees that gold is gold, and glitter only glitter.

I will discuss sketchily the physics of modularity, the treatment of identical particles. I do so because it seems to me that in part, at least, this must lie at the bottom of what we regard as the visual symmetries, the symmetries of perception. Let me describe it in the simplest possible way.

How can we impartially label three electrons? We begin by writing down the obvious labels, e_1:e_2:e_3, in an arbitrary order. Now we rearrange the three labels in all possible ways. One can choose any of the three

numbers (1, 2, 3) to mark the first electron; for each such choice I have
two choices among the two digits that remain, but no more, for after
choosing two digits only one digit remains. For this example, then, there
are six and only six orders for the three particles. Whenever I refer to the
electrons, I must consider all six orders, and take only those descriptions
of measurement which do not discriminate orderings. For example, the
mass of one electron could be written as 1/3 the sum of the mass of each
electron: $m = 1/3 (m_1 + m_2 + m_3)$. It clearly makes no difference in what
order I choose the electrons; the mass so defined is impartial among the
labels. I would calculate averages by taking into account equally all six
orders of the three particles taken together. Then my theory is blind to labels as it must be.

This is no lecture in the mathematics of groups, but the obligation to
deal with the world's marvelous modularity, plunging description of nature into an extraordinary metaphor, a very "mystic rose" of physics.

The rather dull task of labeling among identical objects at once describes something apparently entirely distinct: the geometrical transformations of a magnificently simple geometrical figure in ordinary space.
That connection unites the linguistic and the visual, the algebraic and the
geometrical, perhaps one might say, the left brain hemisphere and the
right one.

Take an equilateral triangle like a capital Greek delta, Δ. Label the apex
1, and continue labeling its corners clockwise with 2 and 3. Now rotate
the triangle about an axis through its center until it is indiscernibly
changed. The peak has rotated one-third of a turn clockwise? Then you
have changed $_3{}^1{}_2$ into $_2{}^3{}_1$. Once more by a third of a turn: $_1{}^2{}_3$. Continue
as you will, always rotating the triangle this way, by one-third of a turn or
any integer multiple whatever. You cannot rotate a triangle of this sort into
any of its indiscernible, symmetrical presentations not described by the
three orders we wrote down. Put them down even more schematically, as
mere triplets of numbers in order: 123, 312, 231. There are no more results of rotation. But what of the six orders of the three digits? Indeed,
here lies dizzy depth. There is a symmetry richer than rotation in the
plane; it goes beyond plane rotation. It is mirroring, reflection, which
turns a triangle into itself again, preserving its full indiscernibility. This
can be thought of as placing a mirror against the paper, held along the
right margin. Then the triangle image is indiscernible from the real triangle apart from the labels we artificially attached. But now they run $_2{}^1{}_3$
(never mind that the digits are unfamiliar in the mirror; we can still make
them out). That order was not contained in the list of products of rotation.
Now rotate the mirror triangle: we have the order of labels 132, 213, 321.

These are all restricted to the results of at least one reflection. But taken all together, the results of pure rotation and reflection plus rotation yield all six orders, all the possible indiscernible positions of the delta, and exactly the results of label exchange. There are no more results possible, as there are no more operations which leave a triangle congruent with itself in Euclidean space: only rotations and reflections, and their combinations. The purely numerical exchanges gave the same results: the relation is called an isomorphism, and it lies at the heart of the theory of groups of symmetries. That theory is both rich and profound; here I have touched only the first lines of any text on the topic.

It is striking to see so deep a connection between Euclidean geometry and particle labels, which do not at all look geometrical. That is what we see on a large scale. If it were not true that one sample of stone is much like another, you couldn't have the sense of symmetry that one pillar is like a second pillar, and like a third pillar . . . in any temple colonnade. I believe at depth that is the same relation. It is a tremendous jump from indiscernible electrons to indiscernible limestone pillars, but it is no stupid jump; if art tries hard to remain pure and precise, cold and exact, it recovers more and more the hidden modularity of the physical world. One needs to carve all the pillars out of the same quarry—you don't want to take one black pillar and one white marble pillar if you want them to look the same—you'd then be emphasizing something different, indeed, you'd be breaking the symmetry, in a striking way. When the Shah Jehan built the Taj Mahal in his wife's memory, he intended to build another mausoleum for himself, much like the one that's on the bank of Jumna now, but across the river, in black instead of white stone, and connected to the tomb of his Queen by a golden bridge! His heirs wouldn't allow him to do it; they put him in jail first. (They didn't have the money, the fate of many magnificent projects!)

I argue that here is a simple but deep level in which we see symmetry arise in the inner nature of matter. Resembling it, mathematically one would say isomorphic to it, we had a relationship in Euclidean 2-space. Take those two examples together, generalize to other dimensions, and you have most of what the world calls symmetry. (So far we have talked little about symmetry *breaking*.)

By the way, to do this with a square, or to talk of *four* identical particles, is much harder. In the first place, there are 24 operations: $4 \times 3 \times 2 \times 1$. They are not isomorphic anymore to rigid geometrical rotation. You have to include the operations upon 3 and upon 2. The simplicity is gone, but still they contain all of the operations on a square, more than you can sort out. By the time you get to 12 particles, probably nobody's ever done it. The permutations grow to be bizarre.

It is remarkable—this is a digression I cannot resist. The physicist has demonstrated that the isomorphism of geometrical and permutation operations is not simple except in special cases. The only schemes that represent all numbers of particles are the very crudest schemes, of the following trivial kind: represent every operation by + 1. If I perform two I get + 1. No matter what I do, I get + 1. So if you call every exchange + 1 you require all permutations to produce no effect at all. That is said to be the symmetric representation. It is of course not one-to-one, therefore they say it is not "true." Nevertheless it is contained in the representations of all numbers of permutations. The other crude possibility is $(-1)^n$, which flips you back and forth between two classes, either the even or the odd value of n. These two are the only ways to represent all of the operations of permutation that can persist, not only for 3 electrons, or 4 electrons, 5 electrons . . . 10^{23} electrons, but any number you happen to have in mind. Now electrons have to live up to one fixed rearrangement of their labeling no matter how many partners with which you require them to rearrange. Otherwise we would see experimentally that once you put many electrons in a box, their symmetry would change. The world is not that way at all. We can conclude that only $(+1)^n$ or $(-1)^n$ are possible descriptions of what happens when you mix labels for any number of electrons. It turns out that relativistic quantum mechanics can show that it must be $(-1)^n$ for electrons, protons, and neutrons, and so on. But it is $(+1)^n$ for mesons and photons; only these two families of particles form the world: a profound proposition. We know of no breaking of this grand class of symmetries. Perhaps here I have said enough to enable you to see a little of how it must be on the level of fundamental particles. Among them we find structures of more complicated symmetry only when we arrange them in space. In the periodic table of atoms, or in the tables of nuclei, we are dealing with just such structures, geometry fused with permutations. I shall not build it up, but you have read about the electron shells, the intricate if approximate models which have some quality of truth. Chemists well know that spatial symmetry is found on the level of molecular bonds because modular symmetry exists on the level of the electrons themselves.

Symmetries of Space and Time

We also have more geometrical symmetries. Perhaps the most important of these is not rotation or mirroring already alluded to, but translation. That is, the fundamental symmetry of a set of columns of the Parthenon, or to invoke less lofty matters, the supports of an elevated expressway. There is symmetry in the sense of an unending iteration in length. You can use two dimensions of three dimensions. Crystals, as everyone knows, reflect this arrangement. Of course, they need not be as simpleminded as that, but yet they do require three unit directions and lengthy iteration of the translation. Notice, however, that such symmetries are necessarily broken in the real world. No real example of an unbroken translational symmetry can exist because the world is not infinite, or least, our experience is not infinite. So you must come at last to an end. A very long colonnade becomes a kind of a nightmare. You never come to the end of it. If instead there is an end to be seen, symmetry is broken by that end. The end is implicit in the finiteness of the structure, probably an important part of the aesthetics of such forms. If you're willing to make the form circular, rotational in nature, you don't have to come to an end because you go back upon yourself, but strict translational symmetry breaks at its inevitable end.

The subtle fabulist from Buenos Aires, Jorge Luis Borges, has told of a mysterious place where columns were built throughout a vast park by some mad prince; I suppose on top of each was a fine capital. They were beautiful, each a scarlet red. The next one you see as you ride along is just like the first. And you pass a dozen more, all about the same, as far as you can see, but if you ride a week you come finally to a white pillar! Yet each one has been like the first without any change. For, says he, the discernible difference between two hues of red that we call the same is, say, 1 percent along the scale of redness. Only after many pillars can you detect one colored a little different from the first. But you never see the two together; you see only a few neighbors, looking all the same. But by riding a week you come to a white pillar, without ever having detected any differences between neighboring examples. This tale is almost real. There is an analogy to the illusion in Escher's drawings, where monks walk up stairs all the time yet never get any higher. In both cases the paradox arises because the global and the local decisions of equality are not the same, a strong point for the idea of symmetry as discernibility.

Leave the fanciful things. Most familiar of all symmetries is human bilateralism, we are the same roughly speaking—our faces look the same, directly viewed or in a mirror. You and your image are one identifiable person. We know, in fact, this is always broken; it is rare to find someone for whom his good friends cannot distinguish a direct view from a mirror image. The flipped photographic print never looks comfortably familiar. It is hard to say what subtle differences allow one to tell. In the same way left hand and right hand are not exactly the same; the left half of the cerebrum and the right half are definitely not the same, one related to speech, the other to geometrical perception. Near the midline, the heart has its strong muscles over on one side. In all but 1 person in 10,000, that is the left side. On the other hand, we know that life reflects an inner molecular handedness; the amino acids in every protein, for example, are the so-called L-form, and their mirror enantiomorphs are not commonly found in nature, except in very special circumstances. You can actually feed bacteria on a symmetrical molecular mixture, to produce a handed residue, because the bacteria won't use up the wrong half of the foodstuff they're given. Here the symmetry, of course, is gone; the indiscernibility is removed for ancient historical reason. We don't really know how, but we have to say that the chain of life is linked, one form always using the molecular parts from another. You couldn't share in the chain of life if you didn't share this asymmetry. As far as we know, nothing whatever prevented life from starting the other way. Nothing would work differently. Had it started the other way, there'd be no necessary distinction except by confronting the two faces. That instead life sticks to one way predominantly is a sign only of its unity, its universal kinship, not a sign of special preference, except maybe at the beginning. Perhaps subtle forces caused the earliest life to choose one direction rather than another? It could have been chance, or it could have been the presence of particular handed crystals in one chance place where it all began. We don't know, it is a goal of active investigation.

There is a strong line of studying symmetry in the domain of fundamental particles, not only the enduring particles, electrons, protons, and so on, but the great zoo of unstable particles on which physicists are working. This is not my topic; I want simply to mention it. But there is, of course, a discernibility known now for more than fifteen years which I can express best in this way. Let me place a disk on an axis so that it can turn around freely, a real metal disk, some carefully-made piece of gyroscopic machinery. Now I paint the surface of this disk with a certain radioactive substance, for example, cobalt-56. If I made this disk well, placed it in a vacuum, and allowed the 50-day half-life of cobalt to pass, in order to allow most of it to decay, it should spit a lot of electrons outward. But inward the electrons find the disk rather thick, so they get caught and remain fixed in the disk. The disk loses preferentially those electrons which started outward. Such a disk will spin spontaneously in a clockwise direction. That's built in to the way that cobalt-56 decays! (The matter is complicated because each atom gives out two particles—neutrinos which are never caught and electrons which can be but always share in such a way that the higher-energy electrons preferentially take off one direction of spin; so do the neutrinos, but they're not getting caught!)

Such an experiment is not practical but is entirely conceivable! Less macroscopic analogies to it have been done with care. This was striking; it was anti-intuitive, and it won the Nobel Prize for Lee and Yang after Wu and others did the experiment in 1957. It was a wonderful result, if not very much more came of it. What was immediately shown was that, to be sure, you could distinguish the right hand from the left hand; there was a direction built into our world. Pauli said, "God was displayed to be a left-hander." But not so. It turns out that if I take other radioactive elements, the disk might not spin the same way. That's all right, one can always recognize cobalt-56 by counting the number of nuclear protons. It seems still perfectly unambiguous. But there's another experiment. Count the protons, everything's correct: I build the same disk, but lo and behold it spins the other way! Yet I've done exactly the same thing, save for one little fact. What is needed to make it spin the other way? A source of anticobalt-56. What is anticobalt-56? It is cobalt-56 in which the protons are replaced by antiprotons, and the neutrons replaced by antineutrons. You can't buy that. But we have made antideuterium. So it is only a question of care and effort; we might in principle be able to go up to cobalt-56. We don't doubt that it would work the same way. But it shows that discernibility is like the mirror. The left and the right hand are exactly the same. I don't know whether a mirror has been interposed or not. In other words, the mirror's image of the left hand is just like the right hand. In the same way, the antimatter image of cobalt-56 changes its handedness. So one does not know whether the sample is from the real world or the antiworld; again there is no way to distinguish left from right. There is no experiment which will enable me to distinguish anticobalt-56 from normal cobalt-56. If I bring the two together, they mutually react. Both disappear in a burst of unstable particles and radiation. It is generally believed that the matter-antimatter mirror is not exact. These are deep questions of broken symmetries, of fundamental interest to the particle physicists of today.

The idea of symmetry, the idea of the indiscernibility of transformations has been fairly straightforward. We change plus to minus, one direction to another, an object to its mirror image, rotation to translation. One profound result can be added. Recall the foundations of classical and, indeed, of quantum physics: energy, momentum, and angular momentum (like spin) are the three great conserved quantities of classical and quantum physics. Everyone knows energy is conserved, momentum too is conserved, and angular momentum is conserved in all processes of which we know. If you gain energy, some other system must have lost it. If you gain momentum, some other system must have lost it. If you gain angular momentum, some other system must have lost it. If you start with none, you can't get any energy at all, because it's just a number, without direction. If you start with no momentum, you can gain momentum, but only by sending away the necessary amount in the opposite direction. If you start with no spin you can gain spin, but only by sending out spin in the opposite direction. Those are the great constancies of physics. We know no exceptions to them; they hold very widely, and a very few others, only a few, like the constancy of electric charge, complement them. It is possible to show that under reasonable conditions the existence of the conservation of energy is essentially a statement that in the development of a system, one time is like another. The conservation of momentum is a statement that in the development of the behavior of a system, one position in space is like another. There's no difference; they are indiscernible. Angular momentum constancy says essentially that one axis direction is like any other. Physicists are used to saying that these are symmetries because there are constants during any process which can be shown inescapably to come with the space-time symmetry of any system. If I take *the whole* of a system, carefully allow nothing to be left out, it doesn't matter at what time I consider that entire system; it's going to work the same way. It's true that if I have an alarm clock, it'll wind down. But if I can deal with the whole system, the winding up and winding down, the effects of air, and so on, if I take all of it into account, I could have done the same thing a million years ago, and it would work the same way. Absolute time makes no difference. On that basis you can show that the abstract quantity we call energy must be conserved. In the same way, if absolute space makes no difference, momentum must be conserved. If absolute direction makes no difference, angular momentum or spin must be conserved. It is probable that all symmetries are a little bit broken because the real world does show some kind of structure in space and time and direction. We can't put our fingers on it yet, but the universe is not the same in all directions. It is more or less the same, but when I point *here,* I do not see the very same galaxy as when I point *there.* On the average, I see the same kind of thing, but when I point at the Andromeda Nebula, it is not the same as M33. Maybe that specificity shows up, but no one has found it yet. People have looked hard, down to a part out of 10^{15} and 10^{20}, and they haven't found it yet. But it might be there—a failure of the conservation of energy or a failure of the conservation of angular momentum.

Symmetry and the Act of Becoming

The next point lies closer to the surface. Examine the changes that go in the world, all kinds of changes: the growing of plants, the froth moving on the waves, the flash of a lightning bolt through space, how each of us ages. Try to give an account of change in some broad, general language. One thing that you might say is that there are certain principles of economy; each natural process spontaneously seeks a minimum of some kind. The philosophers of the eighteenth century put a great deal of weight on that. They thought that here was a great economical principle, Nature seeking to do great things with least means. We have a less broad view of it now because we recognize that these minima are not global, but local. That is to say, you might find a new way of doing something by going along a completely different path, still easier than the way it actually goes, but on a different branch of behavior. For example, you know how light goes from air to water; it breaks toward the normal. It's an easy demonstration: if I know the speed in air and the speed in water, the shortest path from start to finish is the path given by Snell's law. That was a grand principle in the old days because they felt something marvelous in the economy of nature. What this means is that the shortest path compared with all nearby paths is realized. But if I put a mirror in the way, I can get a much longer path, yet still just as natural. It turns out that they are often the *longest* paths that you can take. It doesn't matter to us; it's not the shortness that counts for us, it's the extremal nature. In the near neighborhood of the actual path, there are many neighboring paths almost the same length. Those are the paths that are realized; whereas any path that goes in a crazy excursion across the countryside has nothing comparable to it in its near neighborhood at all. Such paths never add up to anything. The real world works on this "most probable" scheme. We ought not to make a great deal out of minimal (or maximal) properties, but there are many situations where we imagine that a minimum state will prevail. For example, a soap bubble is spherical; clearly the spherical form gives you the minimum surface area for the given volume enclosed. If you ask what is the energy, there's the gas pressure inside the bubble and the surface energy of the film. Minimize the total energy by making a spherical bubble of a certain size for every given pressure. You can see how that would work. All I want to argue is that the fact that it is minimal gives a mathematical target at which to shoot. (Actually, it's only the extremal that really counts, the reinforcing property, near the minimum or sometimes near the maximum, that is important.)

Notice when I talk about soap bubbles, I am far from talking about the domain of atoms. When we think of soap bubbles, there is a volume inside full of gas and the filmy surface of the soap bubble. Of course, it's not a surface; the film, too, is a volume. In the atomic view of the world, there is a collection of certain atoms of high density in one place and others of low density in another; certain forces in one place, other forces in another; there is no distinction in kind. But it is a good approximation to deal as though one were in a true continuum containing various continuous substances; that's what we do. The film is different from the air. That is not a molecular view of the world, but a continuum view of the world, suitable for almost all work in classical physics, or on the scale of art we can perceive. You rarely have to consider molecular structure itself.

Let me describe a situation in which the breaking of symmetry during the act of becoming is made manifest to the eye. There is a wonderful, if expensive, toy which does the job remarkably well. (It is the invention and product of a clever designer in Montréal and is sold under the name *Atomix*. Many a reader will have enjoyed one.)

You see a transparent plastic slab about the size and shape of a paperback book: say five inches square and a little more than an inch thick. It is not, in fact, a solid slab, but a sandwich. The two pieces of "bread" are carefully fastened together with a sealed air space as the "filling" between, a very thin filling, little more than a millimeter thick, over a square somewhat smaller than the two defining plastic pieces of the sandwich. That air space is occupied by some five thousand small steel spheres, similar to the ball of the ball-point pen. They are good commercial spheres, quite alike, indiscernible to the eye. They can very well represent individual atoms as the designer intends. Of course, they are grossly inaccurate on the atomic scale; no two are alike to microscopic examination. You could learn to call each one by name, unlike atoms. Human scale does not permit the modularity of the particle world; after all, each little sphere contains more atoms of iron than the number of snowflakes which fall worldwide per year. They are not identical spheres, but they are indeed similar. They are smooth and free to roll within the air space, where they take up perhaps two-thirds the volume. The space they occupy is so thin that they remain spread out in a single layer however they move; there is no room for one to pile on another along the thickness of the air space.

What a mobile and wonderful layer the small spheres make! They respond to every tilt of the device by rolling downwards under gravity, until they touch the walls or other spheres and can no longer fall. They are spheres, and in the uniform packing which similar spheres can take up, each one touches six neighbors in the plane located symmetrically around the central one. They arrange themselves into such hexagonal patterns over the whole area of their space, obedient to the requirements of seeking a minimum energy. They can't pack more closely, for the steel is rigid and little deformed under the small weight. They fall until they touch, for the force of gravity overcomes the little friction, and they stop, rolling only when another sphere or the plastic wall matches the gravitational pull. They remain in place in their thousands once they have come to the equilibrium of minimum energy, in a striking crystalline array, a plane layer of identical spherical "atoms" under minimum energy, exactly analogous to the structure of crystals, but, of course, on a very different scale, one perceptible to the eye. No plans for design were needed; the orderly crystallinity, the symmetry we see, arises out of the automatic play of force on an array of identical spheres. The spheres are indiscernible; each one acts like its neighbors, and the enforced result is the hexagonal packing clearly visible whenever the device is viewed. Order has appeared fundamentally out of the symmetry of the sphere and the rule of modularity. No designer need intervene.

Yet that order, that symmetry, is not perfect. It is definitely broken, but not in a chaotic way. Here and there a hole occupies a point where a sphere would fit. In rolling there, some spheres by chance so fitted that beneath them a hole formed which could not be reached by a sphere that would have fitted there. A hair more shaking and that hole would have filled. As chance would have it, the place remained empty. In a typical shake, there are dozens of such spaces left. But its neighbors pretend the whole is filled and make their hexagon anyway. It is much better for the energy minimum that they each have five real neighbors than that they act somehow crazily. Rarely there are two holes nearby. More often there are lines of slip and mismatch—quite understandable and quite like those which the physicists have found in real atomic crystals. A whole line slipped out, and two neighbor lines made up the space by a jump of only one ball spacing. Or one region of five hundred balls arranged itself in the hexagon pattern while the balls rolled; so did a region nearby. But there was no way in which one region could signal the orientation of its hexagons. The two meet at a boundary line along which there have to be a few regular spaces. The texture looks different; this is what in real material is called crystal grain. All the most important defects of real crystals are exemplified; but that is not our story, helpful as it is to generalization.

No, the point is that the perfect symmetry is broken. It must be arrived at by local force interactions, but any chance event will modify it slightly. As the order continues on in space, sooner or later it must find some other center of order with its own modification. There a conflict will occur; the final order is partial, regional, incomplete. If with arbitrary slowness we allowed the balls to roll one by one, and avoided all interference, if we intervened to make sure that one hexagon pattern formed with the same orientation as its neighbors, we might make a perfect "crystal." (A patient and dexterous player can, in fact, do this with an *Atomix* model.) But in the real world of finite rates, of noisy interference, of blind forces of assembly and disassembly, perfection is, not inconceivable, but rare to the point of no practical significance. Large systems cannot be perfect unless they are formed at a snail's pace. Big gems grow slowly and are rare. Local symmetry and global repentence is all that a world of atoms can achieve. Symmetry almost always is broken by growth, as by destruction. Most often the system has to satisfy the minimal principle. But if it anywhere doesn't reach minimum, there's going to be a defect. The memory of a defect will force a defect somewhere else. The whole thing has got to work out in the long run.

The symmetry of the array is broken, really because of time. Since no process, certainly no spontaneous process, can go on for infinite time, some irregularity must occur to propagate its mismatch faintly throughout the entire system. It is just as wrong to expect that perfect symmetry to be present as it would be to neglect the symmetry entirely. The first sort of error would be to ignore the whole idea of the energy minimization principle. The other sort of error is to imagine that the whole system is constructed with infinite time available, or, if you like, with every perfection as a goal. But in reality structure is built by local forces. They do the best they can at every point, but they cannot achieve perfection. When once they fail the effect must be made up somewhere in the system as a whole. The spheres give up finding an absolute minimization at every point in favor of a practical solution. In most places the structure can come close to the minimum. With a finite rate of formation, it will do the best it can to adjust. It doesn't wait until somebody comes to make exactly the right geometry. It can't grow that way. It has to grow by local events. The global adjustment must be propagated, yet it can't propagate quickly enough to satisfy the minimal condition everywhere.

This is a metaphor, an analogy, for what happens in all real processes in the material world. We can never minimize energy truly in the presence of real processes that require real times of growth. Everything must grow at some rate. We always trade off error against time; we trade an impossible symmetry for the scarce but possible energy. We have to pay in the coin of energy to be absolved from a need to try hard for exact form. That is the nature of broken symmetries in the world. Order forms spontaneously all the time; the Second Law has nothing against that at all. The Second Law insists only that if you make order, you must pay for it. You must pay in energy. Of course, if you make an improbable arrangement which is yet very orderly, you have had to use some of the energy that otherwise you might have gained. The reason is quite clear: The world is immersed in disturbance. The world does not consist of isolated subsystems sheltered in their growth. When I shake the ball bearing model to try to make them move into place, that corresponds, if you will, to a sort of Brownian motion at a high temperature. If I went in there with tweezers to move each one, carefully enough so my hand doesn't slip, I might get them all arranged perfectly evenly, at a very great price. Any speed of working is going to breed errors; those errors must break the symmetry.

The world is governed by minimal principles which apply locally in every place. They cannot be expected to overcome the constraints imposed by time and topology. It is just as remarkable to see an absolutely perfect crystal as an assembly of the right atoms with no crystallinity at all. Neither could occur. The only way to have an absolutely perfect assemblage would be to have the whole thing formed at the absolute zero temperature—no randomness. But, of course, at true absolute zero the rates of formation would go to zero. Nothing would ever happen. Only an isolated world where nothing ever happens can be perfect. (Is that Nirvana?)

Broken Symmetry and Art

I am not justified in making a moralizing conclusion in aesthetics, but I can state a prejudice. . . .

What we regard as highly satisfying works of art, even many natural things of beauty, contain broken symmetries. The symmetry is made manifest in some form, yet it is not carried out to perfection. The contrast, making visible both sides of the act of becoming, demands appreciation. A soap bubble is beautiful. Somehow everyone would agree that it has a kind of simplicity, a coldness, which bars it from the category of great beauty. In fact, the very reflections and color changes which make it something other than a perfect sphere enhance its beauty. A cube of glass, too, is a beautiful object but no work of high art. If you see the work of a lapidary, a rough crystal, the crystallinity plain on some faces, but hidden in the matrix of others, it is a more satisfying object. I suspect we react to the fundamental thermodynamic quality: an expression of symmetry, yet one not allowed to dominate exclusively, just as it cannot in the real world, for some feature always breaks every macroscopic symmetry in the end.

Looking at the World

To describe the evolutions in the dance of these gods, their juxtapositions and their advances, to tell which came into line and which in opposition, to describe all this without visual models would be labor spent in vain.

—Plato, *Timaeus*

O f all our senses it is vision that most informs the mind. We are versatile diurnal primates with a big visual cortex; we use sunlit color in constant examination of the bright world, though we also can watch by night. Our nocturnal primate cousins mostly remain high in the trees of the forest, patiently hunting insects in the darkness.

It is no great wonder that the instruments of science also favor vision; but they extend it far into new domains of scale, of intensity, and of color. Inaudible and invisible man-made signals now fill every ordinary living room, easily revealed in all their artifice to ear and eye by that not-so-simple instrument, the radio, and the even more complicated TV set. It is very much this path of novelty that science has followed into sensory domains beyond any direct biological perception. There, complex instruments assemble partial images of the three-dimensional space in which we dwell, images rich and detailed although at scales outside the physical limits of visible light.

The images finely perceived by eye and brain in a sense span the scientific knowledge of our times (though it is risky to neglect the hand). The world is displayed by our science in diverse ways, by manifold instruments and by elaborate theories that no single person can claim any longer to master in all detail. The presentation of the whole world we know as though it were a real scene before the eyes remains an attractive goal. It should be evident that no such assemblage could be complete, no picture could be final, nor could any image plumb the depths of what we have come to surmise or to understand. Behind every representation stands much more than can be imaged, including concepts of a subtle and often perplexing kind. Yet it is probably true—truer than the specialists might be willing to admit—that the linked conceptual structures of science are not more central to an overall understanding than the visual models we can prepare.

The Gamut of the Sciences

The world at arm's length—roughly one meter in scale—is the world of most artifacts and of the most familiar of living forms. No single building crosses the kilometer scale; no massive architecture, from pyramid to Pentagon, is so large. A similar limit applies to living forms: The giant trees hardly reach a hundred meters in height, and no animals are or have ever been that large. The smallest individual artifacts we can use and directly appreciate—the elegant letters in some fine manuscript, or the polished eye of a fine needle—perhaps go down to a few tenths of a millimeter. Six orders of magnitude cover the domain of familiarity. Science conducted at these scales is rather implicit: The most salient disciplines are those that address the roots of human behavior.

Let us begin to marshal the furniture of our world according to its physical size. On larger scales, only occasionally does the work of our energetic species show up: a bridge, a wall, a dam, or a highway. These are typically less than fully three-dimensional. They seem long ribbons when occasionally they are caught in aerial views. Only in their collectivity do we see human artifacts that occupy large surface areas (still not three-dimensional) in the ten- to hundred-kilometer range, sometimes even beyond. These are the cultivated plains and terraces, the irrigated lands, the clearings of the ancient forest, the great cities and their environs. Theirs is a history of growth more than one of design. For the rest of life, too, we find a similar display. Blades of grass are small, but grasslands and savannahs, like the dark forests north and south, extend over whole regions, easily up to a thousand kilometers across. It is these regions that make up the visible large-scale landscape. Here the cognizant sciences are those that aim at the nature and use of lands. Perhaps still more germane are the descriptions offered by the historians and geographers of an earlier time, and those offered by the adepts of elaborated practical technologies, from agriculture to forestry and engineering, ancient and modern alike.

Once past the scale of a thousand kilometers, we lose sight of our species. At the global and regional scale, up to ten thousand kilometers, cooler sciences enter in strength. The swift motions of the air, its clouds and ceaseless winds, the slower flow of rivers, ocean currents, glacial ice, and the majestically slow drift of the solid continents themselves lie behind the single views. These occupy the dynamical sciences of meteorology, oceanography, hydrology, and geology. Within this generation, geology has far extended its grasp; until recently, the globe as a whole was hardly a geological topic. Regions were well understood, but no known single process had joined distant shores of wide oceans or ringed the whole globe. All that has changed: Today a geologist may take the Earth for province.

One leaves the Earth, but not yet the domain of humanity, beyond the range of ten thousand kilometers. Out to the moon we have sent intrepid explorers, while the geosynchronous Earth orbit, a ring five Earth radii out into space above the equator, is now a well-exploited natural resource. Satellites orbiting within that gravitational band neither rise nor set, as watched from the spinning Earth, but remain always in view of the artfully aimed fixed dishes; they relay word and image over radio links to and from almost every nation.

It takes a scale six powers of ten larger to reach the boundaries of our solar system, out there among the unseen comets. The sciences of the solar system—the studies of the surfaces and the interiors of the planets, large and small, their satellites, meteorites, the comets, the dispersed dust—are today more than just astronomy. No longer do we merely look from afar; we touch and sample now, at least vicariously with our robot probes. Astronomy proper begins now with the stars; one of them, the Sun, is our own life-giving hearth, the only star close enough to permit detailed study. A great gulf is open between our home region near the Sun and even the second closest star; the steps must cross four or five powers of ten to enter the realm of the stars. It is a remarkable story, first told in our century: the birth, development, and life history of the stars, diverse globes of gas into which most of the visible matter in all the universe is bound. This is the kind of astronomy held in the root of that word itself: the study of the stars. It is mature now, though certainly unfinished.

Let us now look the other way, inward from the submillimeter world of the attentive but unaided eye to the microcosm. First in interest come the intricate machinery of our own bodies and its counterparts in all the larger forms of life. Here we engage anatomy, physiology, histology, cytology— a battery of specialties, ending with the study of the cell itself, the ubiquitous unit of living forms. Three or four further powers of ten span that whole microscopic world of life—microbiology—down to the smaller cells of the most ancient forms of life, to uncover the not quite living parasites, the viruses. But at that level, on the scale of a thousand angstroms or so, we encounter the mechanisms of molecular biology (and its newer emulation in the textured artifacts of microelectronics). These scenes relate form to function: The form is molecular; the functions are among the deepest properties of life, shared by the full web of life during all the time we now know of terrestrial evolution. Here we speak of genetics and of the biochemistry of large molecules and their cycles of interaction. Before long we cross the vague frontier that parts life itself (so much the most subtle of chemical processes) from the chemist's world of random motion and atomic bonds.

Look out again at the celestial scale. There too we cross a real boundary of nature once we leave the Milky Way environs to see galaxies, whole spinning pools of stars bound together over time. The astronomy of the stars first extended to the dilute interstellar medium, the star-forming matter from which new stars are born, and then went on to galactic and extragalactic astronomy. The fascinating forms and varieties of star-pools are strewn throughout space as far as we can see into the distant world.

Again in journey inward from the world of the large molecules, we reach at last the individual atom, at a scale of about an angstrom. Below that scale, all science is physics and chemistry; once we set out to explore the innermost spaces of the atom, we enter a strange domain well beyond any direct imagery. We can represent it only as it has been puzzled out with the powerful tools and concepts of modern physics. Our study has led far, to new fundamental laws, at first paradoxical but now fruitful in explaining the patterned and stable world we know within matter. The modular world of the hundred chemical elements and their larger but still limited variety of nuclear species is a world ruled by a subtle interplay between order and chance.

The two ends of our procession of images, the terminal scales of large size and small, mark the limits only of contemporary knowledge. On the one end, far out where the galaxies appear like a glowing froth in darkness, all our sciences become only one: cosmology. We know of no spatial novelties beyond the billion-light-year point. All the distinct structures we are aware of are safely smaller than that. There is certainly wonderful novelty, but it is expressed not spatially but over time.

The universe has come to be filled with those diverse galaxies formed out of a once bland and uniform stuff. At the other end, for the very small we have again one science only: particle physics. There are even hints that the two ends inform each other: The fiercely hot early universe may once have held only the kind of matter we see transiently now in the particle labs. Ours is a modular world, built out of myriad replications of the simplest structures, structures that we are only now beginning to understand. Within the nucleus is the proton; within the proton, the interacting quarks. Within quarks? The magnetized rings and tubes that are our great accelerators, the ultramicroscope probes of our time, have not given the final answer.

Forty-two powers of ten so far span our firm knowledge; we have only brave hints and conjectures beyond that. We do not yet know, though we can argue about it, whether infinity lies within the real world as it lies within the mind's reach. How far can we continue the excursion in either direction, out to the macrocosm or in to the microcosm? Some day we hope to learn.

The Same and the Different

Over the wide sweep of size, the eye scans all that is made. When the thought of Aristotle turned to such a gamut, he saw the sky-world as wholly different from this Earth on which we dwell a little while. We know his distinctions: up there, the objects are shining, circular, perpetual. Here below the moon, matter is usually dark, and its motion is neither elegant nor enduring. But in our times we have seen these two worlds made one. The satellite we launch is earthy, faulted, marked with the finiteness of the minds that designed it and the hands that wrought it. Yet given that one terrible flaming impulse of the rocket, it enters the celestial regime, into an orbit that can be as purely circular as that of any celestial body, there to shine enduringly, bright as a star in the night sky.

The world is not dual; that below and that above are one and the same. The difference is but distance and motion. Aristotle's theory, which has its insights, was fully superseded by the wider insights of Copernicus three centuries ago. Since then we have come to know well that our Earth is just as celestial as any other planet; it too shines, its motion is circular and perpetual—if epicyclic—once viewed from far enough away, once seen from elsewhere in the Sun's cortege. We dwell on Spaceship Earth. By the same set of experiences, we recognize that the red wandering star we call Mars is not an ineffably luminous body, quite unlike the Earth we know. Our cameras show instead that it looks a good deal like the desert of Arizona. All the rest are places, too, places as physical, as finite, as wonderfully complex as Earth. Of course they are not identical to Earth; our Sun, for example, is a giant fiery ball, shining and hot as no planet can be. But its globe of gas gravitates just as do the rocks of Earth; its nuclear fires, so enduring as to have warmed our Earth since its earliest days and to have nourished somehow out of water and air the thread of life, are now under detailed study. The solar fires are long-lasting but finite; one day we may set up their counterparts here (as we have already done transiently in grim thermonuclear explosions) to run the shafts of industry. Heaven and Earth are not wholly distinct, nor are they one; they differ in their nature, yet they are joined in a wider unity.

So it is with the microcosm. Again the Greek insight was wise. The early philosophers tried to explain how wood could change to fire and ashes, how bread could nourish the hungry, how black iron might rust red. They had the idea that deep down below the size we perceive, matter was a web of small modules—the atoms—whose incessant rearrangements account for all becomings. That profound idea, turned from speculation to a powerful and growing mastery of the hundred elements, has teased out the linked fabric of every substance old and new. It has led to today's view of atomic matter, made clear in the images we build of the world of the small. Again, there is no trivial path to that understanding: The atomic world is not just like the one in which our senses place us. To be sure, it is the same world, for we have found no other anywhere, but it is related to the world of familiar experience through that same curious blend of the marvelous and the homely that we find out there among the planets. The step-by-step examination is best shared by a traveler who is pleased alike by unexpected familiarity and by exotic novelty.

Lilliput and Brobdingnag

Dr. Lemuel Gulliver, acting for the ironic Jonathan Swift, is the most famous of travelers among the powers of ten. (Truth to tell, he visited the two neighboring powers of twelve.) The pettiness of the sectarian wars in Lilliput and the disdain of the magnanimous Brobdingnagian giants for us self-serving little humans are both deftly reported as a new quality of moral judgment arising from size alone. In physical dimensions, these peoples, big and little, differ from us mainly through our perceptions of them. Gulliver sees the Lilliputian world at a distance, say to admire the elegant threading of an invisible needle with invisible thread. The giants he sees as if in a microscope; their most elegant artifacts seem crude, their very bodies gross and marred to his too-close gaze. Exactly this had been the real surprise offered by the microscope: The finest needle points turned jagged and coarse in its magnifying lens, and the most admirable complexion viewed microscopically was a sea of blemishes.

But the effects of scale go well beyond perception in engendering novelty. The world works differently at different scales. Swift well knew the foundations of geometry; he understood that it takes not simply twelve, not even twelve times twelve, but twelve times twelve times twelve six-inch Lilliputians to equal the bulk of one six-foot man. For not only must the height be increased twelvefold but the width and the length must be increased as well. (Eight cubes of sugar—two layers high, each layer a two-by-two square—need to be stacked together to build up a single cube with an edge twice as long as the edge of one of the original cubes.) Thus the Lilliputian planners requisitioned for their prisoner Gulliver $12 \times 12 \times 12$, or 1728, daily rations.

Experience shows that so simple a trust in schoolbook geometry is unrealistic. Consider that a man might get by on a loaf or two of bread a day. But to feed a small animal like a chipmunk, six inches long, on about one part in two thousand of that diet means the active little creature is fed each day a piece of bread no larger than your thumbnail. Starvation rations! The animal will contentedly eat a third to a half of a slice each day.

For the truth is that the world is not ruled wholly by the simplicities of Euclid. A scale model may resemble its counterpart with fidelity to the eye, but, in general, the model cannot work in the same way. It was Galileo himself who first pointed out this result. Put aside the complexities of nutrition, heat loss, and the rest, and consider one single static but important property—structural strength. The learned Salviati asks early on the first day of the dialogue in Galileo's graceful *Discourses Concerning Two New Sciences* (1638), *"For who does not see that a horse falling from a height of six or eight feet will break its bones, while a dog falling from the same height . . . will suffer no harm?"* Later he remarks, *"A little dog might carry on its back two or three dogs of the same size, whereas I doubt if a horse could carry even one horse of the same size."* Indeed, small things are relatively strong, large ones are weak. Great steel bridges cannot support a load equal to their own weight; any wooden plank does much better. It is no special miracle of design that fits a worker ant to drag back to her nest a fly much bigger than she is, nor one that allows a small bird to fly so well, while humans can barely fly on muscle power but must enlist the aid of hot and thirsty metal engines. These facts imply that form follows not function alone but size, especially over large changes of scale. Such is the effect, in the physical world, of "adding another zero."

We cannot follow up the reasons for all the effects of scale change; that would require treading most of the paths of theoretical science. It is enough to see how one case works out: the structural weakness of the great and the strength of the small. Every structure on Earth must support itself against the pull of gravity. Gravity is exigent; it reaches inside every candy box, no matter what the wrapping, to distinguish the full pound from the empty container. Humans, horses, and dogs walk about on their feet; only the foot surface is available to bear the full weight of the whole organism. That surface area, like all areas, increases by 10×10—a hundredfold—for one tenfold step in linear size, say from a small dog to a horse. So the bearing area goes up by one hundred. But the weight of the whole body, the total load to be held by the bearing surface, goes up by $10 \times 10 \times 10$—a thousandfold. The horse is sure to have proportionately less structural reserve than a dog. Whatever its design, a structure enlarged sufficiently must therefore fail. Where the design remains similar in form, the reckoning is accurate. Of course a dog is not in fact built very like a horse. That difference in form partly reflects the distinct differences in behavior of dog and horse, but it partly reflects the adaptation to a change in scale. Large animals tend to thick and sturdy form; small animals are recognizably graceful, agile, subject to chill, restlessly hungry, easily waterlogged. Each of these familiar traits can be given at least rough explanation as a simple demand of scale.

There is an inner reason for the difference. Dog and horse are built of the same materials, of flesh and bone. Deep inside matter, the atoms do not change in size as the size of the creature increases. The scale model is inconsistent. The stamp of intrinsic size is held in the nature of the atom; in another universe than ours, one with differing atoms, these arguments might be circumvented. But here they rule. Nor is it only the elegant engineering of living forms that reflects the decisive result of adding zeros. Human technology naturally must obey the same rules. Moreover, natural forms, quite apart from life and its evolution, are no different; mountains too must bear their own weight. They can reach heights only very small compared to the whole Earth, or even to Mars. Every planet, every star is nearly a sphere. But smaller bodies, like the little moon of Mars called Phobos, can be quite unlike a sphere: Phobos resembles a rather poorly formed potato. The reason is in its size alone. As the size of a body increases (usual materials always assumed), the self-attraction arising from gravitational forces pulls internally on all portions of the object. Let the body become of great size, and no material has sufficient strength to withstand the effects of gravity. The object is self-compressed; it strains to become as compact as possible. Once it is massive enough, it must approximate a sphere. It turns out that the yield limit for rock is reached somewhere near a couple of hundred kilometers; below that size, satellites and asteroids come in all shapes: bars, bricks, lumps. Much above that limit, they are all spherical. Astronomy is thus the regime of the sphere; no such thing as a teacup the diameter of Jupiter is possible in our world.

Let us look at still smaller objects—and at other properties. No one would expect that, if you dipped your hand into a bowl of sugar cubes at the table, you would bring out a cube with it, unless you positioned your fingers to grasp the cube. But if the bowl instead contained granulated sugar, some crystals would surely stick. If the bowl held powdery confectioner's sugar, you would fully expect to withdraw a white-dusted sugary hand from the container. Why? It is all the same material; the adhesion of sugar to skin remains the same. But that adhesion grows with the area of contact between the two surfaces. The total stickiness increases with the area of the sugar sample removed; but the weight of the sugar varies rather with its volume. Normal adhesion cannot lift a sugar cube: too much volume, too little surface. For the powder, the circumstances are reversed. The world under the microscope is dominated by just such surface effects; it is frictional, sticky, clinging. Small bodies do not coast; they show little effect of inertia. Bacteria experience pure water as we would a pool of honey; the water is the same, but the cell's surface is so large by comparison with its tiny volume that even the water we find so yielding fiercely inhibits their motions. We repeatedly encounter phenomena associated with scale change throughout our journey. Stars shine, planets are round, bridges remain geologically rather small, cells divide rapidly, atoms randomly vibrate, electrons disobey Newton, all because of scale.

Invisible Order

No visual model can convey unaided the full content of our scientific understanding, the less if it is restricted to the static. Pictures in a book mostly present a static account of the world, a limitation not imposed upon swift-flying film or videotape because of their vivid fidelity to the world of change. Film and the video processes together constitute the most characteristic form of art in this changeful period of human history. The limitation of the static image is not simply that it lacks the flow that marks our visual perception of motion: Real change in the universe is often too slow or too fast for any responses of the visual system. The deeper lack is one of content. A single take belies the manifold event. Several still images in a sequence chosen to bring out the nature of the change can often do as well cognitively, if not evocatively, as the flow of re-created motion itself. But that sequence is the key: It takes more knowledge to convey change. While our attention is seized by the moving image, we usually forget that it is nothing but a subtle illusion induced in eye and brain by a multiplicity of still pictures. But what a profligate flow it is! A couple of dozen distinct pictures each second are needed to simulate ordinary moving reality.

All that implies how much a visual model needs some additional means by which to present time, to mark change and its rates. For the world is a pageant of events whose relative rates of unfolding in time are as important as the size of the stage upon which they play.

One familiar kind of unfolding can offer a transparent example. The physical world is in motion, whether openly (as in our everyday surroundings) or more hidden (as it is within molecular matter). Sometimes the tempo is so slow that change is concealed from one quick glance, as in the larger cosmos. We measure the speed of motion most directly by noting by how much an object's position in space changes in a given time. The apparent speed of the planets in the sky provides a correct clue about distance that not even the most archaic of naked-eye observers would ignore. The perceived sequence of decreasingly swift motions is familiar: birds, clouds, moon, stars. . . . Since Copernicus, we have been able to translate the observed round of the planets—how many years each requires to return to the same point, taken not with respect to the Earth, but to the central Sun—into their true orbital speeds. For the size of their circuits can be read from a Copernican map, relative not to the meter standard but to the orbit Earth. Jupiter's near-circle is five times the diameter of Earth's. But Jupiter's year is about ten times longer than our year. That result (even then known with accuracy) was in the hands of the Copernicans from the first; it falls at once out of a scaled diagram of nested circles. Copernicus's system is by no means as simple as his drawing might suggest. Copernicus found it necessary, as today's nonastronomical reader does not, to take into account the small deviations of the planets from uniform circular motion, long known to the old star-clerks.

The speeds turned out to be of supreme interest. In the figure, the measured relation of speed to orbit size is plotted in graphic form using the data of today (taking a mean radius to represent the actual ellipse). The regularity from planet to planet is striking; it is a relation first adduced by Kepler and made rational in detail a lifetime later by Isaac Newton. Our point is simply made: Without watching over time, we could not have found this rule. Nor is this merely one detail of order missed by the static onlooker; it is a profoundly diagnostic one. Compound this regularity among planets with a similar rule for the little moons of Jupiter, for those of Saturn, and for our moon itself (its motion related to the fall of an apple at the Earth's surface) and something of the order behind celestial appearances comes through. Go on to artificial satellites, whose motion was predicted by Newton in the seventeenth century, and on to the distant double stars, and the point is universally made. Once time is added to space, once the visibly moving model is augmented with quantitative inferences, the invisible order of the world earns the awe with which the Enlightenment saw it. Their sense is shared by the physicist today, even one who recognizes fully that the world holds essential disorder, too.

The simplicity of the celestial rounds has long made them the paradigm of order in science. But the study of change has gone well beyond the planets. If the eighteenth century admired the Newtonian revolution to excess, our present thought is equally caught up in the physicist's notion of energy. Energy does not mean something close to sunshine, gasoline, or jelly doughnuts; the technical meaning is clear, but altogether more abstract. The concept has been extended successfully beyond the mechanical domain, in which it first arose, to every process of physical change. Energy is a quantity that can be defined for any system; its value can be calculated, but not directly observed, by procedures that demand much detailed analysis of the system. Once the value is found, it holds without change during all the changes any system can undergo: thermal, chemical, biological. . . . Only if the system is in some sense open (that is, if somehow energy can flow along specific paths out into or back from the world outside) can the energy content change. Even for such an open system, it is always possible to seek out the compensating change in the environment that, once included, leaves the total energy content of system and its surroundings again invariant. The aphorism of the nineteenth-century physicist Rudolf Clausius, a pioneer in the wider understanding of the concept, holds today: *"The energy of the world is constant."*

To energy one should add a variety of other examples of constancy, like momentum along a line, momentum of rotation, and electric charge, which so far seem to hold unreservedly. Their utility is wider than that of Newton's laws or of any other of the more detailed statements of chemist or physicist: They turn out to be as true and indeed as useful in the quantum theory of the nucleus as they are in the study of the distant galaxies. For many complicated natural systems, they imply the major quantitative statements we can make. Despite any entrepreneurs who may now be busy selling perpetual motion devices, no physicist fears the failure of these high principles. To be sure, for the universe as a whole—if that somewhat paradoxical concept can be as well defined as cosmologists now think—limits may someday appear even to this class of generalization. Short of that, the invariants work always and everywhere, at every one of the forty-plus orders of magnitude. These invariants (there are more) find a rational connection with plausible properties of space and time; they imply a consistency we tend now to demand of proposed new theories of any kind. Einstein and Schrödinger offered their novel theories, far from Newton's paradigm, but the conservation laws hold in those new theories too in quite recognizable ways. Yet they cannot appear in the model held before the eye, save by implication.

Such invisible regularities can reveal the invisible, even literally. A few star images seem to undulate along the line of motion very slightly as the years go by, following a path like a gently meandering stream. Why? The momentum law explains that such a mass cannot behave so by itself; there must be an unseen companion to share the momentum, now pulling the visible star to one side, then to the other. Disturbances were found in the motion of the planet Uranus in the 1840s; on that evidence, an unseen planet was predicted by John Couch Adams and Urbain LeVerrier, promptly found in the telescope, and named Neptune. The last major planet to be discovered, Pluto, was forecast in a similar way. (Pluto turns out to be so small that we now believe the original data were inadequate, and the final success of the search was a tribute more to patience than to Newton.) What worked for planets worked for atoms too. In 1932, certain cloud-chamber tracks showed protons recoiling—from nothing. No initiating track was seen toward the proton from the novel radioactive source outside. But the inference was clear, exact, and soon verified: James Chadwick disclosed the neutron by directly applying the same notion that had found Neptune. It is this catholicity of scale—planet to nucleus—that warrants the conservation laws as the strongest pillars of the invisible order.

Color is part of the visual experience, like change. But the perception of color is no simple accomplishment of the eye and brain. Rather, it usually includes an unconscious internal comparison over the whole visual field, to allow for the quality of the light that illumines the scene. The basis of color, the physical quality that distinguishes red light from blue, is the energy carried by light in each single interaction with atomic matter, ultimately in the retina. All those forms of energy that are strict analogues to light—from radio waves to gamma rays—can also be ordered by the interaction energy they carry. The variations in energy are large—sixteen or eighteen powers of ten—from the very low energies in radio to the highest-energy gamma rays. (The visible gamut from red to blue spans only a two-fold increase in energy.) These rays all move at the speed of light in empty space; but they interact differently with atomic matter, according to their intrinsic energy. At the low-energy end, the interaction is diffuse, gentle. Energy carried along into matter by radio—though it can be great overall—requires a huge number of single interactions. At the gamma-ray end, the same total energy would express itself through only relatively few atomic interactions.

All these types of radiation are called electromagnetic because they interact solely with electric charge and its magnetic counterparts and with no other property of matter. The images they form are subject to strict geometric limitations. Low-energy radio is incapable of revealing fine structures; its limits are macroscopic. High-energy rays can probe deep within single material particles, but they disturb the receiving particles by their disruptive transfer of energy. The limit for spatial discrimination is proportional to the energy carried. This limit is roughly the wavelength of the radiation, to describe it in language that expresses the wavelike and not the raylike aspect of the radiant energy. Individual atoms of angstrom size can be imaged conveniently by radiation whose size discrimination is at their own scale; that is, the x rays. Quite elegant means have been found to recover the spatial arrangement of a crystal exposed to x rays from the record of the redirection of the radiation as it passes through the repetitive array of atoms. That is the technique that maps the forms of such molecules as DNA. The reconstruction is carried out by computation; it is simply impractical to devise lensing systems that work well enough for the x-ray region.

The fiery color play of the opal is like x-ray diffraction at another scale. An opal is a stacking of tiny balls of silica, each about the size of the minimum discriminating distance of visible light. Its effect on such light is to disperse it into different directions, which depend upon the energy of the interacting light. The different colors are thus separated, and the fire results. A careful record made with light of one sharp energy— one color—would produce, instead of the color play, a geometrical pattern of bright spots in that one color; that pattern could be analyzed to form an image of the opal array in space. The x rays produce the same phenomenon but on the scale of the atom itself.

It is extension of this same set of ideas to incoming beams of particles, electrons and even protons, that has allowed indirect study of the internal forms of the nucleus and of the subnuclear world. The method is there more complex because the results are so much more diverse. If light enters an opal, only light comes out. But if a fast proton enters a nucleus, many new particles may emerge. Yet the ordering principles that depend on the transfer of energy, momentum, charge, and spin still control the events.

The Fine Print in the Almanac

The refinement of astronomical observations is not a new story. The long time during which the motions of the planets have been studied has allowed us to store up a great treasure of data. In a specific city, a solar eclipse can be predicted a century in advance with an error measured in seconds. That precision has led students of the motions of the solar system to a concern with fine points. The simple story of circular orbits in strict obedience to the unmoving Sun is evidently too simple. It describes reality well, but with a broad brush. We can speak of a model of the real system. From Newton's day onward, it was the task of the theorists to rationalize more and more precise observations by more and more attention to detail: improving the model. The orbits are clearly not circles but ellipses. That itself goes a long way toward better agreement. But they are not truly ellipses either. The ellipse is the strictly correct form of an orbit on which only the massive Sun acts. Yet the planets in fact pull upon each other and even on the Sun itself; the full study of the overall motions must take some account of the influences of each of the planets. The slighter the disturbing effect to be accounted for, the smaller the planet and the farther off the source of the disturbance that must be considered. Calculations can include a great many interacting bodies by drawing on the full computing power of these times. The orbits are in reality complicated, slowly evolving, endless rosettes, always near a shifting ellipse but beyond any simple description. It is this step-by-step complication of a first approach that is the chief current task of many a science, though few have the power celestial mechanics possesses. The model of the circles has been inordinately refined.

Why push so hard the improvement of a model? Initial understanding would follow mainly from the circles and the simple curve of the graph. Why the rest, which demands most of the expertise and carries the effort into technical domains beyond the nonspecialist? There is the challenge, like that of the mountaintops. More than that, detail is sometimes a necessity. No one can aim a spacecraft at Mars and expect it to arrive there if the aim has been based on an approximation of circular orbits and the assumption that the only gravity that matters is that of the Sun. So the unknown future is a strong incentive to model improvement. Most of all, no one can know what is hidden in the deviations; there is always some hope of a grand discovery. Indeed, Neptune was hidden in the small deviations of the motions of Uranus; a major planet or two is no small prize. The astronomers of the late Victorian decades looked hard for additional planets near the Sun, because their orbital calculations of Mercury's motion were plainly infected by a small error that would not go away. No innermost planet has yet been found that might be responsible for turning the Mercury rosette just a hair faster to fit the now-certain facts; instead, that discrepancy has been accounted for fully by the new and radical view of gravitation that Albert Einstein presented in 1916. It was the first fruit of his most original accomplishment, the theory of gravitation known as general relativity.

It is the occasional grand consequence of a minor misfit that has formed a good deal of the popular caricature of theoretical change in natural science: a succession of schemes of the universe, each one all-embracing, each to be shattered beyond repair by an ocean of change that floods through a single crack. No one would deny that this view contains some truth, but it is hyperbolic. We no longer need, like the philosopher Kant, to imagine that human thinking itself requires Euclid's geometry and Newton's laws of motion. Nor is the universe seen as a clock of infinite precision, a clock that, once wound up, is fated to tick out its inexorable rhythms. We are more modest today in physics and probably much closer to reality: There is room for strict cause as there is place for a reasoned uncertainty, noise, and chance. The two together play out the intricate universal drama. It is true that Einstein's theory lends itself well to difficult talk of curved spaces and the geometrization of space-time, and those insights are even irresistible to a degree. But, in our visual model of the world, such effects can hardly be seen, for the old descriptions of force and motion on a large scale (even if wanting when pressed too far) neatly account for most of what we see within the larger cosmos, out to a billion light-years and more.

Let it be affirmed that gravitation, in the plain old Newtonian style, is today by no means the effete and elegant subject of precisionists. Their studies saw no grandly new phenomena but only agreement in the last decimal places: fine print in the almanac. One can easily take that inference from the history of celestial mechanics: Nothing very new is to be expected. But that is not true. Gravitation is still full of surprises, though to be sure not in the motions of the major planets. Their orbits have mutually smoothed themselves out in space over the long repetitive span of solar time. But look at the galaxies in collision, their long arms and counterarms telling a strange yet purely gravitational tale. Even the new compact objects, like the neutron stars, are understandable as the outcome of Newtonian pull, no less than the intricate rings of the planets are. The wild comets, never disciplined to the flat plane of the solar system, are occasionally imprisoned for indefinite sentences inside Jupiter's orbit. All these are signs of the versatility of gravitation. In effect, the whole world of the large is one drama of resistance to the inexorable, never saturated, never forgiving pull of gravity, from which its orbit, its high temperature, or its quantum motion may preserve a system, at least for a while.

Quantum Motion: Surprise and Commonplace

If fifteen of the largest powers of ten within our visual grasp are dominated by gravitation and its richness, several scenes at the small end are examples of quantum motion. They follow new laws, not those of Newton; the effects of gravity are all but absent. The domain of the atom—both in its external partnerships, up to the scale of molecular mechanisms, and within it, down to the unending quest for the ultimate in fine structure—is the domain of quantum motion. That motion, represented in our visual model by a few rather striking conventions, is both the most surprising of all the remarkable discoveries of physics and the one that most resembles everyday life. For it is not the almanac that governs much everyday experience, nor is it the more or less accidental orbits of gravitating systems. What we find is material stability: Like begets like; gold always glitters; bread is nourishing. If atoms were so many tiny solar systems, that stability would be impossible. No two planets are alike; yet all electrons are identical, and so are all atoms, taken species by species. No two stars can have identical corteges of planets, even if we discover plenty of solar systems where until now we have seen none. Detailed study of atomic structure and atomic bonding have shown, in analogy to the solar system, that there is a central heavy bonding center with an external coterie of circling electrons, analogues of the planets. Even the Keplerian laws of motion—in a modified form, to be sure—were found true. The atomic forces are enormously stronger, particle for particle; they proved to be electrical, and thus saturable, not insatiable like gravity. The atom wants to gain or lose at most only a few more electrons. But the Sun will pick up additional mass whenever it is offered. Gravity always attracts, but electric forces can repel.

The key is the modular, precisely repetitive, stable form. No large system is like that. Identical building blocks—identical and stable forms kept at an energy minimum and resisting any small disturbance—characterize the atom. They are the signs of quantum motion. The electrons bind to an attracting charge just as they would in large-scale physics; the forces are the same, and the fall from orbit inward to hug the center is prevented by motion. But there the analogy runs out. The quantum motion is of fixed pattern; only certain arrangements in space are open to the moving electrons, and of these only one kind of structure has the lowest energy. Given that state, the electron is present, but it is untrackable from moment to moment. No orbit can be traced step by step; there is only an entire stable pattern. Can an electron never be followed along a circular orbit? It can, but only if given an energy much greater than the minimum possible, denying it a larger volume in which to spread out. The energy present is then by no means sharply defined but spread among many energy states, as no stable atom could allow.

Scale determines. Given a larger particle than an electron (say a dust mote, or even an electron with much more energy, like those that paint the TV image), the motion can be followed as sequentially as any circuit of the moon. Quantum motion fully reduces to that of Newton once given the scale conditions of the macroscopic world. But in the atomic regime, no stopwatch, no light flash, no fast camera is adequate to note the electron in its flight between two points within its usual domain; and the use of any radiation or any material sensors that could function on the appropriate scales of time and space would require energy transfers so great as to break up the atom. You might track your electron, but no longer would it be an electron within a hydrogen atom bonded normally to the carbon atom.

For that reason, our visual representation of quantum motion conceals the particles involved; the individual electrons in the bonded atoms are not drawn. Rather, we show the electron charge cloud that they paint out over time, in a pattern that, on the average, becomes sure but cannot be found to unfold from moment to moment. The convention seems a fair one; the concepts it represents are profound, even paradoxical. But in a way they are not so strange as the haphazard quality of the might-be-anywhere orbits of the large-scale world. We are used to a stable world of substances with fixed if manifold properties; individual atoms in a way have no history. They are identical. A gold coin may contain last year's gold from the mines of the Rand or gold won from the California placers in 1848. The two metals are one in kind. All chemical change is by the reassortment of atoms and electrons, never by their wearing out or their gradual readjustment to some long history of attrition. When atoms heal, their recovery is total. Even nuclear decay is ordered and modular; aging uranium always turns into lead atoms of the right sort at the right rate. It is this stability even during change that marks the world of quantum motion, the world of modular identity.

Human Scale

Between the galaxies and the atoms, amidst color, life, and familiar landscapes. This is the realm recognizable in history. It is better known than that of the universe at large and not so strange, in its subtle unfolding, as the atomic patterns, which seem to obey immutable laws whose history we can hardly conceive. Life too has had a long history: evolution. Its intricate mechanisms and its wide adaptations, its fitness like its beauty, are slowly ripened fruits of time, time long enough for Darwinian editing of the manifold rearrangements of an inner, modular, chemical world. Human artifacts, from wide fields and great cities to tiny electronic chips, undergo a similar though far more rapid pattern of evolution; their changes are entangled with those of the human mind.

ASTRONOMY SHINES OUT

How could our predecessors understand even the fiery sun by day without a physical theory of gravitation, heat and radiation, ions, nuclei? How could the astronomers of the past hope to grasp a world whose color scheme is often painted in spectral bands far beyond the grasp of eye or camera?

Twentieth-century physics and the dazzling range of instrumentation it has inspired have brought a flowering of astronomy not rivaled since Newton's time, and led to a similar blossoming in just about every field science. What good fortune it is to live in the era of N-body simulations, microwave astronomy, neutrino pairs and neutral currents, IR satellites, and CCD spectroscopy! (Of course, it is equally our bad fortune to miss what will come hereafter.)

The pieces here are not up to date, but remain valid guides to today's issues. Ideas that were in gross error have been cast out, a process. as William Blake said, "also part of God's design".

What Is Astronomy?

Everyone knows, as the most commonplace part of any discussion in astronomy, with the man on the street or even in the horoscope next to the funny papers, that astronomy is in some version or other one of the oldest of explicit scientific efforts to understand the world. It offers the oldest sign of order in our universe; day and night entrain us all. Of course, the rhythm of the seasons adds more to that; and nothing except the intimacy of heartbeat and breath can be closer to us than the rhythms of day and night and summer and winter. It must go back a long ways in our species—and I venture to say in our genus—as long as we had any way of formulating ideas, that we recognized that there is something outside there that forms order.

When Isaac Newton at the end of that great century, the seventeenth century, when modern science began, formulated the first really powerful mathematically and experimentally complete "system of the world," he drew heavily on the results of astronomy. The apple was the connection between the moon and the Earth; it was an interpolating device. But the Keplerian results on planets and satellites, the values of periods and radii, were the main facts on which Newton rested his system of the world. The effort to describe the tides, and the pendulum swinging in the laboratory, were important; they went toward universality, but they couldn't quite get there. The real impression was made by the understanding of the events in the sky, then of course kinematically extremely well known and studied for thousands of years. Even with the telescope there were results that had been well established for 60 years or so.

Even 50 years after Newton, the main support for Newtonian physics was astronomical, or at least large-scale. It was Maskelyne, the Royal Astronomer in the middle of the eighteenth century, who first measured the gravitational constant. He did not set up lead balls in the laboratory—that was to come later. He used a mountain in Scotland. By observing the zenith from two sides of that mountain he could see what the mountain did to his local definition of the vertical. So it was large-scale physics (not laboratory physics) that determined the most important results the physicist had in those days. It wasn't till Cavendish, 30 years later, that the physicists left the outdoors world and started to build their laboratories and their tunnels and put large lead balls in place, to do the physics that of course by now has risen to extraordinary power and maturity. Everyone will agree that it was celestial mechanics that gave rise to the remarkable mathematical structures that made contact between the particle physics of Newtonian mechanics and the partial differential equations of continuum theory that dominated nineteenth-century physics, ending in Maxwell.

By the time the 1930s came along, there had already been 70 years of spectroscopy, most of it done with very little understanding of atomic physics. The great success of the old quantum theory, culminating in the work of Saha in the 1920s, finally allowed the great pioneers of modern astronomy, Hertzsprung and Russell and Eddington and Milne and Payne, to do the atomic physics that made it for the first time possible to analyze the stars quantitatively. They discovered that the world, the whole visible cosmos, is made of hydrogen and all the rest we so much count upon.

Then, as everyone knows, came nuclear physics. It has its slow beginnings in World War I, and broke onto the world with an air of omen and power in World War II. In between it acquired that maturity which in the hands of such masters of the subject as Hans Bethe was able to answer a long-voiced query. Nobody who had the slightest feeling for the nature of the world could have failed to ask, "what makes the stars and the Sun shine?" At last this could be answered.

Today we see astronomy as the application of physics to an inaccessible laboratory. It demands the best and the most complex physical understanding that we have. But it is without the parallel support and drive of an important industrial technology that demands laboratory answers for its applications, as, for example, aerodynamics or plasma physics. There we know the physicist is going into those particular interesting applications because they're enormously needed, and well paid for. Because of our ancient human connection with astronomy, and because it is so far yet from any abilities to control, it remains a special case. It is applied physics, where application, yes, is to special systems, not given to us to choose but only to observe from afar.

Other beautiful branches of applied physics have grown up in such a way that they mostly take away a field right out of the physics profession. There are now not very many physicists, only a few, who know a lot about aerodynamics, because aerodynamicists have become a world of their own; they have so much power and so much need and so much desire to enrich one particular subject. So it has gone for quite a few specialties, but not yet in astronomy. Astronomers did that for a while, from the time when the photographic plate came in through a time when the spectroscope was the astronomer's real monopoly.

The new beginning was presaged by Karl Jansky (the pioneer of radio astronomy) before World War II. But it was in the postwar world that physicists were enormously attracted to astronomy. It happened on two sides. On the side of experiment (probably the more important), new channels have been opened into the astronomical world by the techniques of the physicist. It also happens, of course, on the theoretical side, because we now have a pretty powerful command of classical and relativistic mechanics, of plasma, of nuclei, of electromagnetic theory, of kinetic theory, and the rest. These enable us to try to cope with the complex systems and bizarre parameters domains the cosmos is pressing upon us. We look theoretically only because some clever experimental investigation has shown us a new signal: for heaven's sake, what can that be? We try hard and then, like Franco Pacini and Tommy Gold, we may come up with a fundamental answer, like a spinning neutron star. That's about as far as we can get with certainty. We're still struggling. We are of course enriching, deepening, and making more quantitative our whole understanding.

Now, I'd like to draw some conclusion from this. Probably everyone agrees there's a dual investigation, the investigation of new channels—in my view absolutely indispensable—and the investigation with new understanding of complex problems, also requisite. It would be impossible to view modern astronomy without both of these invigorating entries, almost rude door-breakings, from the side of the physics laboratories. That is not to say that the classical discipline of astronomy is not a monument to the effort of the astronomers. Witness, say, the magnificent instrument 50 years old now that is Palomar. Theirs was the social effort and the intellectual effort, to make something that would last and be worthy of the sense of frontier that astronomers have felt for a long time. They represent the human mind struggling with its origins and fate, in the best way we know, by looking at the world outside our own domain, where we look so far and so long ago.

Now what we have found, in the last 30 years or so, is that we live in a dual cosmos. We should have realized that earlier, but for a clear reason we didn't. I'll try to say what that reason is. The reason is very simple.

But let me first sketch what the duality is, as I see it. In the first place, the cosmos, the telescopic view is dominated, like the time on telescopes and the papers in the *Astrophysical Journal,* everything we do is dominated by the stars. That's as it should be, for we are creatures of the stars. We are adapted to a star; we were evolved under the light of a star; we can see stars; we can see deep into their interiors with neutrinos, even if only in a murky and puzzling fashion at the moment, but at least we try! Now the star is an example of something very familiar to humankind, familiar probably in the Rift Valley six or eight hundred thousand years ago, if not earlier, when the first campfires were kindled under the dark sky by those who were our forebears. The hearth is the characteristic word I'd associate with that ancient experience.

If you ask, what is the nature of this as a physical phenomenon, I think it is quite plain to say: The issue in the course of evolution of the astronomical world, dominated on the large scale by gravitation as it is, is the struggle between gravity seeking to crush and anything else that tries to maintain. What maintains the stars, of course, just as what maintains terrestrial objects against the crush of gravity, is the resistance of random motion, whether it be thermal motion, or zero-point energy, the fluctuations of the electrons in the van der Waals forces between atoms. Whatever way you look at it, speaking broadly, I think it is fair to say that it is a randomized energy in which many motions in many directions take part. That's what we're used to. Accompanying this are certain characteristic features: the emission of radiation is unpolarized, because no direction is singled out. Thus the characteristic shape of this astronomical world is a sphere. The Sun is a sphere, and the Earth is a sphere, and Jupiter is a sphere, and Titan is a sphere, and the stars are spheres. Again, this is a sign of the lack of particular direction, the lack of particular motion; the sense of symmetry is characteristic.

Now we understood pretty soon that this is not all there is in our astronomical surrounding. There is something else unchanging; it is like the laws of Medes and Persians. We see it both in the solar system and in the galaxies. It is not only random motion that fights gravity, it is also ordered circular motion. The planets circle around the Sun, and in the galaxies, the stars, like the Sun, circle in the galaxy. It's not now a question of randomness; in fact, we see very well that the solar system and the spiral galaxies are characterized by an axis, by flatness. So we're quite right in saying it's not all spheres, as I described above. But look how stable the orbit is! Look how long we've been circling this little Earth, ever since the Sun was made, in the same orbit, with hardly any changes. There are only slight changes that our savants now tell us might make the Ice Ages, but so small that you can hardly believe it; it's not quite visible. The energy content of the motion of the Sun around the galaxy is about the same as the internal thermal energy of the Sun, and just as important. Yet nothing changes. Why worry about it? It's not a phenomenon of interest; it was made so in the beginning, and "as it was in the beginning, so it will be to the end," until 10^{25} years from now—gravitational radiation will take its ultimate toll. But we can't worry about that. That's the state of affairs in the universe we know and the universe we were born to. But what we have discovered is that beside the glowing hearth, a second completely different world exists in the universe. It is a world in which polarization, lack of sphericity, sudden change, explosion and expulsion play an important role. It is no hearth; rather it is the whirlpool and the fountain. This is the characteristic side of the new cosmos we are now beginning to investigate.

Why did we not see it in the past? Again the answer is very clear: we didn't see it because we looked with the eye. The eye was evolved for the gentle light of Planck radiation at 6,000 K, the big broad curve that gives us white light. Everything else is a small departure from that—the dye-stuff on the tablecloth or what you will. But what we recognize is that sunlight, radiant sunlight, is what we study. When the astronomers work, they study starlight; maybe they push a little bit to infrared or they go a little down near ultraviolet. If you look hard at the stars and have good eyes, you can see that there is some kind of color in Rigel or Betelgeuse, but it isn't very saturated; it's not the kind of vivid color we are talking about. There are hints, of course. One of those hints is held in the aurora that are visible at high latitudes. The aurora is not white; the aurora has not steady; the aurora is not in thermal equilibrium; the aurora has no simple temperature; the aurora is full of mystery, so we hardly understand it yet.

In fact, a similar hint is what led me directly into astronomy from the side of physics. The cosmic-ray physicists were also studying something a little more randomized, but again not temperaturelike, no hearthlike equilibrium. It has that characteristic power-law spectrum that goes with turbulence, and not the absolutely necessary Boltzmann distribution that goes with thermal equilibrium. We all recognized that, so we followed the cosmic rays, first into the laboratories. They took particle physics away from cosmic rays, since they could make mesons at 10 million times the rate of the cosmic-ray beam. But they left us the chance to follow the cosmic rays back to their origins, which we're still looking for far into the sky!

One other thing I want to mention, of course, the comets. Halley's comet has just come and gone. The comets represent a kind of gravitational instability not present in the rest of the solar system. Comets come in, we don't know when and we really don't know whence; they are not in the plane; they are not uniform; they are not combed out, like the planets, into circles in a plane. Slowly a great light dawns upon us—it is blazoned big among the clusters of galaxies—that gravitation, celestial mechanics, is not the quiet subject we all think it is. The almanac is there, and the experts in celestial mechanics, what do they do? They push the numbers one digit farther, to add another hundred years to prediction. That's the view I grew up with. I laughed at the perturbation theory that they taught us in dynamics, because, while it's very nice, elegant, and good for mathematicians (and no doubt osculating orbits are worthwhile, as well as Poisson brackets), there's nothing in this subject: no novelty, no romance, nothing new under the Sun, just the Sun and the circulating planets with their slow orbits. Well, the comets don't quite fit that, and nowadays we learn to our astonishment that it may be that the evolution of life itself owes a great deal to perturbations of comets that, half-expected, bring comets hurtling into the central solar system, to cross all the orbits, destroy regularities, break down the stabilities: no more circles, but plenty of collisions in the solar system. That, of course, is a sign of real trouble.

If you think about that, you realize that the reason we see light with the eye, not X-rays and infrared, and millimeter radiation, the reason we see circular orbits logarithmically spaced out into the solar system in a flat plane is that we come late upon the cosmic scene. We depend for our origins on the very systems we're studying. Of course, gravitation appears to us the rock-bottom safety of circular orbit. But that's not the way it really is. That's only the way it's been combed out to be after five billion years of circling around the Sun. What it really is, is galaxies in collision and accretion disks, fierce fountains bubbling up along the axes of spinning disks in a way that we can barely begin to understand, on the largest scale and on small scale too.

Gravitation is an explosive phenomenon, not an orderly one. It depends on what conditions you have opened it to. The Laplacian notion that you can predict the future with precision, that if only we improve our measurements, we can project predictions further off, depends entirely on dealing with the two-body approximation to the circulating planets. As soon as you start putting in multiple bodies, reduce the degree of symmetry of the system, and ask for the evolution, then you find from Monte Carlo calculations the evaporation of clusters and the tails behind interacting galaxies and a hundred other phenomena, too many to evoke here. In fact, there really isn't any physical system of interest that you can predict with Newtonian precision far into the future. It's only because we have looked hard for those few systems that do work that way that we have the view that classical physics is a simple determinate mechanical structure. That's where we live: at a hearth. Of course, we teach that to the students, because otherwise they can't solve the problems. So for a long, long time separable differential equations and the two-body problem solved by the constants of the motion are everything in the textbooks. The fact is that if you go to three bodies only and generalize, you don't know what is going to happen next. You can study it forever; there is nothing but a finite circle of convergence. No matter how well you measure it, there will come a time when you can't say whether a system will hold together or not. That is even true probably for our solar system. A young colleague of mine at MIT has now built a grand digital orrery with chips to try to push the orbits of Jupiter and the other planets five or ten billion years into the future, just according to the laws of Newton, to see if in fact they will blow up one day. The stability theorem that governs the solar system is simply not adequate to answer this question.

So that is the way it stands now: there are two worlds, the world of the hearth and the world of the whirlpool. We recognize the universe contains both of them; we recognize that we belong to the one, while the other is alien. That comes from history alone, from the fact of our heritage and legacy, our adaptation to this world, the hearth-world of Sun and planets, where life has lived for four billion years, one way or another.

I think we have come to the most important juncture in physical sciences, certainly since the glorious days near 1930 of the discovery of quantum mechanics, which makes possible much of what we do. We find ourselves surrounded by the radiation everyone knows about, the 3° radiation, in which we are clearly not central. Our motion shows very strongly in the map of the sky, so it isn't that. Copernicus was right; we are not central. But there is something strange about it, because we know of no other such perfection on the large scale as we see in the uniformity (to ± 1/1000) of the incoming blackbody radiation. There is no other case where any macroscopic ball has that degree of accuracy, and for reasons that we don't understand. This makes one queasy.

The current proposal to explain this has remarkable implications. Those implications, taken together with what else we know, suggest that we live in a universe here in which we have not only found the whirlpool component we had neglected, but will find one day soon that most of the matter is our universe is alien to us! It is matter we have never seen, matter we have never examined. I'm not trying to say this is a sure prediction, but it is fair to say that that is the most popular and the most intriguing proposal of the day. The mass that holds the orbiting stars of galaxies in a tight grasp, even though the starlight within is not adequate to explain its presence were they normal stars, might be a kind of matter not known to us at all, a kind of matter left over from a previous regime of matter when the protons were broken up and free quarks reigned in the world. This would be remarkable if it were true. It's a challenge, after four centuries, to confront for the first time a failure of that great principle of Copernicus: as here below, so above. The world is really somehow one; it is up to the physicists, the astronomers, to show how. Maybe we're going to find that it is more dual than I ever expected; maybe we will not, but perhaps we will know that soon.

It seems to me it's a glorious time. We have had in the last 35 years a fitful peace, a happy, breathless time, we physicists and we astronomers. I hope we can continue in peace in this troubled world, to try to puzzle out the nature of the cosmos into which we were born.

The Explosive Core

The Latin word nucleus—meaning little nut—denotes the "kernel, core, inner part." Three times it has become the key word for a new science, in each case calling attention to the central core in which decisive events occur. The first use came in the 1830s with the study of the living cell, when a nucleus was recognized and named by the microscopists. The investigation of the cell nucleus constitutes most of modern molecular biology. In 1911, the term was applied to the innermost part of the atom, and that dense, energy-rich nucleus is now the subject of nuclear physics. Finally, beginning 25 years ago, we have singled out a new kind of core. This is the nucleus, not of the smallest entity of life or matter in the universe, but of the largest—an entire galaxy.

A galaxy is a great island of stars and gas, adrift in the archipelago of the universe. Like the islands of the sea, galaxies are many sizes and shapes, from elegant, flat pinwheels to great puffy clouds. Some are smoothly globular, others mere irregular splashes. All of them contain many stars: a billion stars in even a dwarf galaxy and a thousand times as many in a giant galaxy.

Surrounding each galactic island is a sea of space having, at most, some gas and a few runaway stars or star clusters. We live in a galaxy, the Milky Way (the Greek word for milk is gala). Our own starry island belongs to an island group of 15 or 20 galaxies. One member galaxy nearby lies beyond the pattern of foreground stars called the constellation Andromeda. That galaxy is the great Andromeda Nebula. It is 2 million light-years from Earth (a light-year equals about 6 trillion miles), yet visible even to the unaided eye as a spindle of light on a dark autumn evening. The Andromeda Nebula is known to astronomers as Messier-31, or M-31, after the catalog number given it by the French observer Charles Joseph Messier in 1781. The galaxy is similar to the Milky Way in size and form.

In fact, a photograph of the Milky Way taken from M-31 would broadly resemble the Andromeda Nebula that we picture in the telescope.

Photographs that show the geography of a galaxy are, like all photographs, sometimes misleading. The whitest parts of the picture—where the white paper shows through—are not more than 20 to 30 times brighter than the inky black night sky. The real range of brightness is much greater, and there is an easy proof of this. In a severely underexposed photograph, the image of the galaxy as a whole disappears. There is not enough light coming from most of it to mark the plate. Only one sign remains—a point of light at the very center about as big as the image of a single star, only slightly fuzzier.

That bright point is the nucleus of the galaxy. Concentrated there are so many stars that their summed light still shows up. It is thousands of times brighter than any other part of the galaxy.

No individual stars in the nucleus can be seen in the photograph, however. The size of a star image is fixed, not by the size of the star, but by the scattering of light within the telescope and photographic plate, plus, for the earthbound observer, the haphazard motions of the Earth's atmosphere. In the photograph, the star images have all merged smoothly into a single point. In reality, the stars do not overlap. They should appear as distinct, tiny spots, like strewn dust, but that we can never see.

The nucleus of a galaxy is only a few tens of light-years in diameter. Yet, the millions of stars there are still far apart by human standards. They are perhaps a light-month (500 billion miles) or so from each other, instead of the five light-year spacing typical in the region of the Milky Way near the Sun. If we lived in such a place, the night sky would not just sparkle, as it does here, with a few dozen bright stars spread across it. It would be adazzle with thousands of brilliant stars filling the whole sky with jostling constellations. But the stars would still be bright points having no visible structure or shape because they are still enormously distant compared to their own sizes.

During the years of World War II, Carl Seyfert, a young observer using the 100-inch Mount Wilson Observatory telescope in California, searched the nuclear regions of most of the thousand or so galaxies that are near enough to be studied in detail. He took not only a photograph of the nucleus, but also the spectrum of that blurred point of light, which showed him the light's intensity at different wave lengths. The spectrum shows that the light of the nucleus in most galaxies is the combined light of many stars, most of them dimmer, redder, and smaller than our Sun. Just as in the case of the Sun, the nuclear spectrum is bright overall, and a few dark lines cross the spectrum at wave lengths that common atoms absorb strongly. At these dark-line wave lengths, the stars' light from hot inner regions was absorbed by the atoms in the outer, and hence cooler, regions.

But Seyfert was surprised. The nucleus of about 1 in every 50 of the nearby spiral galaxies is disturbed. Bright lines cross the spectrum, light that could not be emitted by stars, but only by thin, intensely hot gas in the spaces between the stars. These nuclei were, overall, brighter, bluer in color, and usually larger than the others. And the bright lines were wide, suggesting violent motions. In these galaxies, now called Seyfert galaxies, the additional light from the tiny nucleus is as bright as 100 million stars like the Sun.

In 1957, Russian astrophysicist Viktor A. Ambartsumian, working at the Byurakan Observatory in the mountains of Soviet Armenia, reached a conclusion, now widely accepted, that something extraordinary is going on in the nucleus of most galaxies. In some, there is a Seyfert pool of hot gas. Others show signs of old explosions, such as gas surging steadily outward. In a few there is even more activity—glowing jets or fountains, thousands of light-years long, of gas and fast electrons. In some pinwheel galaxies, unusual spiral arms, each containing millions of stars, appear to have been flung out of the nucleus. In others, bright patches near the center distort the galaxy. In a few galaxies, violence is carried to the extreme. Nothing can be seen except the abnormal nucleus. All the outside stars are dimmed by the glare of this overwhelming nuclear activity.

Let us look at a few examples. M-31—the Andromeda Nebula—is quiet enough. Its nucleus is bright and dense with stars, but it has no strong radio emission, no hot gas, nothing very strange. It is as calm a nucleus as any big spiral is likely to have.

Our Galaxy has a dense starry nucleus, too, much like that of M-31. But we cannot directly see the stars in the nucleus of the Milky Way by visible light, even though we are 60 times closer to them than we are to M-31. We live right in the dusty central plane of our own immense spiral, and our view of the nucleus—which would be spectacular—is ruined by that galactic smog.

Using penetrating infrared detectors, however, we can see through the dust. The infrared view of the Milky Way nucleus is much like the infrared view of the M-31 nucleus; all looks quiet. Yet, there is something disturbed about the Milky Way's nucleus. There is a radio source exactly at the center, modest, but not to be dismissed, as well as other signs of excitement. An outward flow of turbulent neutral hydrogen gas can be traced by radio emission toward its origin, close enough to the galactic center to implicate an explosive nucleus as the ancient source. Other streamers of hydrogen gas falling back into the plane of the Galaxy have suggested to some radio astronomers that its origin was another, earlier, nuclear explosion.

A strange source of very long wave infrared may also have been found at the very center of our Galaxy in 1968 by William F. Hoffman of the Goddard Institute for Space Studies and in 1969 by Frank Low at the University of Arizona. High energy gamma rays, too, may have been detected coming from the nucleus in 1969 by George Clark of Massachusetts Institute of Technology, Gordon P. Garmire of California Institute of Technology, and William L. Kraushaar of the University of Wisconsin. And there was even an unconfirmed report in 1969 by Joseph Weber of the University of Maryland that gravitational wave pulses of incredible strength comes from the center of the Galaxy every few days. Altogether, these varied signals make a tantalizing, if still uncertain, case for a tumult in the nucleus of the Milky Way.

A great ball of a galaxy, called M-87, lies about as far away as astronomers can clearly see detailed structure—about 50 million light-years. It is in the nearest large cluster of galaxies beyond the stars of the constellation Virgo. More than 100,000 light-years in diameter, it is also a very bright source of radio emission. For 50 years, M-87 has been puzzling, for, in photographic underexposures, a marvelous glowing jet can be seen emerging from the bright nucleus, like the luminous hand of a clock. The nucleus itself is a dense nest of stars, as the dark lines of its spectrum show, but it also holds hot gas, whose bright lines are strong and wide.

The jet, however, is unique: it shines by a bluish light and its spectrum shows neither dark nor bright lines. Furthermore, that light is polarized, particularly in the several bright knots that emit most of the light of the jet. Polarized light means that the jet of M-87 shines by virtue of magnetic fields that circulate a high density of electrons having cosmic-ray energy. By its motion, each such electron possesses perhaps 100 billion times more energy than an electron moving randomly in the gaseous surface of a white-hot star. The mix of cosmic-ray electrons and magnetic fields is like the beam inside a huge particle accelerator. But this natural beam is enormous. The jet, which from its size and properties may be only tens or hundreds of thousands of years old, contains enough cosmic-ray electrons to equal the mass of the Sun. This remarkable jet also copiously emits x rays and infrared light. Some photographs indicate there is, as well, a second, weak counterjet.

A clearly different type of nuclear unrest is seen in a close neighbor galaxy, M-82. This galaxy is some 6 to 8 million light-years from Earth, in a direction not far from the pointer stars of the Big Dipper. It is a dusty, flattened galaxy, a spiral seen edge-on, that has long been recognized as disturbed. What appears to be a great wave of compression in the gas is rushing out in all directions from the center of the galaxy. A symmetrical explosion must have occurred there a couple of million years ago. The compression wave is much less violent and obvious than the great jet of M-87, but, all the same, a marvel. It looks like the explosion of a huge firecracker. The wave is moving out at about 1 per cent of the speed of light, and the leading edge of the disturbance has reached out 15,000 light-years from the core. Most probably, M-82 is a galaxy of the Seyfert class, but seen edge-on through its dust.

There are a number of other results of nuclear disturbances in galaxies. Among the most spectacular are enormous clouds of fast electrons in magnetic fields, each cloud bigger than most galaxies. These clouds lie in pairs, one to either side of a central galaxy that itself looks disturbed. The electrons are not energetic enough to give off visible light, but they produce polarized radiation in the radio part of the spectrum. This suggests they may be the late stage of a double jet: the central explosion flung the telltale fast-electron magnetized plasma off in opposite directions. At first the jets shone intensely. Looking now, many millions of years after the event, we see the emitted clouds, their intensity diminished, large overall, and moved far out.

The clouds of Cygnus A, the brightest of all radio sources, are a good example. And, a nearby southern galaxy and strong radio source, NGC-5128, sometimes called Centaurus A, has at least two pairs of radio clouds in its neighborhood. This quartet suggests that the explosion was repeated, the more ancient pair having had time to move farther away. In Centaurus A, the central galaxy remains somehow awry, but its nucleus does not seem overbright. Rather, telltale emission lines of hot gas show disturbance in the entire galaxy, or at least its main body. It is not difficult to believe, though, that the optical fury and the radio clouds originated in the nucleus.

Finally, we come to the quasars. These enigmatic objects—quasistellar radio sources—look like unimportant stars. Their spectra show a few wide, dark lines and many quasars have some bright emission lines as well. The quasars are radio sources, too, and many of the nearly 200 known show outrider radio-emission regions very much like the twin radio clouds of galaxies. No central galaxy can be seen in any quasar, although a few quasars show a small wisp or jet leading from the starlike center in the direction of a radio-emitting patch. Most remarkably, the continuous optical spectrum (though not generally the bright lines, which add only a small part) often varies considerably in brightness in only weeks or months. In some cases, a quasar's brightness varies by a factor of two or more.

The puzzle deepens when the distance to the quasar is considered. Most astronomers believe the distance is very great. The closest and brightest quasar, 3C-273, is estimated to be from 1.5 to 2 billion light-years away. At such a distance, the starlike quasar looks a thousand times brighter than an entire galaxy would. But its true size, like that of a star, is hidden within the image size. All we can say from telescope observations is that the quasar cannot be larger than the smear of light it makes—a few thousand light-years across. That is only 10 per cent of an ordinary galaxy's diameter, yet all the light that 3C-273 emits must come from a region smaller than that.

But there is more information. Radio astronomers have measured the diameter of the 3C-273 radio source using the cooperative signals of radio dishes in California and in Sweden—almost as though they had a radio telescope the size of the Earth. The measurements show that an important amount of the radio energy, emitted almost 2 billion light-years away, comes from a spot that is, at most, only a few light-years across. Even this dimension is too large, however. The intensity of light from the quasar varies strongly over a period of months or even weeks. Thus, its source must be remarkably small, even smaller than the radio measurements show. It can be no larger than the distance that light travels during the time of the variation in brightness.

Why this is so can be argued best from an analogy. Suppose you are far from a large, crowded stadium. At a cheerleader's signal, the crowd shouts a sharp "Win!" Will you hear a crisp shout? Not at all. Even if every spectator shouts at exactly the same instant, you will hear a long-drawn-out roar. Even though the farthest and the nearest of the cheering crowd shout exactly together, the sound takes longer to reach you from the far side of the stadium than from the near side.

That delay is the time it takes the sound, traveling faster than a thousand feet a second, to cross the stands and the field, perhaps three quarters of a second. We can measure the size of the stadium by timing the sounds of the cheers.

Can the crowd fool you? Suppose the nearer fans delay their shouts just long enough to accompany the sound they hear from the far side. Then, indeed, they would deceive you. But another listener beyond the far side of the stadium, would hear a great delay. The deceit would depend on the direction to the listener.

A quasar signal is light, not sound, but the effect is the same. Thus, depending on our direction, some strange effects inside a quasar could fool us into measuring variations as weeklong when they really are yearlong. But we would certainly not be in the right direction to be fooled by another quasar. Since the rapid variations in light and radio emission are known for half a dozen quasars, we know we can measure a quasar's dimension by the time of its light variation. The emitting region is thus a light-month or two, and possibly smaller.

By both optical and radio measurements, then, quasar radiation comes from a region as small as the nucleus of a galaxy. Even if there is an ordinary galaxy around the quasar, the galaxy would not be seen. Its starlight would be lost in the thousand-times brighter glare omitted by its spectacular nucleus.

The galaxy M-87 has such a small core, too. The widely spaced radio dishes have located a small source of very great energy in that galaxy's heart. Out along the jet there is another one. It is as if the nucleus had flung out a twin to itself that left a jet trail. The twin still shines from the end of the jet.

Indeed, such jets may exist even in galaxies that are not strong radio or optical emitters, but ordinary clouds of stars. Halton C. Arp of the Hale Observatories has pointed out that some galaxies have strange outflung arms of stars, often bearing star-knots at their very ends. Could these knots be similar small nuclei and their progeny, long since shot out, leaving a trail of gas that later condensed into the new, bright stars that mark the arms? The star trail is sometimes curved by the rotation that all the orbiting gas of a spiral galaxy must have.

We are sure now, from all this evidence, that a strangely compact and energetic source of light, radio waves, fast electrons, infrared, and x rays lies deep in the heart of many galaxies. It may be missing, as seems to be the case for M-31; modest, as in our own Galaxy; notable, as in the Seyferts, and, perhaps, in the strange-armed Arp galaxies; conspicuous, as in M-87; or spectacular, as in the quasars. We know that ordinary stars exist having a wide range of power. Is there some sort of larger object that can exist over a similarly wide range—but many millions of times more powerful than any stable star—which accounts for all these nuclear mysteries?

Many hopeful explanations have been put forward, but none is yet known to be right. Most of them, however, cannot be shown to be wrong. Let us look at three theories, to gain some feeling for what may unravel the riddle that sits in the heart of a galaxy.

Every galactic nucleus is a dense swarm of stars, all in rapid motion, orbiting the center of gravity of the galaxy. However, their motion is not in a neat plane as in the solar system. Do some of these millions of crisscrossing stars collide from time to time? Such a collision should tear the stars apart, flinging their hot gases into space, and stretching out the stellar magnetic fields. Other stars will plow through this debris and slow down slightly, and fall more closely together. Some of these stars, in turn, should collide.

A nucleus of colliding stars with one crash per year might look like a quasar from a great distance. If collisions are less frequent, the core would look more like a Seyfert nucleus. And the modest activity of the Milky Way might require a collision only once every few thousand years. But it is hard to see how the jets of M-87 or of 3C-273 could be expelled by this process—a jet more massive than a great many stars. Still, the idea is reasonable. It might explain part of the nuclear riddle.

But suppose the galactic nucleus is much more dense at the very center. When enough matter comes together densely enough, its gravity prevents particles from escaping unless they have sufficient speed. For example, the Earth will hold particles unless they move away at least as fast as the Apollo capsule, some 7 miles per second. A big enough mass will require an escape velocity greater than the velocity of light itself. Thus, nothing can leave it, not even light. It will be utterly black.

Such an object—no one has ever detected one—would create a "black hole," a region of space like a huge bathtub drain. Anything that comes close enough to it—star, particles, a light ray—is in danger of merging with it, of falling down the gravitational drain and leaving behind no sign of its existence other than the gravitational pull its increment of mass adds.

Suppose such a black hole has formed by the condensation of much gas at the center of a galaxy. The hole might be a light-month or less in diameter. Unlucky stars and gas would tend to fall into it. But unless they aimed straight at it, they would have to swirl around it just as water does around the bathtub drain, for the same mechanical reason: nothing with a residue of orbital motion can approach closely to the center. It must orbit the center until it imparts some motion to the surroundings. Then it can enter.

In this picture, the energy source of a brilliant galactic nucleus is the inexorable gravitational pull of a black hole. Gas and stars moving at speeds approaching the velocity of light orbit the hole, forming a whirling disk of colliding matter. Collisions and interactions within this vortex make the radiation we see. As this energy leaves, the particles that supplied it feed the greedy hole, which will in the end swallow everything that ever comes near.

This marvelous conception, worked out most fully by D. Lynden-Bell of the Greenwich Observatory in England in 1969, neither explains the jets nor the vast symmetrical explosion of M-82. Yet, it may prove true somewhere, perhaps for quasars.

A third model of the explosive core, which I proposed in 1969, is the spinar, a whirling, flattened disk of hot gas similar to a huge star. A star loses energy mostly by radiation. But the hypothetical spinar is too large and too cool to radiate much. It loses energy when its spinning, magnetized bulk flings off particles at high speeds. As the particles leave, the spinar shrinks, spins faster, and flings off more gas. It has so much energy in its spin that it easily supplies the energetic electrons for the radio sources associated with active galaxies and quasars. The spinning disk is a light-month across, and has the mass of from a million to a billion stars. I believe a small spinar powers a modest nucleus; a huge spinar powers a quasar.

Sometimes the big disk splits, and sends off smaller spinars—one at a time, as in the jet of M-87. Sometimes two spinars leave at once, torn from opposite sides of the disk by a vibration of the main body. In time, they slow down and become the great double radio clouds of the quasars and Cygnus A. Each such cloud, according to this theory, must hide within it a small, dense spinar.

How did the spinar get there? It condensed from the gas that makes a galaxy. The more the mass condenses, the faster it spins; the more it spins, the closer it comes to becoming a spinar. An ice skater spins a little faster when he brings his arms in close to his body. But a spinar in the nucleus of a galaxy contracts its diameter much more, by a factor of 100,000, and speeds up from one turn in a few hundred million years to one turn per year.

Normally, only large masses can be spinars. Small masses never spin fast enough; they lose their spin too soon by magnetic forces. They never condense far enough. They remain stars, losing energy almost entirely by radiation.

Those newly discovered pulsing stars, the pulsars, are exceptional cases of stars that became spinars. They did it by blowing up first. Within the explosion, a Sun-sized star collapses to a mountain-sized pulsar perhaps 10 miles in diameter. If this theory is right, then quasars, all the other riddles of the nucleus of the galaxy, and the strange stars called pulsars are all spinars of different sizes and strengths.

There is one weak argument favoring this picture. One quasar, 3C-345, does seem to pulse like a pulsar. Of course, its pulse is not once a second as in a tiny pulsar; rather it is about once a year. It is not sunbright, like a typical pulsar, or bright as a thousand Suns, like the brightest pulsar. Its pulse is as bright as 100 billion Suns, and lasts a week or so before its lighthouse beam turns away from our view. What a titan of a beam that is!

We do not know if any of these models explains the strange violence of the galactic nucleus. The central riddle, the "little nut" of the galaxies, is still a riddle. But we have some clues, and the answer should not be long in coming. Perhaps one of the theories will fit the new facts, perhaps none will; possibly all will, for there may be many kinds of galactic nuclei. But it is most likely that we are in for a surprise. For whatever we may predict, the universe is incorrigibly otherwise. The answer will be a great excitement.

Is M82 Really Exploding?

T he "silver sliver" of the galaxy Messier 82 lies not far from the Pointers of Ursa Major, so close to the spiral galaxy M81 that the two are caught in the same low-power telescope field. M81 is a showy spiral in big telescopes, a common textbook illustration, but M82 was called "nondescript" by Edwin P. Hubble. Yet in most textbooks of our day, M82 appears with special billing as the nearest exploding galaxy. How it came to have that raffish reputation, and how it seems more recently to have lost it, is a fine tale, rich in the interplay between ideas and observations.

Hubble himself put M82 in the class of irregular galaxies, which includes the one galaxy in every 40 that shows no sign of rotational symmetry. About half of that class form one rather homogeneous type, *Irr I,* whose prototype is the Magellanic Clouds. They are irregular swarms of highly luminous blue stars, without much sign of nucleus, spiral arms, or other organization. Galaxies like M82—half a dozen others like it are well known—form the other kind of irregular, *Irr II.* Very different from the Magellanic variety, they show an amorphous texture, curdled and dusty, often with no resolved bright stars at all, and no signs of organized structure. M82 is reddish photographically, though its overall spectrum is ordinary enough, like that of a small spiral, more or less dominated by the common stars of spectral class A. These anomalies of the Irr II's led Allan Sandage to comment in his wonderful *Hubble Atlas* of 1961: "Galaxies of type Irr II present a mystery."

The galaxy pair M81 and M82 were early recognized as the central members of a small clump of galaxies, not unlike our Local Group. The distance to one or two of them is not hard to measure, using bright stars and such variables as novae and Cepheids. The members of the M82 group of galaxies—all with similar small redshifts and not very differing magnitudes or diameters—are pretty recognizable, even though they are scattered across 15° of the sky.

The distance assigned to M82 is fairly sure, 10,000,000 light-years or about 3,000,000 parsecs. Thus the spindle of M82 is about 15,000 light-years long, the diameter of its underlying galactic disk. Evidently the galaxy is circular, and appears spindle-shaped only by foreshortening. We see M82 nearly edge on, tilted only about 10° from an edgewise view, though the direction of tilt is not so easy to tell. Since M82 is only about five times as far as M31 in Andromeda, we can therefore expect to study it and its partners in fair detail, as one of the closest neighbors to our Local Group. (In fact, a clumping in Sculptor, near the south galactic pole, is our next-neighbor group, but it is less well placed for study by Northern Hemisphere telescopes.)

The Signs of an Explosion

In the early 1960s, the idea of exploding galaxies was new and exciting. The big double radio sources had been found, and the quasars too. Radio emission was seen as a sign of intense activity. In the celebrated third radio-source catalogue from the University of Cambridge, M82 appeared under the label 3C 231, perhaps the most familiar and nearby galaxy to enter the list, although making it only by a few flux units above the lower intensity limit of the study.

Astronomers were familiar with the beautiful long-exposure photographs of M82 in the *Hubble Atlas*. These very contrasty prints showed a faint, delicate extension of the optical form of M82. With increasing exposure, it grew out in all directions, appearing as a kind of faint corona all around the familiar curdled spindle.

I. S. Shklovskij had made clear how important it was to seek a polarized visual-light continuum like that which marked the smooth magnetic fields in the Crab nebula. With radio emission, polarized light was a strong clue to powerful explosions. The continuum in the outskirts of M82 was indeed polarized, strongly and systematically—another step in the identification of a nearby galaxy explosion.

Aina Elvius made the best study of the polarization. She stoutly, and I believe correctly, maintained all along it was the polarization to be expected from dust scattering around what was so evidently a dusty object. But that seemed rather prosaic in the 1960s for so unusual a galaxy, a radio source at that. Spectroscopy produced a good curve of rotation from the sloping lines in the spectrum of the combined stars in M82, and it was not very unusual. Indeed, we could be pretty sure that the spindle was a disk of stars seen all but edge on, with the stars in orbit around the minor axis of the image at velocities of some 200 kilometers per second, as one might expect for a modest galaxy.

An obvious but difficult task was to study the spectrum of that faint polarized filamentary halo around the main image. This halo could be photographed quite strongly in the red line of hydrogen, H-alpha, whose wavelength was measured with care throughout the region, even though outside the spindle its light is very faint. What appeared was unmistakable—the halo out along the minor axis showed quite a simple motion. To the northwest of the disk, the gas was moving *away* from us in the line of sight, at about 100 kilometers per second, and on the southeast side it was approaching us at about the same speed.

Now, the spectroscope cannot show the true space velocity of any moving gas, only that component along the line of sight. It was not unreasonable to expect that the gas (and any dust it carried along) flowing out from the center of M82 moved along the rotation axis of the galaxy, nearly in the plane of the sky. The observed Doppler shift would have been zero, whatever the speed, if the gas motion were perfectly perpendicular to our viewing direction, but there is no reason to assume that. Rather, the spindle form gave away the tilt of the galaxy. Fitting the spindle shape to the image of a tilted disk gave a clear interpretation. The tilt was about 10° (with the northwest edge closer to us) and the motions were very rapid indeed.

Stuff streaming northwestward and southeastward from the center of M82 was moving out at about 1,000 or 1,100 kilometers per second! The measured 100 kilometers per second was simply the line-of-sight component of this motion. The faint stuff could be followed for several kiloparsecs, seemingly streaming with explosive speed out of the nucleus both ways along the axis of the galaxy. One could estimate from the emitted light the streaming mass, and from the speeds calculate the enormous energy of the flow. It was conjectured that the event might have released the whole rest energy of up to 10,000 suns.

According to this picture, we were living close to a fine example of a galactic explosion, to be sure on a modest scale and rather concealed in a dusty region within the M82 spindle, but plain enough by radio, by the widespread and regular optical polarization, and now by the unprecedented gas motions. Halton C. Arp soon found a remarkable faint ring around one end of M81, M82's bright partner. While no one could be sure, it too presumably had something to do with the explosion. M82 duly took its textbook place as the best domestic example of a galaxy with an active nucleus. Only 10,000,000 light-years away, we could point to it as a remarkable specimen of the growing class, mostly so distant and so poorly seen.

Looking Inside M82

Blue and ultraviolet light are dim from M82, but it is bright in the infrared, especially from the center of the spindle. Two groups used special photographic emulsions to record an image in the near infrared. That was the beginning of step-by-step retracing of the entire case so far, which had pretty surely been an example of expectations shaping conclusions, as of course they always do in some degree.

Those rather fuzzy photographs in dust-penetrating near-infrared light looked deep into the core of M82. Clustered there, bright patches could now be seen, which stood out in contrast more strongly the longer the wavelength used. That meant that they were brighter than the surrounding image within the dust. Had they been merely regions less heavily veiled by dust, they would have become less marked when more penetrating radiation was used.

A small paper by Sidney van den Bergh opened for me the way to understand this puzzle. His discussion of the images of the bright internal sources was supported by line spectra, which made clear the presence of numbers of very hot, blue stars (types O and B). Those patches seen through the fog were in fact super-Orions, outsized clusters of the most luminous and youngest familiar class of stars. There were no such stars in the outer reaches of M82; they would have shown up well on the near-infrared plates. But deep in the heart of the dust were crowds of bright stars, a total luminosity of several billion suns, all hidden from sight and made known to us only by the strong infrared reemission of the illuminated dust, with a little of the powerful glow leaking even through the dust, especially at near-infrared wavelengths.

Yet this starry M82 was still an enigma. Deep within it there had been an epidemic of star formation, not so very long ago. Had the dust something to do with that time? Certainly the detailed infrared work and the high-resolution radio studies which followed showed a consistent picture. There was no single hot compact center in M82. Instead, scattered over 1,000 light-years or more, there were many nonthermal radio sources, infrared, bright star clusters, thermally ionized clouds—all the features of a region vigorous in birthing stars—with no doubt a full share of supernova remnants, runaway stars, traveling shock waves, even star encounters. All this star-birth activity accounted well for the rich radio emission of the galaxy.

But the work made clear that M82 behaved in this respect more like a heightened version of the dense center of a normal galaxy than like a galaxy that harbored a single, compact, powerful, nonthermal central engine of mystery, kin to a quasar. The energy store tapped in M82 was surely nuclear, star-burned fuel, in manageable amounts. The absence of any outstanding single concentration in the mapping made the emission of M82 understandable as an impeded view of its core.

Moving Mirrors

But what about the polarized light all the way to the outskirts? Even more, what about the swift outflow of gas? Star clusters cannot well account for either, it appeared. Were there perhaps two phenomena at work, a rash of new bright stars *and* a single explosive nucleus?

To study the polarized light coming from the halo around M82 required powerful techniques, for the light from the halo fell off rapidly in intensity as the farther portions of the image were viewed. Palomar investigators in the early 1970s mapped the polarization in patches so far out into the fringes that the light was only a few percent of the general night sky background. Natarajan Visvanathan and Sandage, using the 200-inch telescope and a fine photoelectric polarimeter, studied the polarization not only in many spots, but over a range of wavelengths.

Their results struck at the roots of the explosion model. They learned that the light of the hydrogen line was polarized in just the same directions and by the same amount as the broad-band optical continuum measured by previous observers. The physics then leaves little doubt. The synchrotron emission coming from fast electrons in magnetic fields, which is seen as an explosion aftermath throughout the Crab nebula, and in the nuclei of quasars and active galaxies, naturally cannot show any lines of hydrogen. For that process does not involve atoms at all, merely free fast electrons gyrating in magnetic fields. Indeed, it is hard to find any atomic process that emits strongly polarized Balmer lines of hydrogen. Therefore, the M82 halo polarization could not be due to radiation coming from the outlying regions directly; there were no big pools of fast electrons there. The light therefore had to be light redirected toward us from its path into space by some form of scattering. As Visvanathan and Sandage wrote in 1972: "H-alpha polarization appears to have destroyed the . . . explosion model."

What scatters the light? There is dust aplenty in M82; you can see that at once. But the dust is the heavy dust of that galaxy itself, confined to the disk where the stars form a layer hundreds of light-years thick. Only Aina Elvius had been willing to conclude that thin but very extended extragalactic dust was present, strewn tens of thousands of light-years outside the main body of M82.

For a while it seemed possible to save the idea of explosion by postulating a sea of relatively slow-moving free electrons, released by a shock wave from the explosion, to provide the far-flung scatterers. But, among other difficulties, electrons are feeble scatterers of light. Compared to dust grains, you need thousands of times the mass in free electrons, with their separated proton partners, to give enough scattering even for that faint halo glow. It had to be dust, spread throughout the space far outside M82.

But the observed Doppler shifts still suggested that the dust was moving fast—carried with gas, no doubt—outward in both directions along the rotation axis of the galaxy. You were back at an explosion, yet there were no signs of explosive phenomena, save that swift motion. We were caught in our model; the footprint on the sands was our own. For the work of the 1960s had not in fact *observed* any high speeds in the halo, only a modest 100 kilometers per second. It was the geometry of the outflow model—the explosion model itself—that led to imputed speeds tenfold more (and energies that are 100 times larger).

The way out was a new geometrical model. What the spectroscopists record is not light made by atoms in the halo that are moving rapidly outward. It is light which was made in the disk of M82 and then scattered around a corner to our waiting detectors. All you need postulate is a slow drift of gas carrying a little dust, not moving in both directions out of the center, but rather toward M82 from the southeast and away from it to the northwest. This steady drift of "mirrors" of moving dust will simulate the two-way outflow. For, as the dust from the southeast approaches the galaxy's disk, it sees the line sources within it as somewhat blueshifted. In turn, the light sent toward us carries that same blueshift. To the northwest the story is reversed: the dust here flows away from the central light source, the shift is therefore redward, and the light sent to us carries that redshift. If it is interpreted as direct light, what we see is a swift flow both ways outward. If interpreted as light scattered around the corner, what we see is only the smooth, slow drift of a dust-bearing gas into and past M82.

No outrush of gas, no high speeds, no explosion! The center is active enough at radio, infrared, and X-ray wavelengths, but it is unmarked by any single focus of radiation. Its activity is an epidemic of star formation at many places in the central region. That event—under 100 million years ago, judging by the lifetimes for such luminous stars—is hidden in the internal dust, probably of its own making. No massive stars have since formed in M82, except perhaps in that central region, for none are seen in the outer portions.

M82 is a more or less ordinary spiral galaxy, its disk of modest stars in normal rotation, hidden beneath the dust mantle. That mantle is so newly made that there has not been time to draw it out much into a marked spiral pattern, and there are no clusters of luminous new blue stars outside to mark the arms. After all that, we must add that M82 is immersed in a large slowly moving external cloud of gasbearing dust, much more dilute than the heavy dust held within the galaxy.

Such is the present interpretation of the Irr II system, M82. Much more detail has been added by recent work. Using a line of neon in the very deep infrared has made the rotation pattern of its stellar component clearer. Other studies have given evidence that the outer halo is not so tangled and filamentary as the exciting earlier pictures seemed to show. The contrast of the finer structures is not very high, and these may be explained by slight irregularities of indirect lighting and dust. The polarization pattern gives every evidence of coming from a few bright central sources, strong but not unusual, emitting both continuum and line radiation which illuminate a roughly uniform cloud of dust. Very recently, more complex line-splitting has been seen in the faint halo, but the line pattern is still within the competence of the reflection model, if nonuniformities in the dust are allowed for.

Best of all, it appears that M82 is not unique. A whole class of spiral galaxies, active because of their strong central emission lines and optically accented nuclei, has been identified as class 2 Seyferts. The older category of mini-quasars, the Seyfert galaxies, was not long ago split into two by several criteria. Class 1 Seyferts appear to be genuinely subject to nuclear activity, with a strong tendency to show variable central sources, emitting X-rays and so on. The class 2 Seyferts are milder. They show the earmarks of a central epidemic of star formation, plenty of dust and strong infrared, but no strong X-rays nor the rapid variability which is the clearest sign of compactness. As a group, the couple of dozen class 2 Seyferts are likely to be stronger and more distant counterparts of our subject, M82.

If all of this is correct—not everyone agrees, and M82 has fooled us before—the little spindle is not the nearest member of the class of active galaxies, not the seat of a galaxy explosion, but rather the nearest place where a huge burst of stars somehow formed together, a fraction of a galaxy-year ago. And around that starcrossed galaxy is a huge drift of dusty gas.

Influential Companions

It seems pretty clear what has happened. A mass of fresh gas and dust swept slowly past M82 not so long ago. Much of it fell into the heart of that galaxy, disrupting the few gas clouds already present where stars were slowly condensing. But once the new gas fell to near center, becoming denser and denser, it gave rise to a huge spurt of star formation, whose products—from supernova remnants giving radio signals to infrared-bright dust formed in the birth of luminous stars—mark M82 today.

What caused this grand star epidemic in an ordinary spiral? The radio astronomers have probably found the answer from their observations of the 21-cm hydrogen line. They can detect very large cool diffuse clouds of normal hydrogen atoms, far away from any stars. Three studies, first by Morton Roberts at the National Radio Astronomy Observatory, then from Florida State and from Jodrell Bank, have mapped out the mass distribution and the motions of a huge hydrogen cloud which seems to engulf M81 and M82 as well.

Its motions are consistent with the idea that a tidal interaction between M81 and M82 pulled gas out of both of them, and that at the present time much of that gas, bearing a little dilute dust and stars so sparse as to be invisible, is still orbiting in space around both galaxies.

Computer models of the tidal collision suggest that much gas fell at once deep into M82, and that much of it still drifts past that galaxy from southeast to northwest. This provides the great screen on which the bright but hidden center of M82 is projecting the illusion of an internal explosion. Of course we cannot demonstrate the true three-dimensional motions from the Doppler shifts alone, only that component along the line of sight, but theoretical modeling fits very well the motions we do see.

There is certainly more to the story. Another Irr II galaxy, NGC 3077, belongs to the same group as M81 and M82. Probably that galaxy can tell us something, for some of the hydrogen reaches out to it as well. We need to know more about that galaxy. Years ago, Alar and Juri Toomre gave us all a hint: even grand M81 seems a little awry, with the ends of its spiral arms spread out a little too widely. It looks as if it had interacted with another galaxy, and they accused it of collision in the famous words: "Et tu, M81?" They turned out to be right, slow as we have been to see it.

Finally, what about the dark bands straight across the face of M81? Arp showed that those were continued into the remarkable faintly glowing ring around the end of M81 nearest to M82 (as already mentioned). Maybe this is part of the story, too; at the moment, I would guess it a false clue, a real red herring. Sandage has made deep photographs in which that structure does not appear much different from the streaky high-latitude debris in our own galaxy, shining faintly by the reflected light of our own stars. This explanation, too, needs testing. But it is a good place to leave off the rich story of M82.

I hope that many astronomers who have shared the unraveling of the M82 tangle outlined here will not too much miss their names in this account. To have put them all here would have made the article cumbersome, and some may prefer not to be associated with a still-unproven point of view. Those I have mentioned are a sample of the most decisive contributors to my own position, but they are not responsible for what I have drawn from their work. I am very grateful to my former close partners at Massachusetts Institute of Technology, Alan Solinger and Tom Markert, and to our research associate Bella Chiu.

A Whisper from Space

Looking at the starry sky, it's a mere commonplace, it's almost the meaning of bright stars in a dark sky, that the space between the stars is dark. If we look beyond our galaxy to the distant galaxies, patches of light, again they lie against the dark background, the space between the stars, the space between the galaxies, is dark, without light, and yet perhaps the most important discovery of the last 50 years, in our study of the universe, has been the remarkable fact now richly confirmed and elaborated since it was first, in 1965, that if you look in the right way, between the stars, and between the galaxies, in what appears to the eye to be the blackness of space, there's invisible radiation, that radiation is intense, the sky between the stars is glowing more brightly than we could have imagined.

For years, the community has worked to understand the deepest sounding of the universe, a universal background far beyond the stars and even the galaxies which fill the sky: the most energetic and the most ancient signal we have ever seen.

Surely the oldest and the best astronomical instrument is the eye. Supplemented by the telescope, and the photographic film, the eye and the mind behind it, has led us to a pretty good understanding, pretty good understanding of the universe in which we live. But there is also a skyful of invisible radiation, hidden from our inbuilt eyes, quite clear to the cunning instruments which sense radio waves, x-rays, untra-violet and the rest.

The great family of big dishes which focus invisible radio is familiar. They can see what the eye cannot. We can speak of seeing because radio and all the waves we shall discuss are close kin to visible light. All of them are electromagnetic radiation, differing from the colors we can see only in wave-length, as indeed red light differs from blue, or by analogy, for the quite different waves in air we call sound, as a tone of low pitch differs from one that is high.

All that the eye sees, like all that the camera sees, is as though we were restricted on the piano to just those few colored notes. Outside that range, up here in the ultraviolet, and down here in the infrared, are invisible to us, as though we had only one octave of the piano.

Much more than that, 6 feet to my left, well beyond the end of the real keys, there is another range of invisible keys, those represent the spectrum of the radio astronomer, and it's in a small part of that spectrum, maybe one octave, where some years ago, a new detector, looking at the sky in a new way, brought us extraordinary information, which has transformed our knowledge of the cosmos.

This microwave horn, looking like an old-fashioned ear trumpet for a hard-of-hearing giant, sits on its hilltop in Holmdel, New Jersey. Among all the listening ears in the world, it was this one that caught the crucial whisper back in 1965, the lucky start toward today's cosmology.

What they sensed was nothing less than the backdrop of the universal theater.

Front stage, of course, our Sun and its planets, merely one of a myriad of stars which orbit in the Milky Way Galaxy.

Near us, too, the other galaxies of our local group—a couple of million light years away. Plenty of other galaxies in groups and singly crowd the stage.

Behind them, out as far as the telescope can see—the thousand million other galaxies we can dimly sample.

Two radio astronomers, Robert Wilson and Arno Penzias, used the Holmdel horn and found their way by a mix of chance and care to the great discovery.

By 1978, the work begun by Penzias and Wilson has convinced the world of cosmologists that our complex universe is in no steady state, but in constant evolution, cooling steadily from a featureless white-hot past.

This antenna is nimble enough to search the whole of the sky you can see from Holmdel. The horn is carefully designed and built to catch hand-sized waves, microwave signals. That horn feeds the finest of sensitive amplifiers, kept very cold to reduce internal disturbances.

They chose first a seven-centimeter wavelength because they expected little signal there; it was a waveband, people thought, free of most galactic emissions, a quiet baseline for an overall survey.

Here we had purposely picked a portion of the spectrum a wave-length of seven centimetres where we expected nothing or almost nothing, no radiation at all from the sky.

Instead what happened is that we found radiation coming into our antenna from all direction—just flooding in at us and clearly was orders of magnitude more than we expected from the galaxy.

This way, to put it baldly, an embarassment. Was it a false reading? The compact amplifiers were easily tested; maybe something in the big horn antenna was making excess noise.

While Penzias puzzled, another group thirty miles east at Princeton was getting ready to look in the same way. Their independent impetus came from the young theorist Jim Peebles, searching for some hard data in a time of open cosmological speculation, a time when I myself still believed in the steady-state universe.

The Princeton group worked around the redoubtable experimenter and innovator, Professor Bob Dicke. He is famous for insight and skill in settling long theoretical arguments by sensitive and novel experiments. In 1965, he had just persuaded his colleagues that a real clue about the early universe was within experimental grasp.

Dicke saw that the earlier universe would at least do one thing: it would be so hot, that it would endow the universe with plenty of radiation to start with. That radiation would still be around today, and Dicke said it should be searched for. He left Peebles to work out the details.

Peebles' calculations in fact showed there was a chance to pick up the signals. So in the summer of 1965 on the roof of the physics lab at Princeton the experimenters Roll and Wilkinson got ready their smaller horn. They stretched a net for smaller wave lengths than the fish they caught at home. While they were working, Peebles had been invited to talk at Johns Hopkins University about his calculations and the planned experiment; his experimental colleagues did not object.

In fact at Holmdel, only an hour's drive from Princeton, the experiment had already been done, unknowingly but very well. The two groups knew nothing of each other. The connection between Princeton and Holmdel was made by an MIT radio astronomer, Prof. Bernard Burke.

The news was out. Holmdel was seeing far beyond everything else in the sky.

Like so many novelties in science, their discovery had in fact been anticipated. Seventeen years earlier, two cosmologists, Robert Hermann and Ralph Alpher had in print predicted the existance of the very same background radiation.

The two work for large corporations now. Their prophesy of the new cosmology had already been forgotten. They were pioneering the memory, recalled only after the fact.

The time has come to explain how we know what that whisper is saying. The tale can begin in an old fashioned way, in the blacksmith's shop.

For the radiation Holmdel sees is now known to be heat radiation—the very same emissions that come also from the fiery forge, heating the blacksmith's face.

The hot metal makes visible what was long hidden. This physics is now sure to us, turn-of-the-century physics, a foundation stone of the modern theory of matter and energy. How strange and yet how familiar to see mere temperature turn strong iron soft as wax.

Everything which has a temperature radiates heat radiation. The eye tells us so, and warns the craftsmen to use the tongs. The self-luminous chain link evidently glows red hot. But most objects we see do not glow at all like the link, made a while ago, already black. The eye says: no glow left.

Even the cooling link gone black to the eye is still vigorously glowing—now with a glow invisible to our Sun-adapted eyes.

Everything in the universe, at any temperature radiates and even at 3 degrees, the temperature of liquid helium, things radiate, perhaps one set of objects does not radiate. Anything at absolute zero doesn't glow to any detector at all. This might be the definition of absolute zero. Short of that, we should be able to pick up thermal radiation from anything that has a temperature.

It is not easy to accept that objects radiate all the time, particularly when you see a batch of pots and clay pigs loaded into a cold dark pottery kiln. The eye is blind to most radiation; we are partial to sunlight and its imitator.

The early universe itself was a uniform and glowing kiln. Moreover, hidden in this glow there is a quality as unique to the physicist as a signature.

Now, we didn't know that without experiment. In the lab we demonstrate the signature of heat radiation—a signature which bears the mark of the temperature of its source.

We take the coiled filament of an ordinary light bulb, glowing white hot, heat provided by the electric current.

The signature is written in the spectrum: like Newton, we pass the light through a prism which draws out the component colors into the rainbow stripe. Then a suitable detector, one less biased than our eyes will record the intensity color by color, and so we will see an impartial representation of the light coming from a hot filament.

The Holmdel experiment gave us just one point which fitted a temperature curve. Even though it agreed with predictions old and new, one point is far from proof. In a few months Princeton had another point, and then the radio astronomers of Cambridge, England brought in a third one. All three points fitted the same curve, so that one could read off the temperature of this radiation source at very cold, only three degrees above absolute zero.

All these points were far below the peak of the curve they seemed to fit. They had not marked out the hump as we did in the lab. Was the judgement premature?

So, we were guessing, if you like, that all the energy would be there once we measured it, on the basis of three points, far far off the maximum of the curve. That is dangerous, because many curves might look like the real curve. I think we're in the position of somebody (I once said) sleeping in a tent in the desert and noticing a camel rope, beautifully braided rope, lying under the edge of the tent, well, there's no question, that was a camel rope all right. But was there a camel on the other end of it? If you tug gently—well, you feel something on the other end, but maybe just tied to a ten pound weight, or maybe it's tied to a tent peg. I don't think you could say you've caught a camel until you can see the hump, and so the game for me was that they had to do the experiments to see the bulk of that energy, the hump of the signature, not just a little edge, even if it fitted mathematically very nicely.

For the experimenters it was a heavy burden. The millimeter waves at the hump just cannot reach the ground through our watery atmosphere. The experiments would have to leave the ground and go outside the atmosphere.

One brief rocket flight gave a result far above the predicted hump, and delighted us skeptics. The rocket turned out to have looked at some of its own heated parts.

The stratosphere balloon turned out the better choice: it soared high enough to leave most of the air below it, and it stayed up overnight, long enough for some careful checking. Many launches were made by a series of labs.

It was Rainer Weiss' lab at MIT, where our demonstrations were made, which first caught the real peak, though there were earlier hints.

Every measurement meant a couple of years hard work by a skilled group, usually based on some new detecting scheme, some new device for improved reliability at a new waveband. In 1977, it was a University of California group at Berkeley who brought back new strong evidence from up high.

The data appear point after point. Beautifully they march up and over the hump and down again. The sharp disturbance on the smooth temperature curve show the effect of the thin air still above the high flying balloons.

And finally, there was simply no doubt, the curve fitted beautifully the signature of temperature radiation all the way from 20 centimeters right over the hump. It was a quite a remarkable result.

You couldn't deny it any longer, it was a coherent fit, to 3 degree temperature radiation. Moreover, it was absolutely uniform in all directions; like the radiation inside the kiln, detail was washed out. It fitted perfectly the ideal temperature radiation.

But, more than that, once you've found the energy in the hump, you had to reckon with how much energy there was. We see the bright stars, the violent events, the supernovae the quasars, anything you like. They're brilliant, hot, extraordinarily rapidly lose their radiation, but don't forget, in between is all that black space, and it turns out that the quiet, microwave thermal glow of this radiation is 99% of all the radiation there is. What comes from the galaxies and the rest, the stars, is unimportant. From the point of view of the universe, there is only the 3 degree background; no more radiation exists, the rest is unimportant.

This is such a striking result that we must give it a fundamental interpretation; it lies at the bottom of things somehow. The first step to understanding then will be to ask this very difficult question: how can all this energy have come from a source, which we don't see, which is every where, and yet which is giving us radiation at the very low temperature of only 3 degrees above absolute zero, and that's the question we then proceeded to answer.

It was the physicist Christian Doppler of Prague who first pointed out 150 years ago that a change of pitch would be expected whenever a steady source of waves moved with respect to an observer: today we call it the Doppler shift.

Approaching, higher pitch, shorter waves; receding, lower pitch, longer waves.

This baroque experiment was actually first tried out by a Dutch physicist in the flat lands of Holland shortly after Doppler published, puffing Billy, uniformed bandsmen, and all.

The Doppler shift is just about symmetrical. Whether source or listener move, the effect is there. If one moves swiftly past a trumpeter, his note will sound high as you approach him, lower after you pass him by.

We shall soon encounter this result in a grander context. It should come as no surprise—Doppler himself expected it, that the motional effect is about the same for light or radio waves as it is for sound. For a given shift, the speeds must be higher by a million fold, for so fast is the speed of electromagnetic radiation, but given high speeds, the Doppler shift has been crucial for understanding the sky.

Astronomers have always known they would find the same effect in light from the very beginning. And in the early 19th century, they found it, in stars, some approaching, some receding, but about the time of the First World War, and thereafter through the twenties, first Slipher and then Hubble, found something quite remarkable. They looked at galaxies, not individual stars, but great collections of stars, whole archipeligoes, and they found enormous Doppler shifts, always receding, always red shifts.

The analogy between lowered pitch and a red shift is close; red shift means receding sources. Since Hubble, we know as well the faster cosmic objects recede, the more remote they are.

The radiation curve is then the record of an invisible glow of most ancient heat, deep in the universe, farther away, we are told by its profound redward shift, than anything else we see in the universe.

Think of that early universe everywhere, a kind of vast glowing neon tube, filled not with the familiar red neon but the simpler primordial hydrogen.

As the universe expanded, that bland plasma, as it is called, slowly grew cooler and cooler, not much, 4,000 degrees, 3,500 degrees, 3,000 degrees. Somewhere in there a remarkable piece of what I can call chemistry occurred. At that point, the temperature is not high enough to break apart the atoms of hydrogen, and so the atoms form, each electron finding a proton to join in a tight dance as hydrogen atoms. Once that happens, the electron mist is no longer opaque. Hydrogen is transparent.

At that point, the radiation is set free from matter, travels through this transparent stuff as unimpeded as the light from a desk lamp falls on the desk, and it's been glowing separate from matter ever since that time. So now the black space is full of radiation, and the shining points mark the condensation of the matter.

Nature has been generous in showing us the three-degree radiation. Think how remarkable it is. It comes from a single time, the time when the electrons and protons recombined to make transparent hydrogen. That time marks it with a single temperature. It comes uniformly to us in all directions. It is the same everywhere. As it travels through space it is successively reddened by the red shift to longer and longer wavelengths. These longer wavelengths shift the energy peak to lower and lower energies, so the measured temperature becomes smaller and smaller, gradually with time. Had we measured a hundred million years ago, we would have found a little more than three degrees. Were our descendants to measure a hundred million years from now, they will find a little less than three degrees. It's plain that it provides in this way a kind of cosmic clock which every observer, on every galaxy, if there are observers among the distant galaxies, will make the same, and be able to calibrate one with the other, the universal time since the flash emerged.

That moment was maybe fifteen or twenty thousand million years back. (Our own clock is not yet so well calibrated.) The heat radiation came once to the forming Sun, one insignificant new star among many. That was five thousand million years ago. We are still bathed in that radiation, as it continues to red shift, growing steadily redder, and cooler as time passes.

It is the relic of the ancient past, a fossil more spectacular than any dinosaur bone. It deserves study for every little detail, every tiny feature we can find. For it can speak to us of times we cannot otherwise know.

In 1977 and 1978 a new reconnaissance in detail was carried on by a different group in the Radiation Lab at Berkeley. They flew high in the air, not by balloon but in an old U-2 spy jet plane, converted for looking up to the far heavens, not down to grim missile sites below.

Their task is not merely to find the radiation—but to scan the sky, comparing one direction with another to see if the signal shows any sign of directionality. Like the heated kiln, true heat radiation is free of all directional detail. It is seamless and bland, uniform in every direction, the sign of an utterly uniform fireball long ago.

Microwave measurements from balloon and aircraft have found a remarkable result. You and I and this city and the Earth, and the Sun and the whole solar system are moving off about in that direction just over that way, about 250 miles a second. Now all motion is relative, but this motion is related to the grandest reference marker you can imagine. It is related to the average motion of all the matter in the universe, the source of the temperature radiation. We never expected we could find a simple measurement that would give us this, which is the closest thing to absolute motion that we can think of measuring.

Once we allow for the Sun's great orbit as it swings around the Milky Way, we can conclude that the center of our Milky Way system is itself moving at a modest celestial speed. In roadside terms, that is a million miles an hour or so, which somehow our whole galaxy picked up during its formation.

So the temperature radiation has given us a great deal, for example this magnificent benchmark, the motion of all the matter in the universe. But remember the kiln, remember how as it glowed more and more strongly, detail was washed out in the uniformity? That is the veil, the opaque veil, which temperature radiation seems to represent, and it may be that we shall never directly see with electromagnetic radiation farther in to the opaque plasma than we do now, by much. Of course, it's not a wall which is impervious to knowledge. We have other means, we have other radiations, neutrinos, gravitation, who knows what will come. But maybe there is a veil of electromagnetic radiation at that time. On the other hand, if we are close to such a veil, then we have every incentive to study what we can see, the last bit that leaks out, as matter became transparent and matter and radiation went their separate ways. We may be justified in studying in great detail, every little thing that happens at that time. That means looking for small differences, small lumps in the otherwise uniform radiation. Perhaps those lumps will be the sign of where this lumpy material universe formed, for the galaxies and the stars are lumps, not smooth.

The idea that out of a bland, homogeneous background, all the complexity in the world could arise, is not an idea new to microwave astronomy. It is an idea as old as the Sanskrit philosophers. For long ago, in part of the creation myth, they told how Vishnu the preserver with a mighty churn made of a mountain, churned the ocean of milk, and out of this blend and nourishing material came the divine cow, first nourisher of mankind, an eight-headed steed, the moon itself and many other such wonders.

There is a logical similarity between the mythmakers and ourselves who make, if you like, a grand myth out of the substance of science. We're looking for how to make the complex from the simple. Those persons long ago saw their metaphor in the acts of the kitchen every morning, where, with a simple churn, the butter could be made out of the uniform milk. We look at a much grander scale, and a much more impersonal and distant sea, a sea of hydrogen plasma, but in the same way, by forces we are only beginning to understand in some detail, gravitation, rotation, tidal effects, there was formed, out of the bland uniform hydrogen, the lumpy complex mix of galaxies and stars of matter, near one in which we dwell, and of which we are part.

What is going to happen in the future? This expansion continues, nothing can stop it short. It must continue for a long time into the future, during which the galaxies will draw further and further and further apart, the universe will become more and more dilute, the black body radiation drops, slowly in temperature from 3 degrees to 2 and perhaps to 1, and that's as far into the future as we can be sure. We now can conjecture, maybe the speed is so fast that they will fly apart forever and end in a zero temperature bath, of dilute, utterly diluted galaxies. Maybe, instead, it is not flying apart quite that fast, the gravitational forces will win over the flight in the end, and it will all start to fall back together again, gradually to become denser, and finally a homogeneous plasma again in some kind of grand replay of the ocean of milk. We don't know the answer to that question. Many people think they have seen some of the answers, but the numbers are uncertain, it is a quantitative question which depends upon better measurements and better understanding. And what about the deep past?

We don't know the answers at either end, what happened first, or what will happen last, but now, thanks to the temperature radiation, the microwave background, we have a secure platform of tens of billions of years, in which to explore. Standing on that platform, the cosmologists of the future will be able to go into the deeper past and the deeper future and tell us what we want to know, the direction and the end of the universe.

SEARCHING FOR INTERSTELLAR COMMUNICATIONS

When Giuseppe Cocconi and I first called public attention to the unique status of radio astronomy in the possible detection of radio communications across interstellar space, we did not know that young Frank Drake had already started a radio search. After decades of pioneering effort by many radio astronomers around the globe, a sustained and systematic large-scale search was formally begun in October 1992 with the new NASA High Resolution Microwave Survey.

These pieces recall the ebb and flow of ideas over the years. The giant radio dish designed for Arecibo that drew our attention long ago is still—with new surfaces, and new secondary "optics" now under construction—the largest radio dish in the world. But today's computer power for processing multiple signal channels has left our naive forecasts far behind. A penetrating but still patient search throughout the entire signal space is now within our grasp.

That grasp is loosening. The NASA Survey that had begun so hopefully was shut down for good just one year later. That was the work of one determined United States Senator within the present ambiance of a grave Federal budget deficit. But the Survey was no billion-dollar annual NASA effort, but a hundred times smaller, or about 1/200 of 1 percent of the deficit figure. We need to hope for and to seek out more patient, entirely private, support if we are ever to mount a sustained search from the USA.

Life Beyond Earth and the Mind of Man

Many surprises, even the most extraordinary surprises, are possible. It is conceivable that a spherical ship will land in front of the Washington Monument and a figure with four antennas and otherwise looking like a professional football player will rush out and demand to see our leader. But I hope very much that the universe of circumstance is wider than the rather shoddy imaginations of science-fiction writers during the past 30 or 40 years. I am pretty well convinced it is. We have not found their guidance so great in any but the most modest activities, like going to the Moon. Science fiction of a hundred years ago told us how to go to the Moon, and we have done that.

I think, on the contrary, that an enormous distance separates us from the nearest existing group of a similar kind. And it is truly an enormous distance—not the distance to the Moon, not the distance to the planets, not the distance to the nearest stars, but tens or even hundreds or perhaps thousands of times that distance. That means that even by traveling at the speed of light, no round trip is likely to be imaginable, and communications would be extremely difficult. They say, "Hello!" You say, "How are you?" And they say, "Fine." That conversation will take at least centuries. And I really do not think that that is going to bring us into conflict with the problems of the day. It may bring us near problems of some other day, but I am unable to see far enough into the future to notice how our little, not petty, but tragic circumstances of contemporary history are going to affect that.

Nor do I think this communication can be by any other means than light, or its cognate, radio. The universe is simply too wide for other means. The cost of getting enough energy to make physical travel possible is overwhelming, even for civilizations with enormous means, far beyond our own. It is conceivable that after a long time of exchange of knowledge, a ceremonial visit might be made. I can understand that. Everybody makes a great effort and finally come together. But that would not be the initial stage, and would not occur for a long time, until enormous rapport had built up. A time measured not in presidentiads, mind you, but in lifetimes of the republic; that is to say, in spans of 100 or 200 years.

So I do not think we are talking about just a normal enterprise; we are talking about an enterprise more like the development of agriculture than even like the discovery of America. The discovery of America looks to us like a sudden event, but from 1492 until 1605, nobody came to New England. And that is a long time, nearly as long as the time since the War of 1812. So we truncate history. Sometimes we think things happen in a snap, but they do not. They happen very slowly.

I think, therefore, that we will get a message, but it will not be simple. I would like to discuss the message itself for a few minutes to give a feeling of what I think is one possible model. The only way, I think, to achieve success in these matters is to invent models and schemes, not because any of them are necessarily right, but because in this way we mirror what the technical people will have to examine to see how we might get this thing done. Also, others will have to tell us the meaning of it, what will come of it. You have got to ask what might actually happen; then, in the light of that, prepare to meet those circumstances. If you are wrong, as you undoubtedly will be, you will have prepared something quite interesting and flexible, which is probably closer to the real event, although not very close, than you would be in the absence of any preparation.

I think there will be two great phases of this eventual time—which will come (perhaps in ten years, or a hundred, or maybe longer)—when some satisfactory radio-telescope work or something similar will acquire evidence of the deliberate beaming of a protracted message from space. First, the most important issue is the recognition of the message. Just that it is there, not what it says. This is often technically called the "acquisition" of the message. There will, of course, be false starts. There will be many claims. There have already been claims—I think rather facetious ones. They turned out to be wrong. I believe that will happen a number of times—three times, ten times, many times. It will continue until we have got something that cannot be anything else save a message.

I think, on the other hand, that when we do acquire this message, in the right way, it will be unmistakable; for example, it will take a week of verification to make sure that the statements are really true, but once that happens it will be so clear that the signal is something vastly different from the complex natural phenomena we know already that nobody will doubt it. I think this is true, just as nobody doubts if you find a Phidias or a Greek vase in the ground. Maybe you argue about how it got there. Maybe you argue if a pebble was really hammered; but these are only early stages in the making of tools. But the message will not be that way at all. It will be an elaborately planned, very great social effort on the part of some distant society. So I think that it will be easy to authenticate, and, of course, the message will be extraordinarily important.

At first you will know very little of what that message says, save that it exists, and maybe some general information about its source—how far away it is, what kind of a star and where. And then, I think, you will have pouring into the radio telescope's recorders, week after week, month after month, decade after decade, an enormous body of obviously interesting and meaningful postals. You will be able to read them, slowly and fitfully, because they will not be coded but anticoded; that is, the beings who designed them will have thought very carefully how to make the maximum number of mathematical clues so that the meaning will be clear. And it will be a large volume of material; it will not be something that the *New York Times* will publish in its entirety. It will be too voluminous, too technical, too uncertain, too much in need of study.

The closest analogy I can use is the enormous impact on modern thought (post-Renaissance European thought) of the Greek world. As a body of material it can be summed up in about 10,000 books. We have only about 10,000 books written from Hesiod to Hero down to the Alexandrians. Every decade or so someone will perhaps find another one buried in a backyard in Alexandria or in the Middle East, but no substantial number will be added. We can never interrogate any playwright, philosopher, or physicist from Hellenic times, and ask him what he meant by his statements. Yet the body of material from that period has been of great importance to the forming of the whole mind of our time. But not because everybody in the street reads it. It has influence because the people who write the books we do read have read it; and the students in the universities come to grips with it, and it informs and stimulates artists, scientists, poets, historians. With the discovery of the message, such a body of literature and technology and science and question answering and the rest will come, I think, every year. A mass of 10,000 books every year.

But it will come with extraordinary differences, so great we can hardly imagine how great they are. First, the people who send it will be incredibly alien. Even if their biochemistry resembles ours, it will not be all that close, you know. You could not eat their food, very likely, even if you have the same biochemistry, anymore than you can eat the food of a mushroom; it is very difficult to eat that. But meaning will slowly come out of this study, and it will contain, for better or worse, the answer to many questions that we cannot ourselves answer, which we will have to debate and interpret and work on and test.

I think the most important thing the message will bring us, if we can finally understand it, will be a description, if one exists at all, of how these beings were able to fashion a world in which they could live, persevere, and maintain something of worth and beauty for a long period of time. Again, we will not be able to translate it directly and make our institutions like theirs; the circumstance will be too different. But something of it will come through in this way. I think, therefore, that this will be the most important message we could receive. But it will be more of a subtle, long-lasting, complex, debatable effect than a sudden revelation of truth, like letters written in fire in the sky.

So I am neither fearful nor terribly expectant. I am anxious for that first acquisition, to make sure that we are not alone. But once that is gained—it might be gained in my lifetime—then I think we can rest with some patience to see what complexities have turned up on other planets. And if after considerable search we do not find that our counterparts exist somewhere else, I cannot think that would be wrong either, because that would give us even a heavier responsibility to represent intelligence in this extraordinarily large and diverse universe.

Twenty Years After

I first came to think about the promise of gamma-ray astronomy at a chamber music performance in the Cornell Student Center. The idea seemed good, and I am afraid I paid less than due attention to the quartet. By the end of 1958, I had published the first summary of what one might learn from gamma-ray astronomy. Like most theorists, I badly underestimated the experimental difficulties and it was to take almost fifteen years before innovative experiments produced results. But the paper was interesting and made the challenge an inviting one.

One spring day, a few months after its publication in 1959, my ingenious friend Giuseppe Cocconi came into my office, which looked northward out over a small lake to the green hills of Ithaca. Giuseppe posed an unlikely question. Would not gamma rays, he asked, be the very medium of choice for communication between the stars? They would work, that was plain, and my answer was enthusiastic yet cautious. Shouldn't we look at all of the electromagnetic spectrum for its possibilities, and genuinely seek out the best means for such a link?

In those days the big dish at Jodrell Bank was in the news, and the fiasco of Sugar Grove was in the planning stage. I do not recall whether we knew even the mere rumor of the big Arecibo dish, then perhaps little more than a gleam in the eyes of the Cornell radio physicists. Certainly neither of us knew the rudiments of radio astronomy. It took a few weeks of reading and discussion to come to understand that: "The wide radio band from, say, 1 megacycle to 10^4 megacycles per second, remains as the rational choice." Together we wrote and rewrote the short letter *Searching for Interstellar Communications* which from the first we had hoped to publish in the rather speculative pages of *Nature*.

Giuseppe wrote to Sir Bernard Lovell directly, urging him to devote some time for this task at Jodrell Bank, but Lovell was intensely skeptical

(Sir Bernard later wrote that he regretted his indifference to our proposal). Giuseppe and I knew, of course, that our proposal was both unorthodox and improbable, but we held, and we still hold, that its argument compels serious attention. "The probability of success is difficult to estimate; but if we never search, the chance of success is zero."

I sent the letter to *Nature* in London via Professor Blackett, the influential and imaginative physicist then in Imperial College, whom I knew personally, to seek his good offices in having it accepted. He acted promptly and successfully, and the note soon appeared.

By the time of its publication, I had left the United States for a year's stay abroad, a round-the-world sabbatical, and I recall the hints of public interest, as various journalists tried to reach me in far away places in Europe and Asia after the article had come out.

How does the question look now? First of all, we have passed through what I would call the pioneer stage. Following Frank Drake's path-breaking *Ozma*—the first real search—a half-dozen teams of radio astronomers in the USA, in Canada, and in the USSR have devoted serious effort to the search. A few of these have even used specially-built equipment for some time, but on the whole one could characterize their searches as preliminary. None of them prepared to continue, with improvements and expansion of effort, for a protracted search, gaining capabilities commensurate with less optimistic estimates of the task. These pioneers have shown the way and our debt to their ideas and devotion is great, but the absence of positive results so far is not at all unexpected. These are people who have walked past the haystack, picked up a few handfuls of straw, and searched cleverly for a needle. This pioneer phase is not quite over; for me, its end would be marked by an effort rather like that proposed by NASA in the spring of 1978 for funding by Congress: a scheme aiming at some specially-constructed equipment, and its progressive use in a systematic way for some years, on the scale of an important radio astronomy project. The task would be shared among several big dishes, using one only for a portion of the time, rather than devoting one dish to the work exclusively. It is humanly unattractive to demand that individual people spend decades in a search which offers no success, even though the eventual outcome is of great social importance. Much better, it seems, to divide the task, so that no one finds a whole career spent in an important but so far fruitless search.

The ubiquitous rise of computing power has been the most positive change in twenty years. People speak easily now of megachannel spectral analyzers and more; it becomes clear that we can carry out a search even if all the guessed strategies about special channels or directions or times prove wrong. What seems needed is to pass from the pioneer phase to a systematic stage, and to learn how and what to do by serious trials. The most negative change in twenty years has been the slowness of the scientific maturity of the auxiliary questions. We have as yet no clear result and have found no life anywhere else in the solar system. The early claims of dark companions of some stars, pointing to the existence of other planetary systems, now seems less than certain. In any case, those stars thought to have planets were not close analogies to our solar system. They are red dwarfs with apparently rather massive companions. Here is where I hope for the earliest real success: there are several new methods which may enable a much wider search for other "solar systems" within the decade.

For me, of course, the point has never been that we need to know much more about the chances before we search. The point is that only a search can make empirical what is after all a very old speculation. Whether we are alone or not needs to rest on experimental search, not on a string of evolutionary inferences. For I do not believe that our science is yet close to reliable theories of planet formation, of the origin of life, or especially the evolution and duration of communicative societies.

A curious view has grown up, which I may call Malthusian. It is something like the assumption of Malthus, that human society or its counterpart will sustain indefinite growth in numbers and capability, so that we will be pressed to travel and swell among the stars, filling the Galaxy in a geological epoch or less. But a simple look at the Galaxy, which shows no signs of intense colonization, strongly suggests that we here must be the first.

"Where is everybody?" is a question already more than thirty years old. Even this argument is more plausible than the opposite one, which I might attribute to Aristotle, that our planet is the single little blue footstool in all the universe, and no other place can harbor mind—and radio telescopes. But it seems to me neither position is as persuasive as the Copernican view taken in our 1959 letter, which suggests that we are finite beings, with finite capabilities, growing but limited, who are by no means sure either of becoming universal or remaining alone. Perhaps we can find out, and that is what we ought to try to do.

Finally, the task requires patience and a mixed strategy, that is, a plan which does not place every egg in one basket. We face ignorance, and we seek to lessen it. That requires a modest stance, willing to try in many directions, and to learn by doing. Even the microwave choice is not sure, though it seems still to me much the most likely. Indeed, I find myself still of the belief that the 21 centimeter line remains the best guide; of course, not right on top of it, but in its neighborhood, implying a small band search tempered by other knowledge. For example, if ever there are to be found the heroic extragalactic signals—hard to accept because of their demands on power and patience—they are most likely to be emitted at the 21 centimeter line *as viewed in the cosmic rest frame,* the frame defined by the microwave thermal background. This is universally known to every observer, and even now to a part in a thousand. Of course, one should tune to different frequencies as different directions are scanned, for the Sun is moving with respect to that great stationary frame. But the frequency choice is almost without ambiguity.

The many radio lines which have been found since the 21 centimeter discovery seem by the very number to lack equal promise, though they ought not to be disregarded. In the same way, narrow-band acquisition signals still seem the best choice, but other forms of modulation, far less familiar than steady signals or even simple fast pulses, deserve consideration. All of this will follow as the search effort becomes stronger and more systematic.

The problem of interference from satellites is serious and one very much hopes for recognition of the SETI task and for some measure of frequency protection. Every interested person can play some part by urging, through his or her government, protection for the rest of the century.

Just today as I wrote this piece, I realized that the then new 1959 astronomy volume of Joseph Needham's magnificent work, *Science and Civilization in China,* had caught my imagination during the same months that Giuseppe Cocconi and I were working out the *Nature* letter. I do not at all recall noticing it, but it would be agreeable to think that I had read his translation from old Teng Mu (on page 221 of volume 3), ending with the now familiar words:

"How unreasonable it would be to suppose that, besides the heaven and earth which we can see, there are no other heavens and no other earths?"

From *The Lute of Po Ya,* 13th century A.D.

Life in the Universe

Cosmic evolution, has become the myth of our time for the scientist. When I say myth, of course I don't mean falsehood nor do I mean a mere tale. I mean a world view embodying substantial values of deep cultural significance, centered around efforts to explain origins. That is clearly what we begin with. Our myth today differs from those of the past in that we have a much wider series of tests for its acceptance. We expect it to transcend past experience, to go beyond language. That, in a way, distinguishes post-Renaissance myth-making, especially at the cosmogonic level, from what went before. I would hasten to say that I think a good deal of what we have put in the myth will fall before the test, just as would fall the Ramayana if we asked to see the stance from which Hanuman leapt in order to get to Ceylon. We won't find that. We might find something like it, but we won't find the actual footprints.

Our myth is a grand cosmic myth. But there is also a style of great *personal* myth: a hero, his life and times, how he was chosen, what he did in childhood, the difficulties he went through, the triumphs he had, the disasters he endured. The myth becomes over time a set piece, perfect in its symmetry, brooking no changes. In India, if you go to see a dramatic performance or a dance, or hear a song, 8 out of 10 times it is from the Ramayana. You know what is going to happen, but that is not a problem: novelty is not a positive value here. We now have such a personal myth growing in our culture, a very important myth that is only about a century old. Its presence has hovered over all the papers at this meeting, although we have almost never mentioned it explicitly. That is, of course, the life of Charles Darwin.

The story of evolution is the Ramayana of the scientist, and Charles Darwin is Rama, or as close as we can come in our impious and divided times. There is nothing we have done that does not have somewhere behind it the evolutionary model. We have transferred that model from self-replicating and mutating systems to a whole variety of cookery, but in the end it is the myth we are aiming at. And the outlines of the myth were not imaginable even 150–200 years ago.

I won't tell the story of the Darwin Ramayana, though it is a nice one to tell, but I will mention a few events that everybody should know about. There was, for example, the voyage of the *Beagle*, and the letter written by Uncle Josiah to get Charles on the *Beagle*. Charles's father, who was a self-assured, well-to-do man, did not like to be crossed in his opinions. He conceived that it was a very poor idea for a young man who was going to become a clergyman to go off on a trip with all those sailors for several years, to waste his time when he could be sitting in the country trying to avoid conducting too many funeral services. He told his son in no uncertain terms that he had half a dozen reasons for opposing the trip. He said: (a) It is a waste of time; it doesn't lead to your vocation. (b) It was offered to somebody else first; you're second choice; what good is that? (c) Being off with all those rude sailors can be very bad for your settling down in the future. And so on.

Darwin was crushed. Here was a tremendous opportunity that had opened for him in a letter from his professor at Cambridge. Darwin said to his father, "Well, can I appeal in any way? Is there anything I can do to change your mind?" His father was a reasonable man. He was firm, but he believed he was controlled by reasonable persuasion, so he said, "Okay, find any thoughtful man of affairs, who knows the world, who agrees with you and not with me and I'll reconsider the whole thing."

So Darwin went off 10–15 miles to his uncle's house, his Uncle Josiah Wedgwood, who made the pottery. (He was the second or third generation of the pottery owners. By the way, it was Josiah's daughter who became Mrs. Charles Darwin a couple of years after he came back from the *Beagle*. So he was obviously pretty friendly in that house.) He came over there, age 22, and talked to Uncle Josiah. Uncle Josiah wrote a wonderful letter back to the father explaining every point. "Of course, now, it's true sailors are unsettled, but how often do you hear the stories of the sailor who comes back and settles, runs a little farm, and never wants to leave home again. That's an equally likely consequence of going off to sea."

Now suppose that Charles hadn't gotten Uncle Josiah to write the letter. He would not have gone on the *Beagle* and would not have seen the Galápagos Islands. He would instead have been a country clergyman and would have written a brilliant journal of ornithology about the birds of some little parish. I am sure that is quite true. Of course, someone else, some Wallace, would have developed the theory of evolution in a decade or two anyhow. Because A. R. Wallace did write the first paper offered for publication which correctly described the evolution of species by the operation of natural selection. As you know, he sent this to Darwin to find help in getting it published.

Darwin was disturbed, because after 10 or 12 years of closely filling his big notebooks with data on this point, here was a man he knew only slightly who had anticipated the entire story, but without any strong evidence, just a very clear logical statement of the whole story. Wallace himself was away in Indonesia, where he worked as a professional collector of animal skins, birds, eggs, and so. He traveled about the tropics on commission and sent back to the museums and to wealthy private collectors in England the things that he could get. He had malaria in Ternate, Indonesia, which took him out of the field for a few months. During that time he wrote his paper. Darwin was the only naturalist he knew who had been a traveler and was therefore likely to be sympathetic, not a bookish Cambridge professor with dusty disdain for people who traveled about far from libraries. Wallace was not a university man. He wrote to Darwin, "I think you would be interested in this general idea; I know you have some concern for the species problem."

And that was the story. It is, of course, full of myth. I imagine that if we last a thousand years (and I think we will), this story will be told in movies and film, and heaven knows what three-dimensional holographic schemes I can't even foresee, in one way or another, as the great myths of the past have told about Rama, and Sita and the deceiving deer, and the jumping of the monkey army across the sea.

I tell this story to indicate that even in human events there is a kind of convergence. This, of course, argues against one of the objections to the Search for Extraterrestrial Intelligence (SETI) program: that evolution can never get to the same point twice. It is true that an exact retracing of a path is impossible. You cannot come from where you slept last night to this auditorium another time by exactly the same path; certainly not to the resolution of a micron, and I think not even to the resolution of a yard. When you cross the highway, for example, the traffic pattern will be different. But it does not matter if the path is not the same provided the end is the same. This is what is easily forgotten. It is improbable that you will find any second path that is close, but it does not seem so improbable that you will come to the same end, by a very different path in that complex interweaving of world lines that is both human history and the history of biological structures.

I am not going to talk anymore about the principle of convergence, but I think it is worthwhile to bring it forward sharply. Many argue, quite correctly, that many conditions of preadaptation and so on need to exist for any big evolutionary change. When you look at those preconditions in any particular case, you may find them unrelated to the end result, and therefore you say, "How could it be that, since the essential preadaptation has nothing to do with the end result, that same end result could nevertheless happen?" Of course, the answer is just the same. Were the end result to happen, by my hypothesis, along a very different route, you would look back and say, "Aha, this is preadapted in such and such a way and therefore it could never happen any other way."

The particular case I would like to mention here, the preadaptation of human beings for speech, was probably quite real. But I would like to focus on something that has moved me very much in the past few years. One of the great new events in the human sciences relates to the development of sign language. It is now generally recognized by linguistic scholars that the large and randomly scattered deaf community has developed, without any assistance, a flexible, vigorous language of signs and gestures. This community is comprised of totally deaf persons in every country, and in our country in particular—persons who at a very early age before the onset of speech have been made totally deaf. It has taken more than a hundred years and the contributions of millions of people, but a gesture language has been developed which you can now see repeated very rapidly along with the television news. It is gradually becoming salient in the American scene. This language, when examined by the methods of grammatical analysis that, since the second World War, have revolutionized our view of languages, seems to show not only a great unity between the Chinese version and the American version (which have little historical connection, though the French and American have), but also a grammar, highly adaptive and flexible, which violates some of the principal canons of human speech. If this new work is sustained, it will demonstrate that human communication can be achieved on a complex and subtle but distinct level that shares many of the characteristics of normal speech, such as plays on words, words in bad taste, poetry, and song. All of these things have been identified and verified in American sign language in the last few years. There are lots of dirty jokes going around; teenagers giggle over them in the corner, but they won't do them in public. In the same way there are slang expressions that are not good form, but are fast. And there is an analog to song, that is, rhythmic exaggerated gestures that have the quality of music; there is also poetry marked by intricate word play, gesture play in which puns and imitations, one gesture for another, are very important. All this is gradually becoming understood by the academics.

I remark on this because it is, of course, a completely different style of language. It may not connect to Broca's area or to any other of the principal compiler regions of the human cortex, where there exists our facile and rapid instrumentation for speech. One of the papers today said that it was difficult to see how speech could have evolved. Of course, I am not saying that it is just as good to do gestures as to have auditory communications; I honestly do not know. It might turn out to be better in the end. But, in any case, if you evolve with one mode and develop the other as a substitute, this is a tribute to the flexibility of the human mind, to a talent not evolutionarily built in at all; it shows that there is more going on than we imagine. It is another example of the convergence that I have talked about and tried to illuminate by the Darwin-Wallace story.

One of the most important occurrences over the 15 or 20 years that people have been thinking much about SETI has been talked about implicitly here, but not a great deal explicitly. Outside this rather rarefied atmosphere, though, it has had a big effect on public imagination, something we should not forget. The event we thought might happen, but did not, was that another form of life would be discovered on another planet. But what has occurred that we did not foresee, but that supports our general view while complicating our detailed quantitative understanding, is the finding of complex polyatomic carbon compounds in a wide variety of cosmic contexts. From studies of the molecular species in the great condensed galactic clouds and from the putative analysis of some asteroids and maybe satellites (in which we find some indication that carbonaceous compounds are important), we have derived a renewed interest in the carbonaceous chondrites themselves. This may represent a physical sequence or simply a set of analogies. But it also demonstrates a point on which biochemists were already very clear 50 years ago: the flexibility and peculiar subtlety of carbon compounds, as well as the high abundance of carbon, makes them preeminently the source of complex chemistry in the Universe.

I am led to another little story that I think will be familiar to many but perhaps not to all. When we seek the input of the physical sciences—astronomy (dynamical astronomy especially), geology, climatology, aeronomy—we know that their practitioners are masters of a powerful deductive structure, with quantitative possibilities. Of course, they are beset by the necessary complexity of their models. They must try to pin down from some a priori model just what the first 6 or 7 hundred million years of Earth history were like. They search for a necessary prelude for the biologists, something essential to the total picture of cosmic evolution. On the other hand, it can be the other way around. If the biologists were to explain why they needed certain conditions, perhaps it could be determined whether these conditions were at all plausible under some existing model. It is clear to me that strong interchange must go on in this domain. We clearly have to learn a more interdisciplinary way of facing such problems.

A famous 19th-century interdisciplinary dispute illustrates this point. There was a lot of dispute, even full conflict, but no resolution at all; logic was clearly on one side, but it turned out to be wrong. This is a charming story, resolved in print by the distinguished geologist T. C. Chamberlain, of the University of Chicago, about the turn of the century. It was also resolved in a lecture by Ernest Rutherford in the early years of the 20th century at the Royal Society. Many would guess what I am talking about: the famous problem of the time scale available for Darwinian evolution.

The absence of substantial observed changes in speciation in the natural world, as compared with the swift changes produced by domestic hybridization during the course of history, is a strong argument for the slowness of natural speciation. This argument was made even stronger by the fact that paleontology showed change clearly; you had to say that the time available in the geological record was very large. Darwin, while he was a man of extraordinary logical ability and in very simple ways a brilliant experimenter, often cutting right to the heart of the matter, was a bad mathematician. He could not calculate anything. He had a touching faith, however, in instrumental methods. His son writes that he discovered his father making measurements with an old paper ruler and writing these down to high accuracy. The son commented, "Well, you know that ruler probably is not right." So he got a better ruler; sure enough, his paper ruler was stretched and deficient. Darwin just fell into depression; the notion that a calibrated ruler, a thing you trust to measure with, might not be right was a breach of faith he could hardly accept from his ruler-making colleagues! That was his style. He kept saying that his study of geological records suggested that there was a great deal of time indeed. He liked to put it that biology required almost infinite time. He didn't, I think, mean what infinity means mathematically. But what he meant was very sensible. He meant a time quite long compared with all times that had been suggested so far. That is what we usually mean physically by *infinite*. You don't literally mean infinite, you mean a willingness to neglect the reciprocal; that was what he was prepared to do.

Now comes Lord Kelvin, armed with the most powerful physics of the 19th century, who was able to show with hammer blows, one after another, that the time available for Darwin's evolution could not be 100 million years, it could not be 80 million years, it could hardly be 60 million

years, perhaps not even 10 million years. Because, if the Sun burns carbonaceous fuel or any similar chemical fuel, it can only last thousands of
years—a palpably inadequate stretch of time. If it derives its energy from
gravitation, we know its mass and we know its size, and we can show that
it cannot last for more than a few tens of millions of years. The proposal
that it catches and feeds on comets and meteors can also be excluded,
though perhaps it can push us into the 50–100-million-year domain. Thus
there was a direct conflict. Kelvin would come to the biologists and
heckle them terribly by saying, "Tell me what time you want and I'll
show if it can be; I can calculate everything." He calculated cooling and
so on. Of course, they couldn't name a time. They simply said, "Well, we
know you're wrong. We feel it in our bones, in our fossils, but we can't
prove it." So the evolutionists were regarded as people without any quantitative basis for their science, though they were, of course, on to something profound.

That was noticed by T. C. Chamberlain immediately after the discovery
of radioactivity about 1900. By 1905 or so, Rutherford himself gave a famous evening talk. It was a most distinguished and formal lecture to the
Royal Society of London. As a young breaker of rules, a young discov-
erer, he noticed that in the front row sat Lord Kelvin himself, very elderly,
but very stern, trying hard to stay awake and check on this young man
who was going to talk about new forms of energy. (Kelvin didn't like radioactivity either, by the way.) But Rutherford tells us that he wondered
how to avoid offending Lord Kelvin. Perhaps the famous man would get
up and leave when Rutherford talked about the fact that nuclear energy
can keep the Sun going for a good long time. Finally Rutherford thought
of the right formulation. He said that he had been able to solve an old
problem, whose magnitude and importance had been shown by Lord Kelvin himself, when he pointed out that there was no known source of energy capable of keeping the Sun going. At last by experiment they had
been able to find a new source, bringing out fully what Kelvin had shown
all those years ago. Says Rutherford, Kelvin went immediately to sleep
and the whole session was a huge success.

Let us hope that some meeting of minds among physicists and biologists will come to relieve us too of our dilemmas. Of course, there are interesting traps: we don't really know whether we have to have a strongly
reducing Earth environment, whether we can find special microenviron-
ments, or whether we can cross one or another of the bridges set before us
from the physical to the biological side.

Of course, there goes with progress some remarkable inventions, the
inventions of life. It is dangerous to pin our view of what happened

wholly on the success of this or that invention. It is very easy to do and might even be true. But it also might have been that another invention would have occurred had "invention #1" not won the day. That is a question you must always ask. Two inventions created recently I find extremely striking, worked out in detail, very appealing to a physicist. Both of them are, so to speak, structural innovations: the coming of specialized structural proteins and the first appearance of lignin. Both are elements making possible what is fashionably called a quantum jump in the behavior of two great kingdoms, plants and animals, when once they have acquired the ability to maintain large systems. Because the large living systems are showy, they are in the museums; they are like ourselves. The simple laws of scale suggest that large organisms, certainly on dry land and probably also in the ocean environment, must have stiff structures; otherwise they cannot survive easily. Mass goes up with the cube, and strength only with the square of linear dimension. You cannot survive without a pretty strong structure. Of course, we have an internal skeleton and our friends the bugs have something different, but that doesn't matter so much.

The notion that a few chemical inventions made this possible is striking. We begin to see a new power to enter into the details, to try to suggest the biochemical pathways and various correlations. This should not end, it seems to me, until we have spelled out quite a number of such jumps. In population genetics and in evolutionary theories, as I see it, we have little predictive power as yet. We cannot predict rates. We cannot predict proportionalities. We have to take the evidence of change and work backward. It is unclear whether we shall ever have a profound predictive power. But at least our analysis should have a much richer texture than in the past. That is beginning to happen, and I am very encouraged to see it. I think we can expect results in terms of convergences, in terms of interacting biochemical inventions.

It is always fascinating to have the physiologist, the paleontologist, the general biologist working together, not only on the remarkable information-transfer story at the DNA-RNA level, but even at the much grosser level of the engineering inventions that enable living forms to work. I want to mention one invention that is especially striking to me. The internal organization of vertebrates, including ourselves and our kin, brings many implications that do not appear on the surface. Consider one behavioral invention, if you like. It has a quality reminding us of the proteins and the lignin, but it is not at all like them.

Let us look at a strategy: the strategy for catching food as a predator does. We are, after all, predators, both on berries and on bears; we belong to the hunting-foraging creatures. Suppose you were in the position of living on lobsters. That is a nice position to be in: it is achieved by some New Englanders and by all common octopi. Octopi are intelligent invertebrates—in some ways our analogs within the invertebrate kingdom.

Let us approach the octopus from the standpoint of a rational analysis of prey-seeking behavior. What is the situation? Lobsters are not as variable in behavior perhaps as some land animals, but still they are not all that uniform either. It is quite likely that any particular desirable game, like lobsters, appears in fluctuating numbers within the field of action of any carnivore. There is little likelihood of a steady flow of lobsters, one dropping down every hour. That is the Santa Claus or Big Rock Candy Mountain theory of life. Most of us hunters don't find it that easy. You've got to go out and scrabble around a little bit to get what you want.

When caribou are numerous you should of course hunt caribou. When the Eskimo or the Indian hunter, skillful person that he is, finds that caribou are unhappily in short supply, he will simply redouble his efforts, for he is hungry and back home the wife and kids are hungry. The whole situation is serious. The same tendency is found in every hunting-gathering mammalian predator activity, say for the leopard. We explain strange behavior on the part of mammals sometimes in that way: "Well, it was hungry. I forgot to feed my Siamese, so it tried to eat the curtain!"

On the other hand, an octopus has a much purer view. It behaves more like the theory of games. When lobsters become few, an octopus does not seek in a frenzy to find those few lobsters or put up with eating mere crayfish. Heaven forbid! Instead the octopus goes to sleep—a most intelligent thing to do. Every once in awhile it wakes up and looks out. "Any more lobsters around?" No. Back to sleep it goes again.

Such control over impatience, anxiety, and hunger is very hard for us to understand. Our thought is based on our design: namely, we have to generate 100 watts all the time. There is a basal metabolism, roughly 100 watts, that we expend. If we don't keep the machine fueled, we're in irreversible danger. But the octopus has no such base load demand. Cold-blooded, he is willing to relax to the ambient temperature of the warm sea environment in which he lives, provided only that every once in awhile he can scrape up a 100th of a watt, turn the ganglia on a little bit, open an eye. That isn't too hard to do. Once you look at it coolly, you realize that human behavior goes completely against the sound principles by which an organism would adaptively go hunting. Whenever it is hard to hunt, don't continue to hunt with greater frenzy over longer hours, as we all do. On the contrary, take it easy. If conditions are not good, there is not much use in hunting. We can't adopt that principle, though, because for a couple hundred million years conditions were generally good enough so that somehow or other it was worth paying to keep all our subtle electronics going in order to have an opportunity to hunt well. How different evolutionary structure can be! We need a special view of the lives of other creatures. If you now carry this logic over to some distant world, then it gives scope to the issues we are talking about.

The strategic discussions for SETI showed in circumstantial and detailed operational terms that we have already begun to make a clear plan. Certainly, a great deal of hope emerges. Indeed we have a remarkable new opportunity: we seem to be on the threshold of finding other planets. We are setting up apparatus dedicated to the purpose of finding planets, both by interferometry and by astrometry with modern techniques. I very much hope that the entire community will support and applaud this effort, because it seems to be one of the most important auxiliary searches that could be made. It is important even if we never have a chance of getting radio signals. It can give us something else to look at than just the single Sun-planet system to which we are so well adapted.

I will conclude with some remarks of a more philosophical nature. The first is straightforward; it is a distinction I have made before, but I think it bears repeating. We can characterize attitudes toward SETI by invoking the names of two philosophers. (Perhaps they should not have to bear this burden; they are not really responsible. But as often happens, it is convenient here to use the names of famous workers of the past as labels. I ask their forgiveness.) First is the Aristotelian view, which seems quite plain: Earth is the cosmic center; the heavens revolve around it, $1/R$ reaches infinity here, and here is the right place! On this view, of course, the whole outward-looking style is neither necessary nor desirable. Astronomers can hardly accept that view; at least they have not accepted it for several centuries now. They are not going to change, and I am pretty sure that most of the other sciences will follow in turn.

The second point of view I like to attribute to Copernicus. Everyone knows what that implies, though I don't know if he actually said it anywhere: namely, that this green and blue Earth is not all that different from the planets and the Sun and all those other things that circle and shine in the sky. They are themselves earthy or gaseous or whatever, but they are physically real objects. Nowadays men have walked upon one such object and shown that it is not different in kind from the one we inhabit. Since we know Earth is also earthy, then it is clear to us that these are only relative categories. Circular, shining, and perpetual orbits was Aristotle's view; it was Copernicus who recognized that Earth was no less circular, shining, and perpetual. We take a very different view of the cosmos post-

Copernicus. That has been the spirit of SETI. The radical Copernicanism of the very first efforts is still viable, though admittedly we have more judgment about where to look.

I would like to mention another point of view, which seems a little curious to me. It is less developed among the general public than in the scientific community, and it seems to be based on a very grand extrapolation indeed. I would like to call it the Malthusian view. Take a piece of semilog paper and plot the growth of more or less anything in human culture (people, telescopes, cities, motorcycles, whatever you want), extrapolate, and very soon it shoots right off the page. At some point the mass will exceed the mass of Earth, or the volume exceed the volume of the Universe, or the velocity of expansion exceed the speed of light. Therefore, there must be some catastrophe ahead. This is true of every exponential, all but independent of its rate. If you change the rate by a factor of 2, you just double the time: 100 years becomes 200 years. You are talking about the long-term future, so the rate doesn't make that much difference. Therefore, this is only a statement of the transient existence of exponential fits: certainly such a fit is a good thing to notice, but it is never totally realistic.

Ecology has brought us to understand that something always turns up to put a plateau above every exponential, some necessary resource. Californians will guess that it is gasoline that limits the Universe, but if that were not true, it would turn out to be something else.

What happens when we apply exponential arguments to space travel? We have heard several discussions this afternoon of the difficulty of interstellar travel. Several recent papers, however, have taken a longer range view. The most conspicuous of them is in a book by Freeman Dyson, a very able physicist indeed. He points out that we have not reckoned with self-multiplying systems of an artificial kind. (There are plenty of multiplying systems that are living; I am not so sure that artificiality makes a big difference, but I will accept the argument.) If you can make a self-multiplying system, put it in a spaceship, and give it an initial stock of capital resources, off it will go to find a suitable solar system. Eventually, it will be well enough ensconced that it will create replicas of itself, which will then be sent out to find other solar systems to be settled, and so on. If you make some calculations it turns out that you can cross the whole Galaxy with this scheme during a geological epoch, in a few hundred million years.

The argument goes next: since the Galaxy doesn't look as though it is densely occupied that way, since many more than 300 million years have gone by if our calculations are right, then nothing like this has ever happened. The most enthusiastic paper I have seen says that we ourselves will be able to start this process in 100 years. It is plain to that author that we must be the first ever to encounter such a possibility. No use, then, spending a lot of experimental effort on SETI; these calculations make it quite clear. Of course, they are not very robust calculations.

Nevertheless, I find it quite interesting that the mere examination of the sky should suggest that because the stars aren't arranged in tasteful letters, or free form Crescent or Cross, or anything else, that the Universe has not been modified by a gardener of any sort. This seems enough for some to say we should not make our patient search through the channels and through the spatial directions to find signals. I hold that this is a semiserious problem. Serious, that is, only to the extent that it develops within the scientific community an antagonism to what seems to me an already difficult but at least an empirically based scheme. I hope we can come to some meeting of the mind with persons who represent this theoretical pessimism (or is it optimism?). Let me simply suggest a mixed strategy: instead of waiting 300 hundred million years to see if the theory is right, I might try waiting only 300 years to test it by some more active procedure.

A curious situation has arisen under the power of intense modern instrumental specialization (in the broad sense, where *instrument* includes *method*). The instrumental specialization of our science has grown steadily since the 19th century. It is sharply reflected in the institutions of our universities, which are slow to change in the face of it. For example, many universities still have a Botany Department and a Zoology Department, and if you bring in any one of a number of microorganisms, they don't know which department ought to study it! Perhaps it doesn't make any difference. But our research structure is inherited from institutional decisions in Scottish and German universities made 100 to 150 years ago. Sooner or later this will change. For all real large-scale engineering activities, such as NASA has carried out so successfully, we know that that is not the way to do it. Such activities require mission teams and a mix of specialties. The universities are going to have to learn.

There is a narrowness of action, though not of intent, which characterizes university departments, and scientific publications and scientists in general: if it is too popular, it is somehow vulgar and wrong. You can't really speak to those people across the street. I live next to the chemists at MIT, but I never see them. I hardly know who they are, yet between physics and chemistry it is hard to know who should study what molecule. I myself am guilty. We form communities not based on the problems of science, but on quite other things. This is part of the general split between the intelligent informed member of the public and the scientist who speaks in narrow focus. But the great theoretical problems which I believe the world expects will somehow be solved by science, problems close to deep philosophical issues are the very problems that find the least expertise, the least degree of organization, the least institutional support in the scientific institutions of America or indeed of the world.

Two of these, of course, are the great questions, "Are we alone?" and "How did life begin?" These questions are treated now in the elementary textbooks, because of the vigor of a few people over the last 30 years, but they are hardly treated anywhere else. The further you go away from the freshman student, the less likely you are to find a colleague interested in it. This is beginning to change. Five or 10 years ago, the radio astronomers, just to name a group of people I know quite well, were pretty hard to talk to about SETI in any way. It wasn't so much that they would disagree, that's fine: they still do. But they laughed, and that was not very pleasant. Well, now at least they are only smiling; this is a kind of gain.

One cure for this ill, though a difficult one, is the pursuit of a scientific discourse on a more philosophical, more consciously aesthetic, better-illustrated style, one willing to grapple with large problems, even though only small solutions can at present be offered for them. I think that science requires this change. I expect to see an enlarging of the disciplines to form at last an interdisciplinary pool, aware of larger philosophical issues. We need not try to solve them or to prescribe their limits, but we must recognize their human importance, their intellectual existence as an increasing element within scientific thought. If that were the only positive result from the SETI investigation, I think it would still be judged by history to have proved extremely worthwhile.

A Talk with Philip Morrison

From an interview conducted by Charlene Anderson and Louis Friedman of the Planetary Society.

Charlene Anderson: How did you get interested in the possibility of communicating with extraterrestrial intelligence?

Philip Morrison: Giuseppe Cocconi and I were thinking about possible new channels in astronomy as a whole. We first thought that cosmic gamma rays would be interesting to look at, then we realized that we knew how to make gamma rays. We were making them downstairs at Cornell in the synchrotron. So Cocconi said, "Maybe they would be useful to communicate through space." It seemed that they could work, but they weren't very strong. I then said, "We should study the whole spectrum to see which is the best wavelength." We knew very little about radio astronomy. But after we made the study, we realized that radio is much better than anything else. We decided to write the paper because it was so clear that it was a real possibility.

CA: What started you thinking about the possible existence of extraterrestrial intelligence?

PM: That wasn't really the issue. It comes about in a backwards way. It's the typical approach of a scientist. We don't know if they are there or not, but using a communications band is a way to find out. See, you were thinking that in order to call up somebody, you have to have somebody to call. I'm saying that before you call, you have to have a telephone system. We got our initial idea from the telephone system, not from thinking that anyone is there. We don't know how to estimate the probability of extraterrestrial intelligence, and it may not be very high. But we said that if we never try, we'll never find it. We went about it from an instrumental point of view.

One of the hardest things that scientists have to communicate to people who are not doing scientific things is that, in science, you can't do what you can't do. It doesn't often help you to talk about something that is not at least indirectly measurable. At least the consequences (of a radio search) were measurable. So what we asked was not, "Who's out there sending?" but, "Suppose there are thousands of civilizations sending, how would we find that out?" There is a way, with radio. Then we could ask, "How many cultures could be sending radio and where would they be?" But the first thing was to establish the fact that there is a telephone system.

Louis Friedman: Why did you narrow the communications band down to the radio frequencies?

PM: The idea of gamma rays was just an inspiration of the moment because we were making gamma rays that week! So I said, "Let's look at it systematically so that we don't jump to any conclusions. Let's find out what's in the rest of the spectrum." So then we looked in the library, talked to people, made calculations, and a few weeks later we simply had gone through every domain. It was clear that microwave radio is much the best and that the 21 centimeter line distinguishes itself.

LF: Why is microwave better than visible light?

PM: The essential reason is, God knows how to make light but He does not know how to make microwaves. Consider the animals on the surface of Earth. It is very hard to find any animal, or any plant for that matter, which does not respond to the light of the Sun. The light of the Sun is a powerful, dominant thing. But no animals or plants respond to microwave radio; that's an invention of human beings—that's generally true. The stars put out huge amounts of light, but they do not put out much microwaves. The transmitter of the Arecibo dish is 1000 times a stronger source of microwaves because you can concentrate direction and frequency in microwaves and you cannot do as well as that with light. Even if we put all the physicists in the world to work on it, we could not make a white light that could be distinguished from the Sun as seen from a distant star. If you set off 1000 hydrogen bombs, and if I were looking at it from 100 light years, I would not notice the explosion. The Sun's light and its natural flaring variations are too great. But the Sun's radio is nothing compared to even little Arecibo in a narrow beam and band.

Radio technology requires large things which are not biological and at the same time not as large as stars. It's a human scale thing, requiring stationary objects, pieces of antenna. That's not very common in the universe; the universe is made of drifting gases.

CA: What do you think of the searches that have been done so far?

PM: Excellent. I have always called them pioneering searches. This is the age of pioneers and they are sort of taking a covered wagon out into the territory, establishing a homestead, but I am looking for some real occupation with cities and railroads.

CA: What about the search for extrasolar planets?

PM: I think they are sure to succeed or find out that there are no extrasolar planets around the hundred nearest stars. I think it will be done in the next five to ten years, both with ground-based observatories and with the Space Telescope. Both plans are moving forward and look promising to me. Of course, that is a long way from SETI, but it is a related question and very interesting.

CA: The Soviet astrophysicist N. S. Kardashev has speculated about three possible levels of extraterrestrial civilization. A Type I civilization could harness the total energy of a civilization for communications, Type II could harness the power output of a star, and Type III could control the entire energy output of a galaxy. What do you think of this type of speculation?

PM: I don't think the important thing is to speculate on how these societies behave, because we won't be able to get the answer to that problem. It's expecting too much from philosophical speculations. I'm excited by his three levels; it makes a good story. But I don't believe it is true, I don't believe it is false. There is no way to tell.

CA: If you had your choice as to the type of search to be done, would you prefer something like Project Cyclops, which would have been massive in size and expense?

PM: That is going about it too strongly. I don't think people want to spend all that money with no reliable experience. The search being proposed by NASA is the very best kind, a big strategy which combines some guessing at magic frequencies, some searching through frequencies, looking mostly in the galactic plane, but not neglecting high galactic latitudes or other galaxies. I have always said to put your bet on many horses; the biggest bet on the best horse, the next biggest on the second horse. But you don't overlook any horse that makes any kind of sense, because you do not know enough to be sure.

LF: What about interstellar travel? Might that be the easiest way to communicate?

PM: I doubt it very much. The costs in energy alone are so prodigiously different. If people won't invest a modest sum, say $1 million a year to do a radio search, it's hard for me to see that they would do a $100 billion search that would take much more than 100 years.

LF: The argument against interstellar travel in the scientific literature is always that if you wait twenty years the technology is so much better that the interstellar probe that you send out now will catch up with the interstellar probe that you would have sent out earlier, and that is always going to be true.

PM: The speed of light is the speed of light. We can never get news faster than the speed of light. And this is the speed that we are using with radio. Of course, we are expecting that the signal is already present so that we are not even waiting for the round trip—half the trip is made. We could find that out tomorrow; we can't find it out tomorrow by any probe. You could send a space probe out and wait for them to send a message back: "Your probe arrived in good shape and here is the telegram that says so." Long before the probe returned you would have a message from the beings who had seen it. Communication is much better than transportation. In fact, I think transportation will probably never occur except maybe in a ceremonial manner. Transportation is used when you have some good to gain. You send people out to live there or you bring back gold or whatever. There is nothing that will bear the cost and time of setting out through the domain of space as long as radio communication is the cheapest and fastest thing. The only valuable resource is information; everything else is negligible in comparison. I would try not to be negative if someone wants to do it, but I see no signs of preparation to voyage to the stars.

CA: Frank Tipler and some other scientists have argued that if intelligent extraterrestrials existed they would have visited us already, therefore they do not exist and SETI is a waste of time and money. What do you think of their arguments?

PM: I first read Tipler's papers some three years ago. I say that there are three ancient philosophical guides to this issue and they categorize the approaches. I name the first one after Aristotle, and that is: What good is SETI? We are the center of the universe, Earth is God's footstool and we have to clear up the Earth before we find out anything about other worlds. The second philosophy is the Malthusian view of Tipler, Gerard O'Neill and others. It says that life and human beings have a built-in exponential function that is never going to be stopped, swoosh, right up to any power that you like. Therefore, they can easily turn the galaxy into Central Park, and since this does not look like Central Park, we must be the very first. That is based on the very unlikely assumption that we are going to be infinitely powerful in some modest time and I think that is a very over-optimistic assumption. Like all other beings in the world, we are limited in our powers; limited in our energy sources, limited in what we want to do and what we can do. That limitation is perpetual. It will not be a tight, firm limitation but it means that we will go slowly, rising in a curve, plateauing off, then slowly rising in bumps over a long time. I imagine that that has been true of all the others. Of course, those bumps will not be exactly the same, and so we expect some differences. But I don't imagine the beings that we are going to meet would be so different that they would dominate a whole galaxy, as Kardashev says. That is an overly optimistic view based on the fact that we have been through a long, exponential rise in population and technical power. What every physicist knows is that there is no exponential curve that doesn't reach a turn-over point.

I don't think we can say, "Well, look, we've gone to the Moon in only 20 years." Actually, our speed in space is very small; the farthest that we've gone is not very far.

The third view is the one we are taking; that's the Copernican view that we are typical in the universe. We are not the central thing. What's going on here can be expected to go on elsewhere. Growth is limited and power is short here, and probably it's somewhat short and limited elsewhere, too. On the other hand, we're not especially singled out; why should we be the only place where there's life? There should be many worlds broadly like ours, but not exactly like it. That is the modest view of Copernicus.

CA: What about the arguments against SETI from the supposed improbability of life, the claim that the sequence of events leading up to life is so improbable that it wouldn't have happened anywhere else?

PM: Well, again, it's a speculative argument, and I think that it's going to be disproved. I imagine that we will make life here on Earth within the next generation or two. And that will answer that. All these things are based on premises and calculations which are very tenuous. I don't want to sit here and speculate on how likely it is; we have to look.

Speculations go on; there's nothing new. Much the same arguments now used by Michael Hart were published in a 1938 book by Le Comte de Nuöy, who did not invent them, but took them, as he says, from a Swiss philosopher of 20 years before (whose name I forgot). These conjectures rise in the popular literature every generation or so. I doubt that they are right.

CA: How do you respond to people who think that SETI can be postponed indefinitely, that in the current economic climate we should not be spending money on it?

PM: I think that's a plausible position. But on the other hand, it's not plausible to postpone every exploration and every novelty. If you do that, you will never get out of your state. The most depressing and negative stance that you can hold is to say that we can try nothing new and innovative because we're in trouble. New and innovative approaches and attitudes are what we need to solve our problems. I'm not arguing that SETI is more important than any particular problem. We're asking for a very small part of the resources which are disposed of daily by the government, even in the same scientific domain. It makes sense to try it on a small basis. It would be inappropriate to try it on a gigantic, gigadollar-a-year basis, which nobody wants to do anyhow.

CA: What about Bruce Murray's argument that we've got about a 20-year window before Earth-based radio noise becomes so great that we won't be able to hear anything from space?

PM: That is an argument that is somewhat supported by the work that they're doing here at JPL. The Deep Space Network is quite sensitive to that. I'm a little less apprehensive. I suspect that the techniques for discriminating the interference will get better. You'll lose five or ten percent of the sky, but the next year you'll regain ten percent. It's a shifting kind of barrier. It's like the appearance of clouds in a telescope. No astronomer desires it, but they can all cope with it. And I think it's much the same thing.

CA: Why do you think it's important that we do SETI?

PM: Well, I don't know. I take a very permissive attitude towards this. I don't think that it can be done enthusiastically until people decide it's worthwhile to do. When they decide to do it, once they decide to do it, they'll do it. If you're not persuaded by thinking of the consequences and the challenge, then I don't think you should do it. After you are persuaded, you cannot be stopped from doing it.

CA: How would you persuade people that it is important?

PM: You just ask them, "Do you want to know if human beings are alone as intelligent beings in the universe?" If that question doesn't move them, I wouldn't go any further.

LF: Do you think that the consequence of receiving an extraterrestrial signal will only be finding out about "existence?" Might we not see more in it than existence?

PM: I don't believe that it will be like that. Nobody will send out only an acquisition signal. It's extremely easy to attach the acquisition signal to an information channel. As soon as you get this, within five years you'll have a giant wide-band signal pouring in its message, which, to begin with, you won't understand at all. But you'll have all the deciphering people and all the philologists in the world working on it, studying it like cuneiform. It's the same kind of problem, except it's exciting because it's a voice not from the past, but from the future. Archaeology of the future is what it should be called. Archaeology of the past is very interesting because it tells us what we once were. But archaeology of the future is the study of what we're going to become, what we have a chance to become.

CA: So you take the approach that it's like finding our place in the universe?

PM: Yes, it's a missing element in our understanding of the universe which tells us what our future is like, and what our place in the universe is. If there's nobody else out there, that's also quite important to know.

CA: With the wide variety of life forms that you think possible, how would we find our future in their example?

PM: Our future has detached itself from the life forms of Earth. These beings are much closer to our future than are any other animals on Earth, even if they're made of a completely different chemistry, because they're using radio communication unique on Earth to humankind.

CA: So it's the technology. . . .

PM: It's the culture. Technology is culture. It isn't something that is evolved by genetics, it was developed by culture. If these beings have got enough culture so that living near a star somewhere they have built radio transmitters, then they are something like our future. It doesn't matter if they have six arms and are blue and made of silicon, they would still have problems of how to deal with their own technical transformation of the universe. Of course, if they've never made radios and they're just sitting contemplating the world from a mystical point of view, we'll never find out about them. So SETI is a biased affair; it looks for people who have the biases of the Jet Propulsion Laboratory. I admit that. But can you think of any other way to go? And maybe they have found something useful to Earth.

CA: I've heard SETI scientists argue that we're going to blow ourselves up if we don't find somebody who's got an answer to prevent us from doing it.

PM: Yes, this is the gospel reason, the good news. I'm neither so pessimistic nor optimistic as that. I don't think we're likely to blow ourselves up. And I don't think that, if we were, getting radio signals would help us. But it might. That's what Fred Hoyle thought. He built a fascinating scenario that says civilizations grew, developed, and declined, then finally they made contact. That stopped them from blowing themselves up. And gradually this spread over the whole universe.

But who knows, it's all science fiction. I say, and say again, it's one thing to sit here and think of possibilities and write good science fiction; it's another thing to look into the real world.

CA: On that note, do you have anything to add?

PM: It's good that this is a receiving and not a transmitting program. It's much easier and much less expensive to receive than to transmit. And it contains the answer to many reasonable people who say (with, I think, an overly pessimistic view) that if you transmit, you might give away your location to some predatory menace in the galaxy. To which I answer, as Fred Hoyle has already said in his novel, *The Black Cloud,* maybe the best thing that you can do to prevent being preyed upon is make yourself so interesting that they won't eat you. But I don't really believe that. All I'm saying is that to avoid all these difficulties it's much better to receive than to send. This is a program which is designed to listen for a long, long time, and which makes no mark on the universe at all. I don't favor sending or establishing a program to send until, after lots of experience, we decide as a society that's the right thing to do.

The Search for Extraterrestrial Communications

Beginnings

Our initial paper (Cocconi and Morrison 1959)[*] was begun in Ithaca and completed in Geneva, at the brand-new CERN where Cocconi had gone as summer came that year, and where he has lived and worked ever since. We felt it an original, if rather eccentric contribution; Giuseppe wrote privately to Sir Bernard Lovell to suggest the use of the big Jodrell Bank dish in the effort, but Lovell was not persuaded. I sent our paper along to Professor P.M.S. Blackett to seek his good offices in publication. A good friend and a man who liked novelty, he acted promptly and the paper soon appeared. Whatever the astronomers thought of the suggestion, the popular press was much taken with it. I spent that year on leave in a round-the-world set of visits, and the press followed me for comments from London to Rehovoth, Bombay, Kyoto, Denver and home again. Everyone surely knows that Frank Drake, quite independently of our paper, was at NRAO making ready in Green Bank the first real radio search, quite minimal by today's standards, but enough to show what we did not then know: not all main-sequence dwarfs are powerful sources of CW signals at 21 centimeters!

Both Cocconi and I were caught up in the larger currents of physics during the decades before. He is a wonderfully reflective experimenter, I a somewhat experimentally-oriented theorist, both moving out of the pre-war days of nuclear physics, into what had passed for particle physics in those naive years: cosmic rays. (For me wartime had meant the engineer-

[*] Cocconi, G. and Morrison, P. 1959, *Nature,* 184, 844–846.

ing of chain reactions fast and slow; Giuseppe recalled air raids more than wartime projects.) Mesons of more than one sort, then the V-particles and other enigmas were all newly found in cosmic rays. But in the early fifties the cosmic-ray physicists began to see that the new particle physics would be machine-made. The new beams of the engineers were much more useful than the celestial accelerator could provide, in every way save that of energy per particle.

But it became clearer and clearer in the fifties that cosmic rays would no longer dominate the experimental study of fundamental particles. Yet cosmic rays raised other questions that could not be be answered within the accelerator lab. Where did cosmic rays come from? How did they travel? What made them vary year by year, and sometimes even change intensity an order of magnitude within a few hours? These were questions of astronomy, in the solar system and beyond. The interests of many cosmic-ray physicists shifted from the gamut of new particles to the site and nature of new astronomical processes. We had known for a few years that the primary cosmic rays were some sort of a sample of cosmic matter as a whole.

High-energy astronomy became explicit and fascinating for me in an unforgettable meeting of cosmic-ray physicists at Guanajuato in Mexico about 1955. We then learned at first-hand the new Soviet results on polarized optical light from the Crab Nebula. Not so long before we had stared in wonder at the bright blue glow of the optical synchrotron radiation (prudently, we looked at it only in a mirror) produced by the new fast-electron beam in the Cornell synchrotron. It was a kind of light, I recall saying, quite new under the sun, for it arose from transitions in a magnetic field, and had no need for Coulomb interaction. Now we realized that it had lighted explosive stars and even solar flares for a long time. As below, so above: hardly a cosmic novelty.

In 1958 that line of thought led me to publish an account of the originating processes and possible importance of gamma-rays in astronomical contexts. I was far too optimistic; it took the experimenters a dozen years and more to develop the art of measurement. But there could be astronomical gamma-ray sources just as we were familiar with gamma rays from our synchrotron at Cornell. Optical synchrotron light was only a minor by-product of a strong fast-electron source, the possible source of gamma-rays. As below, so above?

Early in 1959 my ingenious friend Cocconi came to me to speculate that perhaps gamma-ray beams could be found in space, generated by other beings just as the Cornell synchrotron had produced them. My paper had made clear that the rays could penetrate the Galaxy. That was the beginning of the new paper, in which we systematically examined the entire electromagnetic (EM) spectrum for a signalling optimum. Still tyros at radio astronomy, we prepared the letter to *Nature* as best we could; microwave radio easily led all the other bands in our simplified signal-to-noise study.

The Dimensions of Search

The search for signals, whether deliberate or mere accidental leakage, is certainly multi-dimensional. The obvious dimensions of direction in space, time, particle type (which for EM signals reduces to two polarizations), and signal intensity all require analysis, but I shall discuss mainly the frequency-time plane.

The overall argument in favor of microwaves rests of course on the spectrum of the natural noise in space. Taken with the strong stellar sources of optical and infrared light, and the quantum fluctuations of any photon-borne signal, the studies agree that somewhere between 1 and 10 GHz is the most suitable frequency domain for energy-efficient signals at interstellar distance. Human technology has already acquired the means of interstellar signalling in these octaves, and only there. Let us accept this time-tried result, always only tentatively.

The next fixed point is the fundamental relation between band width and signal duration, $\Delta v \Delta t = 1$. The search for sharp pulses, with small Δt, is one attractive limit, early pursued in the USSR (Troitskii *et al.* 1979).[*] Plainly one goes to a pulse width such that $\Delta v/v$ is not very small, say only one or a few powers of ten below unity. For the microwave domain, that means pulses in the sub-microsecond range. (A physical lower limit to pulse length is in fractional nanoseconds.) The receivers now can be broad-band; a search to match receiver time to the pulse time is really implied, parallel to the ordinary radio search over frequency, but a pulse might be quite powerful and need little time averaging. Such was the early Soviet search mentioned; what it found was instructive, if rather to be expected. Plenty of phenomena much simpler than radio astronomers are able to generate sharp pulses. Geophysical pulses, their exact origin not known (at least to me), showed up in receivers spaced apart ten thousand kilometers. The sharp pulse, a limiting type of signal quite useful by mutual design, does not appear easy to employ between partners with no

---

[*] Troitskii, V. S., Starodubstev, A. M., and Bondar, L. N. 1979, 'Search for Radio Emissions from Extraterrestrial Civilizations,' *Acta Astronautica*, 6, 81–94.

chance to prearrange the situation. They are too commonplace.

The other limit is clear: narrow-band carriers, with Δv small. Then Δt is large; these signals can vary only slowly. Most familiar radio circuits are of that kind, to be sure often for the frugal utilization of a crowded spectrum. It looks as though energy savings would be greater the smaller the bandwidth used. Moreover, we never encounter any natural celestial signal at all that could be called narrow-band in engineering terms. Since the first signal acquisition, if not the transmission of much intelligence, could be made on very narrow bands indeed, it is very important to recognize, as Drake and Helou 1976[*] first did, that in the Galaxy a physical lower limit to bandwidth is set by the dispersion imposed by plasma drifts in interstellar space. There is no gain in going below about a hundredth of a Hz for Δv. It is a search for this wonderful limiting case, the extreme narrow-band approach, standard for human signals, unknown in the skies, even in masers, which was realized by Paul Horowitz of Harvard.

There is much more choice for signal structure than the two extremes of narrow-band CW and wide-band sharp pulse; the whole domain invites eventual study. But the limiting cases are visibly attractive for our search.

Inspired (?) Guesses

Systematic search over all the combinations of parameters that define the multi-dimensional possibilities is daunting. From the first, anyone looking at the problem is tempted to make an inspired guess, to invoke a piece of magic, that is, an attractive assumption that all at once sets a preferred value of some parameter, noticing a clue or hunch that may appeal to all reasonable searchers, at either end of the hoped-for signal link. For direction, one invokes evolutionary symmetry; we look for stars more or less like our own. There are certainly plenty of those. Or perhaps one looks at groups of stars all together, a cluster. Most likely one looks in all possible directions using a finite antenna beam. For frequency, even in our first paper the 21-centimeter line, the most abundant photon in the universe, stood out as a tempting spectral landmark. That choice still appeals to most of the radio astronomers who have attended to the problem. Plainly this is no objective matter; one can only argue plausibility, and then depend upon hope.

* Drake, F. D. and Helou, G. 1976. Report 76, National Astronomy and Ionosphere Center (Cornell University, Ithaca, New York).

It is of some value to recall that during the first twenty years of search we of earth, investigators in nine countries, tried three dozen pioneering searches. Nearly all used the broad microwave band. Meanwhile our farthest space probe has sped out from the sun perhaps 25 or 30 astronomical units, almost ludicrously short of the stars. If that rudimentary experience is any guide, it confirms the old judgment that electromagnetic search is far easier: easily-created bosons handily win over conserved and costly fermions. (The almost uncoupled neutrinos have few supporters.)

Change

Since 1959 receivers have gotten better, and there are fine new radio dishes. None is larger than the dish at Arecibo, in design way back then. Nor do the receiver improvements amount to an order of magnitude. What has changed by an unforeseeable factor, many orders of magnitude, is computing power, made possible by microelectronics. It would have been the state of the art to invoke a receiver capable of integrating the signal in 1024 channels at once. Right now Horowitz has two multi-channel receivers in everyday reliable operation integrating over a tenth of a megachannel all at once, and multi-channel devices, are under construction that listen to ten megachannels at a time. The use of that informal unit, the megachannel, makes clearer than anything else I can say what a new epoch has arrived. The search over wavebands has been empowered beyond the dreams of 1959.

A second novelty has come from astronomy. It is nothing so general as the rise of microchips, but it offers a certain elegance. It bears on the old guess that the 21-centimeter line is the place to look.

The trouble with a guessed frequency is of course that the natural line is not only a place of high noise power, but is also poorly-defined in frequency position. The natural line is megahertz wide; searching such a band in tiny nibbles of a hundredth of a hertz for every direction in space is a chore, even for megachannel receivers. Can we narrow our guess?

The natural width is a consequence of Doppler shifts; gases in space always move. Transmitters and receivers move too, so the signal can be given a precisely-assigned frequency only if there is some natural frame of reference that stays at rest, known both to transmitter and receiver. There is no place in the Galaxy that fits the bill, save its center. Since we move more or less transverse to that center, we might use it. But there is now known a much grander rest frame, that of the thermal background radiation, the mean motion of all the matter in the universe. Surely we can expect that all observers know that frame; we ourselves know it fairly well now. In a few years we can expect to fix it within a part in ten thousand or so. That implies the opportunity to tune to a 21-centimeter signal within a matter of a hundred kilohertz, a search within ten million minimal-width channels. We will soon be able to do that in one integration time. Of course we must correct away all the motions of sun and earth, as the transmitter must do for its part. But the microwave background, observable everywhere, renders the procedure entirely a matter of care, independently and mutually practical at both ends of the link.

That seems the happiest of hopeful guesses to me: the most abundant photon of all, frequency fixed in the grandest of all rest frames, agreement a matter of independent measurement. The scheme works for any signal path within the Local Group of galaxies. If we ever extend our search across tens of millions of light-years to more distant galaxies, the same frame offers other possibilities for a frequency choice that can correct for the velocity change with time that we call the red shift; that has been proposed by J.R. Gott of Princeton (Gott 1979).*

Criticism

Most, though not all, of those who have looked into the matter agree that it is less useful now to estimate the chances of success from various a priori points of view than it is to mount as powerful a search as possible. Certainly that is the lesson from astronomy as a whole. All the same, it is useful to engage two particularly well-expressed criticisms of present efforts.

The most direct is a position that argues that no signal is likely to exist (Bates 1978)** because no conceivable society would signal into the unknown for a time long enough to have a chance of an answer, perhaps thousands of years or even more. How could any government, Professor Bates asks, ever be concerned about posterity to such an extent? One important rejoinder is this: the notion of protracted search is not a necessary part of the decision. It is enough for each generation to begin the task anew, or by improvement upon the past. There is no requirement for abso-

* Gott, J. R. 1982, Cosmology and Life in the Universe, in Extraterrestrials—Where are They?, Hart, M. H. and Zuckerman, B. eds. (Pergamon: New York), 122–134.

** Bates, D. R. 1978, Astrophys. and Sp. Sci., 55, 7–13.

lute continuity; the game is a probabilistic one. Only the duty cycle really matters, how much of total time is spent in a signalling mode. Each new technique, each new insight, may bring a wave of activity. If the wave does not succeed, very likely it will die out, until the next impetus. That is the lesson of our own science; the Greek philosophers were the forerunners of the particle physicists of today. The enterprise of the ultimate analysis of matter is older than any government, university, or laboratory. If it dwindles for a while, it can grow again. Only a final end to all investigation can kill it. Without making any claim to certainty, I will at least argue the possibility of such a course for the searchers after signals from afar, whether in the cheap receiving mode or in the more advanced state of deliberate transmission. Leakage detection is also a minor possibility.

Much better-known is a stance I have called Malthusian. It rests on the idea of the exponential improvement in all technology. That must entail, the proponents say, an eventual physical tour by crews or by mere automata, of the whole Galaxy. Using various estimates, all of which really exploit mainly the unbounded quality of an exponential rise, these critics say that we must be the first ever to think of beings afar. Otherwise they would have been here already; and where are they? This argument is modified to include deliberate concealment, the "zoo" idea, in which we humans have been preserved as unwitting pristine specimens, and more. All of this is interesting, but it shares the defect of Malthus. In the real world there are no unlimited exponentials. Something limits every growth. We are surely finite beings, and it is likely that we will remain finite forever, unless in finite time we disappear into zero. I do not know what the upper bound of our grasp can be; I suspect it will fall short of making a rose garden out of the Galaxy. With that finiteness the power of the argument fades; it all becomes a discussion over the values of limiting parameters that none of us know.

No, the best means to seek the unknown is by experiment, absent a tested theory able to exclude the result we seek—and even then, be careful about the truncation of experience. Consult any textbook for examples.

Mixed Strategy

The task before us is to organize the best search we can, and to improve it steadily. That certainly means that we do not dogmatically optimize all that we do according to the most popular assumptions. There is a place in every particular scheme of search for a logical design optimizing the search that follows from the assumptions. That is good engineering. But where we lack any real theory of what we seek we must take a more modest and more resilient stand. We ought to search with some effort in all the modes and under all the assumptions that we can convince ourselves remain possible. We can well spend most effort on the most widely-held views, but it is a mistake to close tight every other door. Unlikely strategies deserve, not equal time, but time and effort roughly proportioned to their judged probability. It would be wrong to bet everything on 21-centimeter signals from solar-type stars, even though that appeals most to me and to many others. For if the best-supported view turns out to be wrong, even if that is unlikely, and one has done nothing else, failure is very much more probable than with a prudently mixed, or hedged, strategy. Gamblers who do not hedge are not reasonable. Thus the NASA study wisely includes both narrow- and wideband microwave search, and looks both at random star fields and at nearby target stars like the sun. There is and will be room for every plausible stratagem; let a hundred schools contend, the least accepted perhaps contending on a shoestring. Overall strategy can be controlled over the long run, as always, only by the engaged community as a whole. Fads and follies will occur, but they will run their course.

In Conclusion

Unlike most of science, this topic extends beyond the test of a well-framed hypothesis; here we try to test an entire view of the world, incomplete and vulnerable in a thousand ways. That has a proud name in the history of thought as well; it is called exploration. We are joined in the early ingenuous stages of a daring exploration, become real only during recent years.

It is a voyage whose end we do not know, like that of science itself. We seek a call, not from person-to-person, but from culture to culture, a call whose content we cannot well foresee. It is only the first step in opening such a channel, the process that communication specialists call the acquisition of a signal, that is now the focus of our work. Let us try as best we can; there is no other way.

PART II

ON LEARNING AND TEACHING

Just as I was happily taught science, and physics in particular, by example, speech, and print, so I hold myself obliged and delighted to teach it to others in turn. Over the years this obligation has extended to include teaching the general public, even via the video tube, and to teaching young children.

The public often differ from the physicists in what they find interesting, but children largely respond quite as the physicists do. (See the Exploratorium in San Francisco.) College freshmen may fall between the two.

The bond between young children and physicists is epistemological; they join in wanting to know how you know. Authority and precept tell you crisp answers, a pallid sort of knowing, with the reflected tinge of the blue book. In primary grades, even in pre-school, both kids and physicists feel more at home. The nub there is not language, but action; the center there is not the knower, but what is to be known. How to nourish the fruits of that style within the staider institutions of a working world is the challenge not yet fully met.

Less May Be More

The Role of Design

I wonder how many people have seen the mime, Marcel Marceau, in films or on stage? When I saw Marceau, who is of course a recognized genius, I was struck by one of his activities, which seemed to me in a certain sense to represent the kind of attitude which at least the theoretical physicist is required to take.

Marceau produces on a bare horizontal plane, that is on the plane of the empty stage, the marvelous effect of a man climbing up seven stories of a spiral staircase to the apartment where he lives somewhere in the Quartier Latin. He does this in exactly the style dynamically appropriate for this endeavor. He starts up. You see him leaping across the stage three steps at a time, and then two steps, and then one step. And now the railing appears. An imposing sturdy railing which plainly goes around the corners carefully, and then finally, at the very end, he has reached the top; now quite tired, quite slow. It seems he has really raised himself in mime 80 feet while you watch—all on the flat and empty floor.

Now one might think of other models—the model of the craftsman who carves some shape, for example. I think this is not terribly appropriate, because we know the instinctive quality of such work; that is, the reflex-trained quality of this work. The craftsman carves, and he knows the consequences of the strokes of his blade, but does not usually think about them, In fact, probably better not. He just feels and knows step by step, through long experience and taste, that this is the right thing to do. A kind of unselfconscious concern for what's going to happen is, I think, his principal internal state when he works.

But in my opinion, there is no question that Marceau had to analyze with some care, probably on the verbal level, certainly on the conscious level, what were the dynamical, muscular, interactions of somebody climbing stairs. He didn't do this impromptu. He went often up the stairs, he watched other people going up, he looked at himself, he considered what he was doing, and then he deferred the execution of this for a very long time. Until now, year after year, he can put it on the stage in New York, Tokyo, wherever he may go. This sort of distinction between an analytical approach to the external world with its presentation, to be sure not in the mathematical or verbal symbols characteristic of the physicist, but in a quite different way, by the means of a kind of dance, seems to me to offer a rather interesting model. The kind of understanding of the world which the physicist tries to gain is far removed from the view that is usually taken of it when it appears in cold dead type; in its growth, it is closer to Marceau.

I want to talk about the problem of inducing this selfconscious analysis of experience. The experience here is a limited experience but a very important one. The experience we discuss is generally speaking not interpersonal, but between man and the physical world. The problem is to induce an analytical approach to this experience, a notion that it has structure, that it consists of a sequence of events which can be separated and ordered. Then one can recreate in some fashion a model, a description; one has a new power of synthesis. Indeed, I think that this power of synthesis is the other side of the Marceau example, a side I want to stress very strongly.

Of course, this is very different from what the physicist does. The physicist might really carry out much of the same preliminary work which Marceau did for this particular game on the stage. Certainly I can easily imagine physicists working with, say, the group of Gray at Cambridge in animal locomotion doing so. What are the muscle tones, what are the postures, what are the leverages, what are the force chains that go on in this kind of locomotion? Then, instead of synthesizing the whole thing again as Marceau does, into a reconstructed performance and remembering it, probably both reflexly in the muscles as well as consciously, the scientist would write it down. Or perhaps he would photograph it, or maybe graph it.[*]

This is another place in which I think that the model of Marceau helps, especially at the introductory level, especially for those students whose introduction is also a formal terminus in some way (that's most of the class of students we are talking about) it is very important that the analytical does not wholly displace the synthetic use of the materials given. And I feel a serious concern that all of the courses we have had heretofore, very largely press the analysis so far and extend it so widely that in the

[*] See, e.g., James Gray, *How Animals Move* (Pelican Books, Baltimore, Maryland, 1959).

time available there is no chance for a synthetic use of the same materials—or very, very little chance. I think that what people call, broadly speaking, design—experimental design, graphical design, theoretical design—is a major portion of the role which science plays; often not for an individual, but for the people who use it, for the world as a whole.

I heard an extreme version of this from Fred Hoyle. We were together in the audience at a meeting and in response to some point made on the platform, he nudged me, and said it was wrong. The speaker had said, "These are questions which we can not answer, don't know how to answer." Hoyle whispered, and you will of course realize the faults as well as the strength of his view, that in science answers are not important, it is the questions that are important. Every sound question is putatively answered, not answered perhaps by the man who asks the question, or even by the generation that asks the question. But no question, no real question in science, will fail to be answered in the species, says Hoyle. Therefore, he said, the real thing is to innovate, to suggest, to ask questions. Because the answers are sure—somebody will make the answers, says Hoyle, only we don't know just who. I am not trying to press his point, but I do think it is a part of my story.

Analysis, that is, the presentation of why something works, how it works, the generalization, the conceptual framework in which it sits, this is invaluable; it is certainly the essence of what we mean by science. In fact, that's what our archives hold. But I think it would be a mistake for physicists, chemists, biologists—the only people who can really handle it—to leave it at that, because it gives a false impression of what is the nature of scientific enterprise when looked at a little less parochially than from the point of view of those who have to publish papers in the *Physical Review*. It is the subsequent utility of these materials in the synthesis of something new; new questions, new experiments, designed things, the stuff of the world, that is equally important. And we leave this out when we have only analysis. This is my essential point of view.

I think most particularly of the future elementary school teacher as representing a particularly large group of liberal arts students, not science-degree headed.

The Character of the Students

Perhaps a few comments on the characteristics of this group as I see them will be worth making.

Size: $0.3\text{--}0.5 \times 10^6$/year

This is an interesting number because it is the number of liberal arts students that one might expect to have in physics and chemistry. And it is also the size, roughly speaking, of each of the high-school physics and chemistry student groups. This means that the scale of the effort might be at most comparable to the large secondary-school physics and chemistry programs. I think, however, that this is an overestimate, for (a) those programs have already been done, which means it is a little bit easier to do things a second time; (b) the innovative possibilities in this much more heterogeneous group of schools are much greater. The high schools, diverse as they are, are much more homogeneous than the very wide spread of colleges that we have. You know that no package is likely to spread over the entire domain; what is probably worthwhile is to initiate a couple of nucleating schemes and see how they go. Then the scale is probably one order of magnitude less, which is helpful; because instead of the 1000-man-year effort of the PSSC, I should think one might do with 100.

Now, second, what sort of students do we see?

(a) Mechanical/mathematical interest
(b) High literary ability
(c) High interest in interpersonal relations

For want of a better word, I am going to say that there exist what I call mechanical or mathematical interests. These are not strictly parallel, but I think they span an area which you will understand. The people in this group (a)—and vocational choices are made rather early—are those who, by and large, become the majors in science, engineering, architecture, and a few other subjects in our schools. You can tell it by talking to them, or looking at their rooms, their cars, or the contents of their pockets and handbags.

Ordinarily you will find that their early experience has already given you a clue. There will not be many people at the age of 18 for whom you can not, in five minutes, determine membership in these groups. I may be wrong, but it is my experience that this is widely true. Of course, you might find someone who thinks he wants to be a physician and who turns out to be a chemist, or somebody who thinks he wants to be an architect and who turns out to be a psychologist. But all of these persons belong to group (a). Few such people are included in the group whose education we are now discussing; it is extremely important to recognize it. A selection mechanism of great power is at work to remove such people from the group we are talking about; if one overlooks this, he has overlooked the foundation of the problem.

I should say in parenthesis I think this is only temporary, at least I hope so! I hope that when we have adequate elementary school instruction in the modern way—which will come in a generation—that this strong dichotomy will weaken. You won't be able to classify people so easily by talking with them, because they will all have had a much richer experience. That's the hope, though it is only a hope. Therefore, I would say that the program which we are talking about now is a remedial program. It's a program to stand in the way between the kinds of schools we have now, inherited perhaps from Aristotle, and the kind of schools we are going to have in the future.

That one group, group (a) is not involved in this account. The people whom I discuss will *not* have made—or even touched—model airplanes, have *not* wired up circuits, have *not* lit the gas flame very often. I do not exaggerate; they haven't done that very often. You find one in twenty who has never struck a match to light a flame.

Then group (b), high literary ability. In the Departments of English and Literature, in the Language Departments, in Philosophy, one finds students of very high linguistic ability who have, from an early age, been reading and writing, and have had a lot to do with paper and ink. This is another important group of the good students of a university; mostly these people are also, for the moment, *not* in the group we are talking about. I would like to bring more of them in, but most of them are still outside.

Finally, the subject is group (c), high interest in interpersonal relations. I am now talking primarily of the preteacher; again I have to use a very cold, a curious term. I think, however, that this does represent reasonably well the sort of person who tends to become an elementary school teacher, or even any kind of a school teacher. They have strength, a developed strength of some sort for interpersonal relations, showing, I am sure, high concerns of responsibility, and of altruism, of dependence on friends. Very often there is deep concern for the family; this is very important to them.

Given these three groups, my view is that if I draw a graph it is going to look something like Fig. 1.

My division lines are probably not quite right, and the tails of the curves may be wider, and the distribution is probably not Gaussian. But in the zeroth approximation, this is the kind of pool we are talking about, distributed among these three parameters. The group (c) is roughly that which falls between the horizontal lines. This is not a statement about intellectual attainments of any kind. This is a statement which has to do with a more complex thing which I have not measured, but I try to base the statement on fairly wide experience. Of course, like all such subjective estimates, it might be wrong. If any social psychologist knows how to

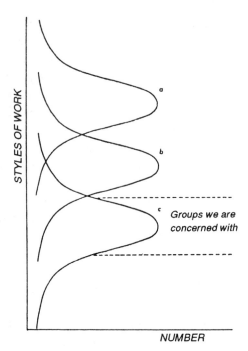

Figure 1. Approximate distribution of styles of work in college student population.

measure these things properly, it would be interesting to know about it. It would not, I think, be very hard.

I have taught in colleges and universities from San Francisco State College to Cornell; and I have worked a lot in the secondary-school context, and for three years, quite a bit in elementary-school contexts. That is, I have worked with elementary-school teachers, and with their trainers— talking to them. I also tried hard to organize at Cornell a small group of about seventy-five students per term who come to a course which was advertised as being for "innocents." Physics for innocents, physics for the naive, physics for people who never had physics and do not want to know anything about it, or something like that. I've formed very many of my judgments from my experiences with them; they turned out to be just the sort that we are talking about, and on a rather high level of ability. It was an elective course, and evidently was not something to press on the academically marginal students who are not doing very well. They wouldn't have the time to take it. But the students who felt they could afford to spend three hours and were somehow talked into it, came in. The characteristic student might be a junior girl, a French Literature major. She had taken high-school biology—her one science—five or six years before, and didn't remember it. Such a girl would be a typical member of this class.

I should at this point make another remark which may not be very popular, but which I think is important. What is the student input into each category looked at from some broad social and sociopersonal point of view? One thing that is extremely important, and probably the rule, will make it clear. People come in two brands genetically, and we have in group (a) 5% women. Group (c) is at the moment 60% or more women. I think we have done a poor job (and when I say we, I mean all teachers in all scientific professions). Everyone has done a poor job in instructing half of the human race in this kind of activity. The reasons for it should be looked at consciously. I don't think you should just say, you know, it's something that's beyond our control. It's not beyond our control. It will be beyond control if we pay no attention to it. Unless we do pay attention to it, we don't have the right to say it is beyond control.

The Style of a New Approach

Now, what are some of the reasons? I am no specialist on this matter, but I can by empathy try to find out what is wrong. I can see some things that are wrong. The characteristic flavor of scientific activity is as far as possible from—well, let me say what it is, rather than what it's far from. I write it down: austere, sober, and unornamented.

In Vesalius' *Anatomy* (you have seen the great plates, they are one of the glorious 17th-century traditions in the foundation of modern science) the cadaver is hero. The lay figure of a man appears and is slowly dissected to the bare bones as the plates go by. And beyond these brilliantly drawn anatomical figures, there is the scene of the Tuscan plain. While the body is there, the intact body, the Tuscan plain is in full summer. As you dissect down, down, down, the seasons turn cold, until at last the wintry, bare-treed, empty Tuscan plain appears behind the skeleton. The engraver who did those plates was a great artist. This sort of thing, which is a contextual approach, away from the purely analytical, is something which the 17th-century scientists felt very keenly. They were not foolish men, as we must agree. They were the founders of contemporary science.

I am not trying to speak for the rich and baroque tone in everything science does, because I know that part of its beauty is its very clarity and austerity. But I am strongly of the opinion that 19th-century utilitarianism induced people to think that scientific laboratories oughtn't ever to be well-painted, that they ought to have lots of heavy engraved brass dials, that the *Physical Review* ought never to crack a smile (certainly it never does, because if it looked at itself, it would die laughing!), that the writing in the archives and journals should always be spare and jargon-like. Perhaps such things are justifiable in terms of busy people and the limitations of space. But that our introductory textbooks should copy this style as closely as possible is clearly wrong.

We know that enormous success attends people who try to add a different tone to the whole thing. For example, Professor Gamow. His attempts may not always be the best, but one cannot deny the fact that he has made a genuine impact upon the public mind, even his little drawings with jokes in them. Don't misunderstand me; it is not at all my idea that the solution is telling jokes, or that we have to moderate the severity of our analysis and its logic. All I say is that we have a one-sided monotonous view of the whole task. A proper mix is, in my opinion, a precondition for appeal to an audience, largely women, with interests based on interpersonal relations, who just will not accept the bare abstractions, the lack of ornament, the lack of contextual reference, the oversimplifications of the presentation of science; which is, by the way, not its true nature, as you know. You cannot walk into a synchroton laboratory without finding its tone very different from the tone of the *Physical Review* or of the early textbooks. Most textbooks are even more sober-sided than the research conferences and the best laboratories. That's unfair!

I learned graduate dynamics from a here unnamable expert. We weren't allowed to ask questions. We sat through 40 lectures; we were never allowed to ask questions. One man once asked a question, whereupon the lecturer stopped took off his glasses, wiped them, put them back on again, trembled a bit, and said in a very quiet voice, "This is a difficult course. There are many theorems to be introduced. If you ask questions we shall never get through." And after thirty seconds to let the vibration dampen, he began again; "The infinitesimal displacements must always be compatible with the constraints. Theorem 7."

His style, which we inherit from Euclid, made rather worse in the process, is, of course, a caricature. Even at that time it was recognized not to be the best kind of teaching. It represents, I think, a polar position; near that polar position, if not on it, is concentrated most of what we do. I think this is one of the reasons, one of the main reasons, why we fail to get many people (and especially in our present culture, a large fraction of them are young women) interested in science.

I'd like to remind you of one last thing. The most important tool of thought, the most important instructional item in any curriculum is the mother tongue—the language. Everyone knows very well, as the Berlitz ads tells us, that a little boy this high can speak Ewe—why can't you? And all those Chinese children who learn Chinese! It causes one to think. How is it that the difficult, subtle, definitive subject—language—is learned by everybody who has to learn it. There are very few exceptions; there is a very high incidence of success.

The student-teacher ratio is, of course, very good, as we know. The social motivation, the functional utility of language is also very high. So you have these two to go on, and I am sure they are major. But another thing that I would argue, is that language instruction, even when it is made in a formal way, for example in preschool, in kindergarten, in the early grades, or even in the middle grades, is carried out on a very wide front. I don't think this only because people have consciously thought it all through, but because the language is so useful, it is used in so many different ways, that it naturally has a large front of exposure. For example, do you think that they ever teach little kids to speak without using games, and puns, and jokes and all sorts of examples with an extremely wide range of emotional approach? No, they certainly do not. This is a very large part of the curricular activity even if disguised.

I think the standard way of teaching science which we employ is rather like trying to teach the language out of a book on contract law. You get a book on contract law. Of course, it is a very big, thick book, so you can't do all of it with children. So you take the 35th page of it, a very simple page, a simple contract: "I, the party of the first part, undertake to buy an ice cream cone for the party of the second part." There's the first day's lesson in language instruction. By the end of five years you have learned an awful lot about contracts, and you are now parsing the sentences, making codicils, and whereases; and so on, and so on in a very interesting way. That's it. That's how you learn English. I venture that the number of mutes would become very great! But I don't think it's an unfair analogy of how we teach science. It's true, I am sure, that if you can once learn contract law extremely well, you will have a very clear, precise, powerful understanding of English sentences, and pronouns, and referents and the lot. But I will bet nobody will listen to you talk. And I doubt if you'll write anything very good or do anything very much with it, except make more contracts. This is the style in which science is taught, by and large. I'll say no more.

Now, to try and be more specific. What can one do? What might one do? I am going to be most tentative only to indicate the sorts of things that should be considered as desirable.

Right-Handed	Appeal to experiment and observation, which is abstracted from the real or given. Mathematical models exist which are metric and qualitative.
Left-Handed	Playfulness, novelty and invention, synthesis/analysis. Diversity in people and analysis.

I would feel that these are the modes which belong in a course in physical science. They are very modest objectives. First, that the investigation of the world of the scientist is on the basis of appeal to experiment and observation (one ought to include both, definitely *not* one without the other) which is *not* the real or given, but which is abstracted from the real or given. Second, mathematical models can exist which are useful, powerful, and isomorphic, in many cases, with the real world. Such models are both metric and qualitative. I believe that to leave out either one or the other would be absurd. To leave out the latter is to be over formal, it is to leave out very largely an important part of the mathematical models, something which textbooks again tend to forget. I remind you only of group theory. You can do group theory with the abstract group tables, or you can do group theory with matrix representation. You have plenty of choices. Some are metric, but there is something beyond it which is not, strictly speaking, metrical.

I call all these matters right-handed, appealing to the powerful figures of speech of Professor J. Bruner. The *right-handed* things we do rather better than anything else. This is the right-handed, the analytic, the only format in which I can test and prove and reason with. But it would be absent, I believe, if somewhere somebody had not carried out the *left-handed* tasks: playfulness, novelty, innovation, synthesis with analysis, and aimed at a diversity of people and types of understanding. This diversity is again never clear on the surface of our books—never. In the nature of things it couldn't be, because the book doesn't have innovation in it. If it's in the book, it's not an innovation! I think no one speaks well for those physical-sciences-encapsulated surveys for the schools of education in which you have geology, physics, astronomy, and chemistry; the origin of the universe, the origin of the earth, the rise of life, all done in six hasty chapters with underlined sentences to memorize at the end of each chapter. Nothing is left to innovate. What you had better learn are the good words that are in this book.

In my view we should teach primarily a nonverbal subject. We should use a mixed strategy, with the nonverbal elements very large. The words, which must be plentiful and diverse, are like the words of other courses;

but there's something else in this course that the other courses mostly cannot have. The lefthandedness is something of great importance exactly for the social input that I was talking about, that is largely of women and of people who are not interested in mechanical things. By and large what such people have not realized, because they have no way to realize it, is that the uncontrolled element which enters into so many human relations—I don't want to call it the creative element, because by now that's a trivial word—but such left-handed elements are found deep within all real science; It's just that they are not found in science courses.

Science, Art, and Letters

It's quite clear that the sensory channels rise to great importance in the presentation of material. At the conference I mentioned earlier that I attended with Hoyle, this became conspicuous. The conference was held in a 19th-century women's college building, say Alumni Hall, which had a proscenium stage. The room was well filled, nearly all of the students women; the lectures very well done. Now, what was the characteristic difficulty in the presentation side? You know what it was. You've been to the same sort of meeting. You couldn't see the slides, you couldn't see the blackboard, and surely you could not hear the speakers at all whenever they wrote on the blackboard. Efforts were made to overcome this, and yet it never really was overcome. You've been through a hundred such meetings.

Now, why does this occur? It's worth thinking about. I asked them, did they ever use this hall before? Oh yes, every year, without exception, from the year One, they have such a symposium in this hall, but this was the first time they ever had a symposium on science. I said to them, I'll make another statement to you—you haven't ever had a symposium on art. They said no, that was right. They have them on Latin-American history, they have them on philosophy, they have them on the novels of anybody, but they had never had them on either art or science. I regard this as somewhat significant, because it is perfectly clear that if you invite an art historian or an artist he has to show you something. He can't just talk. If he does talk only, you oughtn't to go. (If you buy Dr. Panofsky's book, which I heartily recommend, you not only get 500 pages of superb analysis, you get 120 plates of Dürer's thrown in! It's a bargain.)

This is the first place where I began to feel strongly a few years ago that the join between what we call vaguely, the arts and the sciences is a very close join, distinct from the join with the letters.

The college of arts and sciences makes good sense. The college of arts and letters, which some universities have, is a curious join. I am not at all attacking literary scholars, the humanistic scholars. I tried to say in the beginning—and I think everyone will agree who knows me—that I have a great respect for the word and the language. This is our major tool. Linguistic activity is what first distinguishes man from non-man. But that it is *only* the verbal channel is a terrible heritage of the past, now washed out by the technical possibility of transferring information through other channels, and by the enormous and powerful growth of craft, of industry, and of laboratory science. It was always accompanied, in all cultures, by the parallel existence of people who contributed richly to the culture, but who didn't do it linguistically. They did it by presenting to the other senses some very remarkable synthesized objects which we recognize—from the rock paintings to the pots, the arrows, or the temples. This join between the arts and the sciences is an extremely important one; there are many places in the universities where one touches upon it, but no place where it has ever really been grasped. Here may be one of the missing clues. I am strong for reaching people through other sense channels than just what I am myself doing now.

The one thing that I can say for physicists, probably mathematicians as well, and for chemists too; they don't feel happy about talking without a blackboard. So far I have written only words on this one, I am sorry to say; I really ought to draw something, and I will—perhaps soon. At least we have the notion that it is not only letters and the words, but it is the hand and the eye that must be involved. This is not to speak against the words, but to insist upon more than the words. If we make it appear that science is primarily words, or even primarily a demonstrated act, far away—even film and TV, which are important tools, have that trouble—then we are on the wrong path.

One more general remark: I suspect that appealing notion called the core, the "core curriculum." Well, you know, who is against the core . . . ? The core contains the seeds of the apple. If you want to make more apples, you've jolly well got to have the core. On the other hand, if you want to eat the fruit, the core isn't very good. This metaphor goes deeper than I thought. It's so easy to ship apples, if you take off all that bulk and just send the core. Of course it's easier and it would work well, and you could probably get apples indefinitely that way, if you're good at it, but you might have some doubt about why! I think this is the trouble that I really see. If you want to present the reason for doing something, and the satisfaction of doing it, as well as the ability to do it, you've got to do more than just the core. You must include some of the core because the core has the seeds, the core has the power—but you must also include the fruit. The effort to achieve high intellectual quality by stripping everything down to the conceptual minimum is appealing, but dangerous.

The Content of a Course

What can we do? I'd like to discuss a year's work, probably on a somewhat demanding basis. We should try not to settle for a three-hour-a-week total exposure. These are perhaps administrative details, but what I'm speaking for is a laboratory involvement which may be painfully slow, which "doesn't get anywhere." You don't "cover the material," but you spend a good many hours of the week doing something. That's what I'd like to see. I have not tried all these schemes, and I put them forward only as an idea. I hope that there are designers and architects who will find some connection with this. We might start with the free use of simple materials. What kind of materials? Well, paper, thread, wire, clay, splints, wood, simple woodworking tools, colored materials, use of lights, and so forth. You recognize the kindergarten and architectural-industrial design-school materials. These are the characteristic materials of the Bauhaus tradition, to name one particular rather narrow use of them. People come in, they're given razor blades, pliers, a wild variety of paper and splints and sticks and clay. They are set design problems: to produce out of these relatively homogeneous features of the world—pieces of paper and so on—some remarkable assemblages which they must synthesize out of the world.

Now this trains architects and designers; it probably doesn't train scientists. It is not my view that you should do only this, because it would not lead people to understand science; but it lies close to that end, and it is plainly a freeing experience. The ability to see that in a piece of paper are contained extraordinarily complex manifolds which can be realized by folding the paper, to see that every line has all the mathematical possibilities which are inherent in it, this kind of thing comes out very well. Of course I would not like to stop here, because if I did I would be teaching architects, and I don't know how to do that. People who do that do a good job on the whole except that they make their buildings rather indifferent to the lives of the people who live in them. But they are beautiful. However, that is something that need not concern us here.

So the free use of simple materials, united by a big plus sign to some degree of *analysis* of the use of these materials. How one divides the time is something I have not decided. Probably the course should begin with the paper and the wire, and the polyhedra, and the magical surfaces, and everything else that you normally make in this kind of a course. Months may go by that way, depending on how interested people are. I know of one colleague who teaches school teachers, a mathematician now working for a Ph.D. at Harvard, who has just entered such a design course, the standard Harvard three-dimensional design course for freshmen graduate-school architects. This person writes with overwhelming enthusiasm. She says, "The trouble is my friends feel that I have left town because I spend all of my time at the visual arts center." It's quite possible that we have here a person who really should have been a designer, not a mathematician; but I suggest that this combination of art and science which seems unusual today is in fact a very natural pedagogical device.

What would you do to add analysis to it? I know of two directions which are absolutely natural. One has been worked out in part for children by Neil Mitchell of the Harvard School of Design. It's the use of the synthesis of weight-maintaining structures in the constant gravitational field as a device to begin the mechanical analysis of the world. The number of degrees of freedom, of course, is prodigious. I know of perhaps half a dozen classes based on this approach. It was very well done; my criticism of it is one criticism only, which shows up in what I said earlier about people who live in modern architects' buildings. That is, the architects have not in general been as strong in the analytic side as they have in the synthetic. They have elevated the unmistakable beauty of their designs into an idol. I am not in favor of destroying the idol; I am no iconoclast, rather I'd like to add another idol to the pantheon: namely, that we should understand that some of what we do may have constraints as well as purposes. It is quite clear that structural beginnings of this sort lead you to the notion that there is a purpose to design, and there is a way to achieve it.

You can see the rich connections this has with the skeletal structures of animals of all kinds. One can connect it also with the scale and unit problem, as D'Arcy Thompson does in his marvellous chapter on magnitude, or as J. B. S. Haldane does in his short essay. This leads to a reasonable basis for the beginning of metrical understanding and the beginning of metrical analysis as well. Structures thus go over into scale, to units, to measurement, and also to invariant theory. It certainly also goes over (and I would include this because it is important to a wide context) not into biology *per se* but to see biological structures as representing mechanical ideas. I call attention to the many books which illustrate this sort of thing. They are marvellous works which delight especially the unprejudiced elementary school children, but most physicists and most architects as well. When Mitchell, working with a third grade class using free play with plasticene to make tall structures, started his process of analysis, he made the first analysis into what he calls bridges and towers. And after a couple of hours they went around the school buildings, outside and inside, and looked at everything. They saw bridges and towers in all the structures of the school yard, the trees, the railings, and the bicycles, in everything that was around them. It was very real because it was there. Thus they began analysis.

It's clear that there are many more things you might do; you can start with force concepts, you can become more quantitative. Certainly you can discuss scaling. From scaling you get clearly the notion of the very wide range of magnitudes, and you then have something like the book, *The Universe in Forty Jumps*. I would hesitate before plunging into the ultrasensory, far into the microcosm and the macrocosm. That I would hesitate doesn't mean that I wouldn't do it. But I would recognize that here I need the instrumental extensions in order to understand. But in the domain of light plus the microscope and telescope you have already twelve or fourteen orders of magnitude. I would certainly not stop here, but I would like to hesitate and discuss the problem.

This is a branched program, not a single line that everybody follows. In the nature of things that would be stultifying. It is a rich collection from which people can choose according to their tastes, abilities, available apparatus, the season of the year, and so forth. That's a far more reasonable way of doing it than making a definite core, but it also has its core. To give you some idea of the kinds of things I think people ought to handle, here are a few types of instruments which I think ought to be involved, activities which ought to be involved in this kind of course. The slide rule, the desk computer evidently. These are related to the mathematical model of the world. Then the microscope and the telescope as instrumental extensions, which alone make the macrocosm and microcosm meaningful. Next, the Simpson meter, the multimeter—this I find the hardest to put in. I think it would be extremely valuable to develop a whole culture of the Simpson meter. How does one get it in the hands of people like the elementary school teachers? I do not wish to make them TV circuit analysts, they are not going to be that. Rather I want to give them the notion that electrical energy is under quantitative control, to let them measure a few simple circuits and let them create circuits symbolic and real that function. The Simpson meter will be behind there, if the circuits are going to work at all. These should be very, very simple things, but I think the ability to handle them will do a great deal to break down the distinction between the engineering world and the world that we enter in the schools and the colleges. Then probably something optically interesting like gratings, polaroid, and so forth. This is the kind of thing that I'd like to see fitted in next to the papers, splints, thread, wire and clay.

We come now to another very natural take-off, though I shouldn't say very much about it as I really haven't done this myself. I feel, however, that it is most hopeful; Phylis Singer and Alan Holden in particular have shown some ways in which this can be done. I am talking about the exploitation of symmetry, the exploitation of infinite rapport, the conditions imposed by infinite periodicity, say, upon the symmetry circumstances at a point. It's rather easy to do experimentally, and it is wondrously rich from the point of view of design and of aesthetics. You can make all the patterns in the world! You can produce powerful, general qualitative mathematical results; for example, the absence of five-rotational symmetry, of five fold axes in the physical world, and their surprising frequency in the world of biology, at our fingertips! This fantastically beautiful Pythagorean kind of result comes from such a study. I think you could very well make this two weeks' work with no trouble at all. If you get no other thing out of it but just that now people understand why this could be, you'd have enough result for me.

This, of course, is very close to group theory and I would really expect that the group-theoretical results are produced in some way, too. Some notion of the group and of the notion of invariance, for example. Let me make one general remark. When I talk this way, I talk, of course, a little bit as though I'm taking a standard course in physics and diluting it enormously. What I mean, however, is not dilution, but the taking out of the course very many topics we now include. And I firmly believe we should do that. This doesn't mean that I believe you reduce the intellectual quality of the things you do do. We don't have group theory in most of our freshman courses. Yet group theory is powerful, perhaps the single most general and most powerful model for the whole world. I don't think that you are reducing the intellectual level if you can manage to bring it in. Of course, it has to be done subtly; I'm not saying you should do Schur's lemma first. I am talking about a sensible way of doing it in which you show people (and they will find it out for themselves) that there are certain conditions of this sort that work. If you spend even a long time on this, talking about the regular polyhedra or about rotations, or the absence of fivefold symmetry, or the lattices, or whatever you want, you can connect it with crystals, you can connect it with biological organisms. You can connect it with the law of the lever and Archimedes. All of these things you can certainly do; they are historically right. If you do this, you have done a lot which replaces, in my view, a much more systematic examination, let's say of circuitry, or field theory, or Bohr atoms, or statics, or whatever it is that you would like to put in an ordinary course. I speak at this point not for selling short the intellectual quality of what you do, but for letting it come very, very slowly, always based upon an ascending degree of familiarity. One should not substitute the quick results, so easy to do, for the kind of understanding that people need. My notion is that this already describes a term's work.

Now what else could you do? As I said, you can write down a number of topics I think you might include, and very well can.

> Geometrical Optics
> Diffraction
> Dynamics
> Energy
> Acoustics and Vibrations
> Astronomy
> Gases and Gas Laws
> Dynamical and Statistical Laws
> Radioactivity
> Circuitry

Now evidently I have written down much more than a good year's work.

But I want to indicate a kind of level of detail and a level of exploratory alternative examination which you have to do. Another remark I would make is that I think our present instruction has been single-pathed. You are in a forest, you walk carefully along some path, and reach the chest of doubloons on the other side, and solve the problem! And that is the way—I too—teach physics. But the kids that try it get lost at each turning of the path. The trouble is that they think there is only one safe path, that they have to stick as close to that as they can, and they are afraid to go off into the deep woods. I think that the only way to teach pathfinding is to make them get lost many, many times, to make all the false starts, to try out all the alternatives. Finally, of course, you can't learn very many paths that way, *but you learn a way of going down a path.* Then, if somebody gives you another start, you might be able to find your way for yourself, hopefully, some other time.

This is a bad metaphor, but I think that the single-pathedness of our teaching is one of our troubles. I see it very often. I have taught a recitation section which belongs to an excellent general course for liberal arts students, nonscience majors, at Cornell, using Jay Orear's book. Jay's book is an up-to-date, competent, lively book. There is no question about that. When you finish it, you have heard about the Fermi surface, isotopic spin, whatever there is you have heard about it. You can even see that, carefully done, it is *not* a survey course. It writes down the formula; it gives you the results. The kids have learned most of the formulas, but too often they don't understand them. How can they understand them? You know, I don't understand them all so very well, and I have spent twenty years trying to understand some of the questions. Yes, of course, I can do most of the pages in Jay's book all right, but that isn't enough; the reason is that there is never any alternative suggested. You couldn't possibly have one. If you took any other alternative path to understand the Fermi surface in copper at the level of the first-year physics students with no calculus, you are not going to get very far. If you follow carefully his sentence-by-sentence argument, each sentence does imply the next and you make a reasonable argument out of it. (I'm not sure of that, but maybe you could.) But I'll guarantee that you couldn't do it for the first time. I guarantee that you cannot discover the Fermi surface by writing the book that way. This is not a criticism, because everybody writes textbooks that way. We never discover anything we put in textbooks, for the most part. But it's very wrong to take people who don't know science and give them this notion of walking a precarious tightrope in which every step is exactly right. This isn't the way in which it was first done, and it couldn't have been done in this way. The cross sections of the subject are more twisted than that. If you make students take one fiber and go along it sufficiently far, in the end you will have covered the whole of physics. A course of the kind I am talking about will just not cover the whole of physics. At the end the students probably will have ignored something really desperately important. They will never have heard of it. Say, the Fermi energy. I just can't help it. I think this is a decision one has to make.

To go back to our list, geometrical optics, I still think, is very beautiful; and it should carry on through lens systems, if you do it. (I am not saying that you have to, I am writing down a number of alternatives from which, in my opinion, a good course could choose.) Hopefully when we finish, there would be twenty topics from which three or four might be drawn each term. Some can be dropped each year, the old ones can be thrown away, and new ones added that are more up-to-date, that are more lively, that have something to do with what goes on in the world. This we can do as we learn how. If I did geometrical optics I would not like to do it without the principle of least time, a mathematical model which I think works; not just to say "Snell's law is the law of refraction, and now you draw your diagrams." I would rather draw fewer diagrams, use simpler systems, but try to show that there is some deep meaning unifying the whole subject. Here again is a place where less means more, where you have something not put in the elementary physics books. The woman I mentioned earlier who is taking the designer's course at Harvard, in fact invented a scheme, which third or fourth grade children could carry out, in the way of working out the paths that go by other paths than the true optical path. They do this by walking on the floor; they can measure the paths and graph them, and soon they see that the stationary one is the one the light is obeying. She even produced a fable, a fairytale to demonstrate the Feynman argument. How does it go? You want to send messages via messengers from a king to a captive princess, and the messengers have to go a dangerous route where the dragons are waiting. The dragonmasters capture these messengers and interrogate them. You want your messages to be safe from disclosure; that's a very hard problem. So you give the messengers all sorts of messages. Not very many are captured, just a very small sample. But the message that you want to be real, you repeat many times. Of course, this is a stationary condition; if you want you can make a nice analogy, and actually, if one does look at the stationary condition, one can say that the number of paths that go near the real path are numerous, the reinforcement is great—all the others cancel out—so that what comes through is the right one. This is a new kind of approach to geometrical optics.

Some clever devices were developed by the people working on this. For example, you know that this theory tells you very nicely how to make an air lens, and how to predict what an air lens will do in water. You can set that as an examination problem, you see, to discover how this works, using a couple of watch glasses and plasticene and a goldfish bowl. I get the notion that this is good by trying it on physicists who always find it good. Perhaps this is the wrong measure, because they are not the kind of people we are after. It turns out that if you get one of these hard workers into the office, and show him these things, he is really pleased to see that he can put such lenses together, and he follows the argument. The air lens is an unmagnifying glass, though it looks like a magnifying glass. (If you make it look like an unmagnifying glass, then it magnifies.) The theory is right there for you; it tells you why that is so.

If you can ever go to diffraction—to ripple tanks, verniers, to moire patterns—then you can go to x-ray analysis. This I would take as another equally large topic. Why do I like this one? The chemist will understand. Diffraction is one of the most powerful methods for obtaining our notions of the microstructure of the world, on the atomic scale. If we understand diffraction as, for example, through the ESI film on matter waves, we can begin to see how we can get patterns which tell us that the molecules are arranged in ways which mirror the visible symmetry. From such relations of structures, the analysis of matter comes.

The next topic is dynamics—(not *next* really, the order of the listing is not significant). What can you do here? My own prejudice, which has worked fairly well with similar groups, is to talk about displacement, velocity, and acceleration, plus mass, defined and discussed in a rather wide context. Then to weight, gravity, weightlessness, which is, of course, a topic of great interest in this decade, and to orbits. Of course, quite possibly the whole thing could be revolutionized if we can make acceleration not a kinematical concept, but a measured dynamical one by the use of the accelerometer. This is an idea I owe to William Walton, though I don't think he's gone very far with it yet. It may be the beginning of something quite important. You recognize that this puts the principle of equivalence as the foundation of dynamics; you can't do anything about it. The accelerometer goes wrong as soon as you tilt it. You can't tell whether you are moving or only tilting; then you begin to think about this. It is a profound way of doing things. It is all one complex; the whole complex of mass, acceleration, Newton's laws, orbiting, the substitution of gravitational fields for acceleration, and so forth. Again it is a place where the intellectual level comes high, even though the technical proficiency will be small.

I would like to see every such course include the Huygens-Newton result mv^2/r for uniform motion in a circle. It is not too hard to do, and it is worth doing. It is powerful, it gives you satellites; it gives you quite a lot about solar systems; everybody does this anyway. I wouldn't go any farther, absolutely no farther. I would not mention the ellipse until it was dragged out of me by somebody. I am prejudiced; maybe other people know how to draw ellipses better than I do.

With energy we come to a serious, important and difficult problem which I have not solved successfully. I know what I would like to do with it. Having worked up energy as an abstract mathematical quantity—which it is, part of a mathematical model of the world—you use it powerfully to introduce the kinetic or atomic theory. I think that is its real use and function. Mechanical energy you can blunder your way through, but the stuff doesn't stay. Then you make this marvelous assertion that if the same abstract quantity is buried internally, it will stay. Then, of course, you have heat and all the rest. This is familiar, but it is the kind of thing we can certainly try to do. Now with atoms and heat, you can do something with chemistry.

An alternative way of doing energy is to do it with pendulums. That is, you can use the repetitive motions and the coupling of repetitive motions, which one sees in those beautiful film loops of Alan Holden and Phylis

Singer, as a way to discuss this abstract transference. I think that this is also very good, though I don't know how it will go in detail. It's another equally attractive point of departure. I don't think you can do it without the notions of velocity, but you might define that, include the dynamics, but now stress the conservation laws more. These are two parts of, let's say, Newtonian and Hamiltonian theory, to use fancy language for it. These two parts of dynamics could receive emphasis one way or the other for this group of students.

Having motion, atoms and heat, you can go on to chemistry. I was trained by that fearless generation of physicists who invented quantum mechanics. I was not one of these people, but I learned from them the precept—which is not true—that you can do anything if you try hard enough. So I tried chemistry, not very successfully. A physicist can do anything that he tries—but not very well. What I tried was biochemistry, expecially catalysis. What I wanted to discuss most was the existence of specific reactions, each with some reaction kinetics, with kinetics that depend, not upon the reaction, but upon something else, upon another member of the reaction than the principal reactants; and then I tried to fit this into some sort of a molecular model—but not very far. It is a very hard job to do. It would be extremely valuable for the chemists to think about this problem. I would like to propose using a biochemical approach with a few elements only, rather than a more thorough discussion of the periodic system and the inorganic elements. Maybe that's wrong, but I'd say that this is one unit, a biochemical unit. The other unit is a periodic system unit using the metals.

You can do a lot with a simple system. We did iron and catalase-catalysed reactions of the decomposition of H_2O_2. This is not too hard to do if you keep your vessels clean. It is rather striking because you have million-fold increases in the reaction rate. You also have poisons, the poisoning of a catalyst, as a paradigm for the sensitivity of biological organisms to poison and generally to small changes in the environment. We thought it terribly powerful, and very appealing to the students. This whole subject will tie in very well with another year, which I'll not describe, which is a biology year; mainly on microbiology, plus structures and morphology, the two sides. I haven't done enough thinking to say much about that.

The question as to whether you can make models of kinetic theory, whether you can discuss catalysis and life and a little bit of the standard enzymatic biochemistry, that's a question that I don't press here at the moment. But some part of this I think you really ought to try to do; it's significant. If you do it in the context of energy and atoms, which we are trying to get into here, you have reached the hardest part of the work.

I don't know how to do atoms well. Maybe diffraction will help. But here is the hardest topic. Indirect arguments are so difficult, so long to pursue, as you know. The monolayer and the Mueller field-emission tube are the closest approaches we have to making it easy; they are not very easy even then, but they are perhaps a start. Finally, two more prejudices that I will mention. One is sensory physiology. You notice the strongly biological flavor that all of physics has taken on. I am sorry about that, but that is the way it is. My argument is that the context is the important thing. I would like chemistry to be biochemistry, and I'd like to have chemistry intimately related to the energetics of the physicists, but also for chemistry to develop on its own. Here I think the biggest problem will be to develop an atomic chemistry at some reasonable level, and not just to parrot what it says in the textbooks. But we have not yet succeeded in doing that. Brownian motion, too, may help.

Now by sensory physiology what do I mean? I mean simple experiments on the limitations of behavior of the senses, the Weber law, and the introduction to logarithmic scales, a very reasonable approach to why physicists are interested in quantity. The problems of threshold, problems of signal-to-noise, perhaps done with the touch sense, problems of dilution done with taste, and a few things of this sort, are worth doing. This is a very brief unit and can be an elegant one, but again I leave this to somebody else.

Whether we should put in acoustics and the analysis of vibration in acoustics, coming out of the study of pendulums, that's another open place which I think quite possibly should be done. This question is of great importance in music, which I have tended to underrate, although as I have said, I shouldn't do that. Music is of great importance and the tie to the pendulum is quite nice. There are many ways of making this bridge which I have not worked out.

I would also find it very difficult, since I have talked so much about the micro world, not to say anything about the macro world, that is, of course, astronomy. There are two ways to do astronomy. One way, which we'll identify with the name of Walton, is—what shall I call it?—"astronomy from the ground up," and it is an excellent kind. Once you have seen some of it, you will recognize its appeal. It's a kind of theory of the construction of sundials, and how to navigate, but not done in such an instrumental way. It is done in a way which suggests that you should find out how these things go and reach the mathematical model. The geometrical models proceed throughout the entire system. It is astronomy before Galileo.

I have tried something else, quite different in style, with great success in two classes. I just want to mention it as a kind of modern version so that you can see how diverse you can make these things. I have done an astronomy of plates, that is an astronomy of studying photographs. You can buy a sky atlas, not black dots on white paper as drawn by somebody, but real photographs taken with a 5-in. lens, of stars down to the fifteenth magnitude, of the whole sky visible from California. These photographs are contact prints of the original exposed photographic plates, so that they are a map, they are isomorphic to the instrumental response to the universe. If an airplane came along, you will see airplanes on that plate. It's not like the star map, and that's what's good about it. You give these kids a couple of almanacs. You give them all this, don't say too much about it. You let them read and discuss a good book, say, like Hubble's *Realm of the Galaxies,* which describes the discovery of the expansion from the cosmological principle; it tells just how it was done. You let them work on the plates; they can do it pretty well. You set them little problems like finding minor planets. They can find asteroids and Neptune on the plates. They can count galaxies and estimate distances. You have to have magnifying glasses, bright lights, and slide rules, small maps. Some students stay in the lab until 7 or 8 o'clock in the evening poring over the prints, counting galaxies, measuring their sizes, all down to the fifteenth magnitude.

It can be quite remarkable. I told them something which I honestly believe, that it is not impossible that they could find something that nobody else knows. The fact that the prints are contact prints gives an immediacy which is not replaceable by any kind of half-tone. You can magnify them and when you magnify sufficiently, of course, you hit a limitation, but this limitation is not imposed by some schoolmaster's trick of giving them halftones; it's imposed by the instruments, the atmosphere, the grain of the photographic plate, and so on. They find that out soon. You give them some standards, some measure, you talk about inherent signal-to-noise, you talk about certain stars, and you talk about triangulation and parallax. That's where this would join on to "astronomy from the ground up." Perhaps this is an astronomy "from the plates down," or something like that. It is harder to do, and I'm not sure it's right, but I would certainly like to see it much more developed. Maybe again this is a prejudice, but I like these plates very much. Even better, I would like the *Palomar Sky Atlas* which, however, is too expensive and it is out of print. It's of course, a marvelous thing, simply marvelous. If one could find ways of making cheap reproductions of it that retained near-photographic resolving power, you would be impressed. (Has everyone seen the *Palomar Sky Atlas?* The *Palomar Sky Atlas* consists of photographs of all parts of the sky visible from southern California, taken on the 48-in. Schmidt camera. What you get are paper contact prints from the original glass plates. They are about 14 × 17 in., two of them for each part of the sky, one in red light, one in blue light. This gives you the beginning of spectroscopy *and* the beginning of time variation, because they did not take both at the same time. So that you can see everything—just everything. They are really beautiful.) You just get lost; if you take a magnifying glass and look at these things, you see galaxy upon galaxy upon galaxy. The sense of the immensity of the whole cosmos absolutely comes through. This is what you often see in *Scientific American,* but instead of having just a few square inches of half-tone, you've got the whole sky displayed.

It's really possible to do research here, because quite a few astrophysicists and astronomers of good reputation make their living by studying these plates all day, every day, no other data. That's what many do in Europe. They have no big telescopes, but they can study these plates, which are very good plates. You can do extragalactic astronomy extremely well. You *can* find super-novae; *we* have even found super-novae on such plates. When somebody comes out and says 3C293 has been discovered, you can rush to the plate and there it is . . . 3C293, an inconspicuous star. This is another way, galaxy counting is a way to do it.

What else have I left out? Quite a few things, which I am just going to note, which are all very accessible in this same style, but just haven't been worked out. For example, gases and the gas laws. The density and reality of gases, demonstrated by density measurements, by displacement of volume, hopefully even by refraction through Schlieren technique. These are things which I think might be worth looking into at this level, but we haven't done it.

Radioactivity . . . I put it in with some misgivings, but I have to put it down because I realize I left out one extremely important thing. And that I want to put up in big letters. When we are talking about mathematical models, I should have added dynamical and statistical laws. That is something else I would like to produce; it's a fundamental result, of course. You see, with the dynamical laws, with scaling, and mechanics, and all the optics, we have all just the same view. The only place that chance enters is in the errors of the observer. I think this is a very old-fashioned view, and I am against it. I would like to show that statistical law is equally orderly in some sense, and that the play of the world is between these two. The best time to do it, I suppose, is about at this point.

I've done a little bit of this with diffusion theory. One plays the Monte Carlo game, you know, you play out random walks. You can play them out very nicely, using for your random numbers telephone books, or the expansion of the number π in 100,000 decimal digits—which you can find published now—or you can use Tippett's tables, you can spin coins, you can do whatever you want. Of course, they all give lovely random walks. They all satisfy the diffusion equations, and the class easily produces various examples in the course of a few hours. Then you can (but we have not done it very successfully yet), you can use real diffusion of dyes in jelly, in agar, or in wet paper towels, or Brownian motion under the microscope, and you get the same results. You can throw chains of paper clips to illustrate polymer behavior, and you can get the mean length of the chain. You can tie this to the thermodynamics of rubber in a semiquantitative way, though only semiquantitative. You can show the very great difference between metals and rubber this way, and it's striking. The famous experiment of measuring the temperature change occurring on the stretching of a rubber band by placing it against the lip—if you do this after the random walks and the chain, you can give a reason why this might be so.

These are the kind of things I would like to do. I'd like to insist upon the orderly behavior of large numbers of chance events. Of course, radioactivity can give you a completely different approach, such as has been tried at Bryn Mawr, which is yet exactly the same. Tossing the coin, tossing the tetrahedral die, produces another exactly similar Monte Carlo approach to the whole subject.

One very serious problem has to be done, and I don't know how to do it. I think something has to be added because electricity is given very short shrift in all that I've said. The reason is fairly clear historically; it's a hard subject and you have to go a long way before you understand it, in some way. The gap between what you can understand and the tape recorder or the television set, which everybody sees all the time, is very great. I can't bridge that gap in any very meaningful way; I have never succeeded in doing it. Here is a real problem. However, I do think that especially for the kind of student we are talking about, the "Simpson Meter" experience will be extremely valuable; maybe even the cathode-ray tube experience, too, if you can work that hard.

I found successful some free play with a very expensive device, a megacycle counter for timing which counts off microseconds, that is, a very fast clock. Such a thing is very useful. You begin to analyze some of the circuitry, perhaps by the use of the hydraulic analogies that people have made—I don't mean the old one about pressure and voltage—but rather the sort of thing described in the *Scientific American.*[*] I am anxious

---

[*] Sci. Am. 207, No. 2, 128–138, 1962.

to see this worked out farther. This high-school student used tubes in which water flowed, used jets, and so on, to simulate, that is, to represent in another isomorphic form, various electronic systems. He made oscillators, amplifiers, pulse circuitry and the rest. For example, you have a "U" tube, which represents the tank circuit. There it is, a real tank with water in it! For the oscillation of the water in the "U" tube, of course, there is an actual period. He excites this "circuit" by water drops falling into it; when the tube slops over, the water drop is fed back; and soon the whole device builds up its oscillations.

How far this can be done is still an open question. The whole problem is that of the use of electricity, so apparent in daily life, and so remote from the things that one does in the textbooks. The textbooks traditionally try to handle circuitry by describing a tube with a grid, and a plate, and a cathode in a envelope; and some glib remarks and that's it. Heaven forbid that you should try to design a circuit from what you learn from these books! Exactly in the same way, the description of how the airplane flies is almost invariably wrong, because they almost never mention the great theorem which alone enables it to fly, you know, the theorem of conservation of circulation. The books just give an "as if" description. When we were doing the PSSC book we honestly didn't try to face this problem. We said that when the kinematics is difficult, the theory is harder still. The kinematics of airplane flight—look at the wind tunnel—are terribly difficult. Those eddies are not going to teach us easily how to fly. The behavior of the nonlinear circuitry . . . I doubt that we should try to understand it . . . I don't. But I think that in some way we ought to try to break into this area, I am just not exactly sure how to do it.

The connection that I think is most missing in all of the work we have done so far in curriculum building, is the connection to technology. One way I've tried, very briefly, is by making an effort at what I call systems analysis. It might be too hard. What I did was to bring in a clock, in fact, several "clocks"—an hour glass, a spring-wound clock, a grandfather's clock, and a megacycle timer, four different kinds of clocks. We spent a couple of days on this, first making a block diagram of each of the clocks. Each one had a display device, an energy source, a timekeeping device, an escapement. There was a kind of functional quality about the analysis that one could gradually elicit. Then we talked about energy flows, and about information display, and then that was the end of that week; we couldn't handle all the mechanics of how to make the energy do the right things. But we could see that there was a common structure to all.

Maybe that kind of approach towards circuitry, with some building of circuitry, would be valuable. Perhaps people would build small motors, and to use them perhaps in connection with original work, to design something that jiggles or moves, something that repeats, something that makes quasirandom choices. This might be a very valuable addition to the sort of static structure which we tend to put in to the architecture courses.

That's the story of how I think this kind of approach can go. It is a fresher approach than that of the standard good books in physics or in chemistry. It gives frequent opportunity to involve a lighter context, as well as a wider one. There is much it doesn't have. It leaves out major things. It's very hard to put in quantum mechanics. It's very hard to put in relativity. You just have to give up many topics if you take this point of view. But in my opinion, it represents an experiment of such plausible importance that it ought to be worked on seriously.

Ice that Sinks

T
he chain of arguments which leads to the picture of the nuclear atom, and then to the domain of nuclear physics itself, is a long and abstract one. The mushroom cloud is real enough, but the understanding which preceded it is not very easy to make real, especially in the modest laboratory of the high school. The clicking counter does demonstrate radioactivity, but the onlooker must already have entered freely the world of the physicist if he is to grasp how the curious radiations of that little foil could have led step by step to mastery of the source of energy of the sun itself.

In a way, the concept of an isotope is a key to nuclear physics. Radioactive carbon faithfully traces normal carbon through the labyrinth of biochemistry; fissionable uranium 235 looks no different from, but costs a thousand times more than, the normal uranium of which it forms one atom in each 140. Most substances, carefully purified by the chemist, and visibly homogeneous, may contain one or more species of atoms called elements and also mixtures of subspecies, the isotopes, which mimic very well each other's chemistry, and yet differ at heart in the most important ways. This note undertakes to suggest a simple, safe, not very costly, experiment, which makes graphic the existence of isotopes, perceptible directly to everyone who will watch.

Isotopes are of course nuclear species with one and the same electron cortege, but with distinct nuclei, differing by one or more mass units. For years it has been known that the lightest element, hydrogen, bearing a single atomic electron, had two stable isotopes: "light hydrogen," sometimes called protium, of mass number 1, and "heavy hydrogen," or deuterium, mass number 2. Deuterium is to nuclear physics what the hydrogen atom has been to atomic physics, for deuterium is the simplest possible stable combination of the fundamental nucleons, neutron and proton. It has one each of these fundamental building blocks, and in it their interaction is most simply studied. Deuterium is the basis of fusion in the multimegaton explosions of today and in the extreme-temperature plasma reactors of tomorrow. It is of all nuclear species the one which is industrially separated in the greatest quantities from the normal random mixture nature presents to us. In this country, the AEC produces and holds it for use by the tens of tons in the great Savannah River plants. It is generally separated by a repetitive partition of H^1- from H^2-bearing molecules between two compounds of hydrogen, H_2S and ordinary water. Natural hydrogen in all forms normally has about one atom of deuterium for each 6500 of protium. The chemical exchange separation (it is not the purpose of this note to explain this operation) is practical because the *relative* difference between proton and deuteron is so great, one weighing nearly double the other.

Suppose before you there sat an ice cube made of pure "heavy water," whose chemical nature is made clear enough by writing D_2O for H_2O, where D is a fairly familiar chemists' abbreviation for deuterium. It is cold, hard, clear; it looks, tastes, and feels like any other ice cube. But if you place it in a beaker of ordinary iced water, in which float and clink the normal ice cubes which are its counterparts, the heavy ice cube sinks straightway to the bottom, weighed down by one extra neutron in every atom. This simply performed experiment makes plain that isotopes exist, that they are chemically similar, but that they differ in atomic mass. I have done the experiment myself as a demonstration, and the general audience, like the physicists who watched, found it fascinating. Good faith was demonstrated by letting a member of the audience taste the heavy cube, to assure us all it was mere fresh water, and by letting it melt in its storage jar in full view of all, to display that it was not loaded with a sinker of any visible kind.

A physicist trained as I was in the pre-war world cannot do this experiment without some emotion, for heavy water was once more precious than rubies, and letting it come in contact with the normal stuff was an abhorrent crime of the laboratory. But consider that a proper ice cube bulks about 40 cc, and that the quantity price of high-grade heavy water, with more than 99% D_2O, is relatively inexpensive. Moreover, there is no need at all to let the whole cube melt wastefully away in the beaker.

A minute or so of the show is long enough, and in well-iced water very little of the cube melts. We dried off the cube carefully with a clean dry towel and allowed it to melt in its own sealed jar. The exchange of normal hydrogen with deuterium is fast when both are in water molecules, but diffusion into the solid ice is very slow, as is all solid diffusion. One repetition and an estimate leads me to believe that the experiment could be repeated dozens of times before the dilution and loss of the heavy water sample would spoil the display.

A few practical hints: the value of heavy water diminishes as ordinary water vapor from the air mixes with it. (Naturally, no one will mix any visible drop of liquid normal water with his sample.) It should be frozen in a polyethylene individual ice-cube cup, and while in the refrigerator, ought to be tightly wrapped in a well-sealed double layer of a vapor-proof plastic film, like polyethylene or polyvinyl. Long-time storage is best done in an all-glass ampoule, sealed by melting the glass. It is simpler and would perhaps be as satisfactory to use a well-stoppered bottle wrapped and sealed again by vapor-proof plastic film. (But a screw-top jar would not be tight enough.)

A brief comment on the theory of the experiment is worth adding, though I prefer to leave a more detailed study for students. Let us make the first approximate computation in this way. The masses of the atoms involved are tabulated (on the new unified scale, with $C^{12} = 12$ atomic mass units) as follows:

natural	H	1.007967 amu
	H^2	2.014094 amu
natural	O	15.9993 amu

Then the molecular weight of ordinary natural water is 18.015 g/mole, while pure D_2O is 20.03, where I have rounded off to keep an accuracy of about one part in 1000. If we assume that the volumes occupied by each D_2O molecule in any form is exactly the same as for H_2O (which is of course not true), the density of heavy ice would be (20.03/18.015) times the density of normal ice. On this basis the heavy ice cube would have a density of 1.019 grams/cc at 0° C and would sink in normal water even at the temperature of maximum density of water, 4° C. Of course pure heavy water is expensive and rare; the 99.5% stuff costs about 40% more than the 99% grade. But for each percent of normal water impurity, the density will fall off by only about 1/9%. So heavy ice of deuterium content down to about 82 atomic percent will still sink, though it is a near thing once below 90%, and the approximations would have to be improved to decide for sure.

Here are a few points which deserve thought even before plunging into the purchase. The volume of heavy ice is in fact about 1.5 parts per thousand greater than that of normal ice per mole at 0° C. Ice contracts when cooled, and heavy ice melts at about 3.82° C under one atmosphere, not at 0° C. What is the thermal bulk expansion of ice at these temperatures? What is that of water? In the practical case, the heavy ice cube is kept solidly frozen before it is immersed, and is rather cooler than the iced water. Dissolved air in the iced water might slightly aid sinking, while many air bubbles in the heavy ice cube must be avoided. The cube should be frozen so that it is nearly clear. What are the true densities under the real conditions of experiment? It is well worth seeing how these points do not spoil the effect—they in fact do not—by working out the whole experiment on paper, and only then deciding how low a grade of heavy water it is safe to buy. The sample you get may be used also for nuclear purposes; it makes photo-neutrons under the gamma rays of a radio-antimony source, for example, and its neutron moderation properties can probably be exhibited even with a hundred-gram sample. Small animals or microorganisms could be studied with only a fraction of the sample. Many other properties of heavy water are interesting to measure.

The heavy water sample will justify its purchase once it makes real the presence of the mass deep in the heart of the atom. The ice sinks, because all those neutrons weigh it down, and yet it is simply heavy ice, nothing else. The concept of an isotope is thus made plain to the watching eye.

The New General Physics

The PSSC (Physical Science Study Committee) first drew our attention in a serious vocational way, beyond the hard job of the daily life of our colleges. This was, I think, an effort to make an intellectually deeper physics. But it was, perhaps wisely, set at a level that did not require a profound change in the educational system. The name itself rapidly became a misnomer, representing the utter failure of the system to allow unification of the high-school level chemistry and physics courses. That had been the dream of the namers; it could never get going because of the wide differences in preparation, social context, laboratory equipment, and training of high school physics and chemistry teachers, and indeed of the professional research practitioners. You can see what its limits were. Led by buoyant opportunities—some clear accomplishments of that effort—some of us went off, especially with Francis Friedman before his untimely death, to take on a still more difficult task, which we called the Elementary Science Study. This set itself the goal of deepening the intellectual level at which science is faced by the majority of the population in their elementary school experience. But as it turned out, the logic of that demand required that we begin the task—which seems desirable to many people in every generation and so far has never succeeded in any generation—of making a profound change in the entire system of the public schools: a change not only in curriculum materials, not only in the tone and style of the apparatus brought in, but in the lasting confrontations between teacher and student, among the students, and with the community. All those things are involved once you take a serious look. What can you do to try to make young children engage the world a little more in the model of the scientist, and a little less in the model of the authoritarian assertions of their elders? But the latter has traditionally been the role of elementary schooling. I think this task has one of the most powerful intellectual appeals of any activity I've ever engaged in. It is fascinating to meet the people who graduated from its schools and fought its battles. The campaign has had only a limited impact; it is carried away now in a great flood of systematic impingements upon the schools.

To show you that these things are perpetual, that we are dealing, as in all real problems, with profoundly opposing positive and negative energy terms, tension and compression members which cannot easily be segregated completely from the system, I would like to recall my very first formal conference on such issues. It was a pre-Sputnik conference, as we like to say. The discussion centered on what should be the look and nature of a high school curriculum in physics. Most participants were quite devoted, level-headed colleagues in physics; a few were in addition knowledgeable about the public schools, though they kept rather quiet for a while. I remember very well the development of a kind of polar discussion, two sides of a big question. It happened that I was rather volubly on one side; the other side was upheld by a redoubtable champion, namely Professor I. I. Rabi. (You know it's very difficult to prevail in such an argument! Naturally I didn't prevail, but I hope we parted with a sense of mutual respect.) The argument was pitched on this point: physics has two aspects. One is quite clear (Rabi's view): the intellectual elegance and power of even a single well-formed, well-justified proposition has consequences so rich that every special case is an exemplification of the general. He said, "Let's take a simple example. To go through Newtonian mechanics is difficult, many things are involved in it. Let us take Snell's law. We then have all of geometrical optics in front of us, from only one simple generalization." I had to admit this was powerful. On the other hand, I said—with such great energy that I was then put to work writing pieces for the textbook—"That's not the real thing. The real thing is that the sovereign rule of physics holds everywhere from the distant stars to the tiniest nuclei. You don't have to know very much more about it than to know that it's there, you have only to know the order of magnitude and a few such simple things. To neglect that enormous reach and look at only a single example, however strong, would be a grave mistake." After all was said and done (of course, it will always go that way) both of these polar opposites appeared in the textbook. The general stuff was in the beginning, somewhat vague, now known to everybody but still interesting, and in those days rather novel. Right after that, the first time one came to deductive consequences, you were given Snell's law and geometrical optics. That was justified with the use of ripple tanks and so on, and mechanics was pushed way back into the sophisticated part, the third or fourth month of study by high school students. That's the way it stood for some time.

But I notice now with only a little dismay, recognizing the real forces in this world, that neither of these decisions remains in the excellent fifth edition. It turns out that once again it starts with kinematical discussions of how to measure position, velocity, acceleration; you don't see Snell's law until page 397, and you never do see the general dimensions of things, because they were left out, maybe quite wisely left to an earlier general course. I emphasize this only to show there are strong forces in society. They may in fact be justified in their stand by great invariants of human nature or maybe not. I don't want to enter upon that problem now. But they are certainly powerful, and they change with ponderous slowness.

About one-tenth of the high school population engages itself in a serious way, a full period during the school day or over one school year, with a chemistry or a physics course. This is the representative population with whom we have some kind of explicit contact. They are either those who come to our university courses, or those who occupy the high school courses. Of course, I feel very strongly that, even though the complex life of a citizen in this interacting world supports the fact that not everyone can possibly be engaged at the cognitive level with all these tough intellectual issues, a tenth is too few. I would like to raise the question of whether there are means and ways of changing what we do and how we look at it, so that in future we would be able to find a wider appeal. We have made pretty successful inroads. There are the many people who take physics or chemistry in high school. You see them finally in vocational life, manning the enormously complicated economic structure of this country, and coming in contact there with a whole variety of technical activities. The system would not work without them, they are an indispensable part of the community, but they are by no means the majority.

What are we to do? Some of the same things again, with a little more sense of experience, to be sure.

First of all, I think it is clear that the sensory and emotional surroundings of learning, of instruction, what I would like to call its tone, cannot always be the same if you are to spread a wide net for personalities and interests. In particular, the tone ought not always be cognitive. It is not always true that the logically most compact and economical procedure— very often it is that of taking an unrealistic simple limit, reasoning from the limit, and then gradually complicating the account—is the way to catch most people up into the net of thought. That device can work, but it by no means goes without saying that that is the *only* thing to do. I have an ancient example which I have heard testified to out of direct experience over fifty years of personal memories. I think everyone who has taught mechanics will recognize the problem; I can offer only a few directions for solution. That is the famous proposition, dear to the formulations of Newtonian mechanics and very important for it, that the Third Law has to be part of the force analysis of any system, in or out of equilibrium. Now, the general context of our understanding of the world suggests immediately that passive agents cannot be said to act in the same way as those that initiate change. It is very difficult to become convinced that the inner wall, seen as infinitely rigid and unyielding, is pushing back in exactly the same way as the straining horse pulling the rope, or the man leaning hard against the wall. I have found individuals who over two or three generations of schooling have come to one and the same barrier at this point. They reckoned that physics is a subject so interested in minutiae, so free from contextual understanding, that it is willing to accept as a fundamental principle something that is manifestly wrong in nearly all experience! There is an answer to this puzzle which is quite plain: there is no rigid body. Bodies only approximate rigidity; what we really mean in this model is that we can neglect the internal strains. You can find cases for which what I call "carpenter's mechanics" is worthwhile. Use a six-foot 2 by 4 and you know it's not going to strain much when you put a glass of water on it. You don't have to worry about that. But anyone who tries to build a bridge or a geodesic dome or almost anything else bigger cannot deal in that abstract approximation; one must take into account the fact that there are reactions visible in the material. No inert unmoved reactor is present. Whether this is to be done by talking about strain first, or by giving a molecular view, which is nowadays rather popular, or by introducing the notion of the flow of momentum as an extension of the notion of force, or by some other means, I am not sure. But I am convinced we must recognize that many intellectual strategies are necessary if we are to find a wider set of people who will understand us a little, enough to be tolerant of physics. There is a lesson to be drawn from this particular example.

Second, and this is an old and by now probably rather well-won battle, the effort to make a comprehensive introduction is, I think, futile. We know too much to have comprehensive courses: one textbook cannot be the full answer. There is no way to make this work. It does not mean that breadth is overlooked. It doesn't mean that the details are overlooked. But it has to be done by some kind of ingenious episodic treatment which gives due attention to both of these issues. When I took elementary college physics with a book which was unfortunately known as "Dull's Physics" (it was not at all a dull book, but quite a good book), there were actually fine-print paragraphs on internal conical refraction. I was much impressed by such sophistication in physical optics, but I must say I don't understand it very well even yet; I think it could have been well left out of our physics course. Nowadays most texts would deny that they are comprehensive, but I think that the tone in which they appear episodic is also important. I am not even convinced that a single textbook, one single authoritative beginning, one compendium, is the right way to go. In most other courses, not so much nowadays as five years ago, the sense of a many-authored approach, of a many-voiced approach, has been stronger in the schools and stronger in the bookstores. That seems a better way to reach the aims I have: engagement, not authority.

I have now to cite an old metaphor, an important one. It comes from the obvious distinction between analysis and synthesis. These two profound intellectual activities are very often separated in the doer, separated in theoretical structure, separated in the social circumstances amidst which they occur, but never really strongly distinct once you look into the details of what actually happens. Certainly in any corporate activity, any social activity, both are almost always present. We physicists, and perhaps it is our nature, we very much emphasize analysis and tend to reduce the role of synthesis. But Purcell's "Widely Applied Physics" course was much more synthetic than usual, in the sense that he set a problem whose answer had to be sought wherever it could be found. That is quite different from the usual procedure, where you know that Chapter Seven is going to be dealt with in Chapter Seven's examination, Chapter Nine in Chapter Nine's examination; it's very unlikely you'll have anything else on the final! This is the general strategy for passing; it is not the way to make experimental physics, the physics of looking into the real world, the basis of what we study. This merger is a most important task, not easily achieved.

Finally, I think I would judge the comprehensiveness of a change in our approach by one simple measure in our society, not a unique measure, but surely a significant one. We are not now making much progress with it. We still have few courses in physics, or little popular work in physics, which can appeal widely to interested and talented women students, in spite of their obvious large number. This is of course not our fault alone. There are somehow social barriers whose nature we poorly understand, but I regard it as a challenge to reasoned educational design to try to overcome those social barriers by all the invention we can bring to bear. Here of course the vision of vocational shift is equally important: more women in labs, electronics shops, and faculty slots.

I still feel that the principal interest for many people in the assertions of physics is not for their practical consequences in war and peace, but is in what I will call, for want of a better term, the philosophical questions which underlie and shape the aims of our science. These are Ionian questions: whence do we come? what is within? what is change? and so on. Astronomy, cosmology, particle physics remain central subjects even though they might or might not be remote from particular practical consequences. The effort to understand them has involved us in very many interesting byways. The remoteness of the problem is not itself a guarantee of its lack of interest; this is not to be forgotten. We do ourselves grave injustice if we forget these points. We do our fellow citizens an injustice by thinking that they do not have, and do not share with us, the ancient human concerns about these problems, to which physics can make trenchant contributions. Of course I don't wish to say that everything has to be done in such a high-flown way—of course not. Simple experience is most important. I see very wide interest in everyday experience. I need mention only sports, weather, the structures of everything around us, from organisms to buildings. Here lie many valuable and well-exploited topics. Again the strategy must be mixed, dissidence is important, for you should not take only the simple-minded things that you know work. You must try to show where something strange appears for the gain of a new point of view and the daring that transcends everyday experience. Like sport, such an attitude attracts because it allows the introduction of conflict and of mastery in a limited domain.

Frank Oppenheimer has made in San Francisco a wonderful prototype of concern for human perception. This set of physical phenomena, the marvelous physical instruments with which we are born, the physical data-processing which we learn to use in applying these inborn instruments, is shared by us who try to understand it cognitively with the artists who try to exploit it more directly; as you know, the famous Exploratorium in San Francisco manages quite successfully to merge these two lines. So far it has had too little echo. Not only is the eye of the artist and the physicist equally present in that museum, but also the machine shop and the electronics shop are as visible as the exhibits, and play as important a role. People walk in off the streets to try to learn how to do things! This sense of building novelty is very close to what the physicist has in mind.

There is in every kind of instruction the need for some sort of mastery, some sort of gain. This can be deferred; people talk a great deal about the measure of the degree of deferment that people will tolerate, but some gain must at least be promised. Nobody who has taught in a university in the last decade can be anything but enormously impressed by the degree of mastery the computer has given to a very large number of students. Their capabilities in that limited and logical world are great, quickly reinforced by their experiences. The result is not always good; that world, like the gambling world which in some ways it resembles, has a curious finiteness, a curious closed quality. Nevertheless, that success is a lesson for all of us. Amateurism in physics has lost a great deal, though it remains vigorous in radio, in various sky studies, perhaps in lasers and a few other places. Extension to other domains is possible, I think, because we allow the amateur practitioner some entry into this activity through computers. Maybe if the particle desert really appears our particle physics will no longer be held by the accelerators alone, with their enlargement of Rutherford over and over again, ever more elaborated. If that field takes a wider cut, a look at the natural world for decay of the proton, superheavies, non-integer charges, and so on, maybe this will begin to redress the balance, to give the amateur some basis for appreciation easier than the perpetual beam experiments of the accelerator labs.

It is clear that a kind of alienation from intellectual activity as a whole, and from science-based activity in particular, is, if not a dominant phenomenon of our time, at least a very prominent one. It is prominent especially on the campuses, in the centers of learning. We need to be much concerned with that. I offer no great wisdom here except to muster the scale of instruction, the scale of contact. It runs from one minute, which is the time of a TV commercial, up to a thousand minutes, which is the time of a TV series, and on to what I estimate as a million minutes, the time of a Ph.D. Now learning has very different consequences over these six orders of magnitude: we don't exploit a wide enough band in that spectrum and we need to do so.

Now for a truism: we live in a world which is ever more specialized. Intellectually it is wildly specialized; it is hopeless to attack this totality, but we must recognize it. I found an extreme case some years ago when I reviewed a successful international work on nutrition and its requirements, a book devoted primarily to the food calorie, taking that calorie apart in many different ways, quite a substantial work. The comment on the international definition for the calorie (better, kilocalorie) was written by one of the learned authors of the book who, I assure you, plainly did not understand its mechanical basis at all. He had its origin quite wrong, but that didn't matter. Once you've got the calorie exemplified in some standard sample of sugar or the like, from then on the nutritional world is self contained. So long as he had tables of how many calories were spent in walking, pumping, and so on, it didn't matter to him that it was mistakenly given in terms of force and distance. He never knew it, because he never came across a dissonance to show that he had the definition wrong within the mechanical domain. You see, specialization is something strong, and it's a mistake to overlook it; specialization is not in itself the enemy; in short, the adroit use of specialization must be our game.

One way we might judge a success even wider than among our students is in the enrichment of the language, the common speech. Many terms in physics, many unit names and so on, are freely bruited about, often with very little understanding. It is still true that the majority of journalists, judged by the published stories in such dailies of high reputation as the *Boston Globe* and the *New York Times,* do not understand the significance of the simple distinction between the kilowatt hour and the kilowatt. Now these are able people, concerned with reporting technical developments correctly, and yet they get it hopelessly mixed up. The idea of a ratio, the idea of a rate, the crude idea of an integral expressed by the kilowatt-hour or man-hour or what you will, is no trivial idea. It is a powerful idea. It needs better language, better expression, and better metaphorical connections, and better use. Until we've succeeded in that, I think we have made little headway toward a broad public education in science. This, for me, is the goal, not so much a true analysis of powerful public issues, nor even public understanding of the up-to-date concerns of scientists. We need to place science, and physics in particular, a little closer, as John Dewey put it, to the subsoil of the mind. Some of the most elementary notions should occur as axioms, as real as language. In language we know *big, bigger,* and *biggest.* There are better scales with more information, like the sequence sand, gravel, pebble, cobble, boulder. But we don't have enough of those, and we see little understanding beyond the simple change of magnitude. We need much more. Here is a domain in which physicists must play some role, of course no less than many other people, writers, teachers in general, and the makers of telemedia.

I must admit a considerable dismay, though I know I am not fully consistent, at efforts to popularize science, physics and astronomy especially, among the wide population, which are based (and I say it very crudely now) upon propositions beginning with the phrase, "Scientists say . . ." or "Scientists find . . ." Well, they do, you know, but I would throw that out of court; it usually means maybe they do and maybe they don't. At least some kind of judgment must be produced; evidence is at least as necessary in science as elsewhere. An appeal to authority, however excellent, still remains an appeal to authority; it cannot be verified, one cannot do anything more than look up the credentials of the authority. That isn't enough. I suspect this is the hardest task we face, because of course it's necessary in all life to take many things on authority. I'm not prepared to draw the subtle distinctions among them. When once I tried to make a TV program, which I think came out pretty well (though it was a program for a small audience fraction), I insisted on showing the Doppler shift, and not just saying "scientists find" that a receding light source gets redder and that's the Doppler shift. I showed the darn thing, in its historical context, which was the real historical context. You could hear it! It was fun to do and pleasant every way around. And instead of saying a black body is an enclosure where there's no gradient of temperature, we made a real black body, a "Hohlraum." Actually, the ceramicists did it for us—they do it every day in the kiln—and you saw the objects disappear as the glow became universal. It took a lot of time, a little bit of money, and it was of course a difficult program. But I think it was a good program. Some demanding programs are necessary. But the very able man (not the producer who worked with us in making the program, but his boss, the executive producer back there in London) who knows more about television and its science audience than I do by a great deal, was hard to persuade to bring this in. He said, "Our program deals with new phenomena and new points of view; here you tell me this Doppler experiment was done in 1842, why should we put that on the air?" We had to say, "Well, some things really are right. They were right in 1842 and they'll be right in 2042 and if you don't know it, you ought to know why we say it's right." I think we have to say that sometimes.

In confronting the image tide, what can we do? We know how much floods in all the time; what is missing in this enormous flow of bits that we must sort out from screen and press? The 3D, real-time experience is less frequent and varied than it was when we were farmers, lower than it was when we were hoping to work mostly in physical factories; now we have many secondary and tertiary industries which deal in bits. The world behind the screen and the print-out is farther and farther away. I am clearly concerned with television—I regard myself as a TV fan of some sort—but I still think that I would trade quite a lot of excellent TV programs on science if only over America a few hundred people would for the first time take a small permanent magnet and put it on the screen of their television set, and watch it cut a hole in the image, and think about all that that means for three minutes. Even if it left some funny jiggle from magnetized elements of their color screens, it's worth it! If I ever have a chance I'm going to do that on television; somebody will be sued!

But physics alone, even science alone, is not a handle to grasp the whole world. In Professor Gerald Holton's eloquent and learned Oersted lecture the two maps on the wall, the political map that changes as the flow of war and nationalism crosses Europe, and the periodic table which never changes in a century, or changes only consistently, do in fact interact. Power comes from truth, but power can be applied in the service either of the true or of the false, or in that pervasive domain where human interests conflict. We will not get out of that bind; we will not solve that problem ourselves. All we can hope to do is to take up our share of it by insisting, where we can, that we bring some sense of structure to the world in which all are immersed, because we are all part of one world of mass and time and particles.

It is not easy to be self-appraising, but if there is a flaw in our community, it is the flaw of overconfidence, and our sin, confidence carried too far, is the sin of arrogance, of pride. It's a natural flaw. You cannot cope with the terrible problems of understanding the depths of the proton without some sense of arrogance about the accomplishments of the past. If you know three people who are in that business, you'll find it not far beneath the surface in one at least. That is hard to mend, because it has its roots in an everyday struggle of the human mind, and of the social structures we have made, to understand the cryptic universe. But that doesn't mean we can accept arrogance without trying to mend it, without trying to ameliorate it, without trying to encompass it within other attitudes and other concerns. We need to do this, we need to do this all the time. We probably suffer now from a half-conscious recognition of this throughout the world. I admire the beautiful American Institute of Physics (AIP) calendar in which a page shows Einstein in Pittsburgh in 1934 being quizzed by reporters. He was pretty fresh to the East Coast, and hadn't travelled much in this country. He had not yet understood the galvanized attention that followed him wherever he travelled. I much wanted to be at that meeting, I was very anxious to see the old gentleman—he seemed then very old to me, but of course he was not as old then as I am now—but you couldn't get in. No student had a prayer of getting in. The tickets were bought up and scalped; they weren't even for sale, but still there was a tremendous trade. I had no chance. But there was one weakness in the plan, because the meeting was held in the theater of the Carnegie Institute which adjoined my school, the Carnegie Institute of Technology. More than that, there were students at Carnegie Tech who were theater students, who rehearsed in that theater often, and of course they knew the ropes. Since I knew them, I too could learn the ropes, and so on the day of the great man's lecture, we climbed to the very top of the theater, behind the scenes hours before the doors were closed. Poking our noses through a little slot, we looked down beyond the chandelier to see the crowded audience below. We could sometimes catch a glimpse of the top of Einstein's head from a zenithal vantage point above him. We couldn't hear a word, but it was a fascinating experience. The cognitive is not everything!

I would like to close with his remark, one that took the reporters by surprise. When they had the press conference the AIP calendar reports, in a facsimile of the story for the Pittsburgh *Post Gazette* of that day, in reply to several questions by his interviewers, the great Professor Einstein "did not hesitate to say, 'I don't know.' "

The Full and Open Classroom

In the left hemisphere of the brain there are two distinct temporal regions. They are not symmetrical; they are only in the left hemisphere. They are known as the region of Broca and the region of Wernicke, after their nineteenth-century neuroanatomical discoverers. I shall read a couple of simplified protocols—accounts of interviews and discussions with patients who, coming to autopsy, were demonstrated to have seriously impairing lesions of these particular parts of the brain.

First, loss of the region of Broca:

> . . . the patient clearly fails to produce correct English sentences. Characteristically the small grammatical words and endings are omitted. This failure persists despite urging by the examiner, and even when the patient attempts to repeat the correct sentence as produced by the examiner. These patients may show a surprising capacity to find single words. Thus, asked about the weather, the patient might say, "Overcast." Urged to produce a sentence he may say, "Weather . . . overcast." These patients invariably show a comparable disorder in their written output, but they may comprehend spoken and written language normally. In striking contrast to these performances, the patient may retain his musical capacities. It is a common but most dramatic finding to observe a patient who produces single substantive words with great effort and poor articulation and yet sings a melody correctly and even elegantly.

The Wernicke's patient contrasts sharply with the Broca's type:

> The patient usually has no paralysis of the opposite side, a fact which reflects the difference in the anatomical localization of his lesion. [The motor regions of the brain are more forward and closer to Broca's area than to the

Wernicke's.] The speech output can be rapid and effortless, and in many cases the rate of production of words exceeds the normal. The output has the rhythm and melody of normal speech, but it is remarkably empty and conveys little or no information. The patient uses many filler words, and the speech is filled with circumlocutions. There may be many errors in word usage, which are called paraphasias. These may take the form of the well-articulated replacement of single sounds (so-called literal or phonemic paraphasias), such as "spoot" for "spoon," or the replacement of one word for another (verbal paraphasias), such as "fork" for "spoon." A typical production might be, "I was over in the other one, and then after they had been in the department, I was in this one." The grammatical skeleton appears to be preserved, but there is a remarkable lack of words with specific denotation.

Such deficiencies in the brain are of interest both to the surgeon and to the analyst of brain circuitry, because they point up the spatial localizations, but from my point of view it is because of the extraordinary degree to which they imply a clear structure of the most elementary kinds of higher functions, which we normally imagine are a single whole. Not at all: it looks as though grammar and content are distinct abilities.

Even a more remarkable case was described a year ago, and has been worked on by Geschwind, Quadfasel and Segarra at the Neurological Service of the Massachusetts General Hospital in Boston:

We studied our patient for nearly 9 years after an episode of carbon monoxide poisoning. During this period she showed no evidence of language comprehension in the ordinary sense, and never uttered a sentence of propositional speech. She was totally helpless and required complete nursing care. In striking contrast to this state were her language performances in certain special areas. She would repeat perfectly, with normal articulation, sentences said to her by the examiner. She would, however, go beyond mere repetition, since she would complete phrases spoken by the examiner. For example, if he said, "Roses are red," she would say, "violets are blue, sugar is sweet, and so are you." Even more surprising, it was found that she was still capable of verbal learning. Songs which did not exist before her illness were played to her several times. Eventually, when the record player was started she would begin to sing. If the record player was then turned off she would continue singing the words and music correctly to the end, despite the lack of a model. Postmortem examination by Segarra showed a remarkable lesion, which was essentially symmetrical. The classical speech area, including Wernicke's area, Broca's area, and the connections between them, was intact, as were the auditory inflow pathways and the motor outflow pathways for the speech organs. In the regions surrounding the speech area either the cortex or the underlying white matter was destroyed. The speech area was indeed isolated. The patient's failure to comprehend presumably resulted from the fact that the language inputs could arouse no associations elsewhere in the brain, and since information from other portions of the brain could not reach the speech areas, there was no propositional speech. On the other hand the intactness of the speech region and its internal connections insured correct repetition. The preservation of verbal learning is particularly interesting. In addition to the speech area, the hippocampal region, which is involved in learning, was also preserved, and this probably accounts for her remarkable ability to carry on the memorizing of verbal material."

Finally, the third is a report from a different group under Roger Sperry at Caltech, doing one of the most remarkable sequences of experiments in the last decade. The patient in question is a much happier person, a housewife, perfectly normal and functioning well in society, who is a recoverer from a very heavy operation carried on to relieve her absolutely disabling epilepsy by a group of surgeons in Southern California who have operated radically on a number of patients in the last decade with excellent therapeutic results. But there appears so profound a modification of the normal person under experimental situations, though not in normal life, that the experience is most revealing. These are so-called split-brain patients in whom a radical destruction is achieved of all those upper brain portions, above the hippocampus and the callosus, which provide the connection that transfers information from one hemisphere of the cortex to the other. These paths are severely cut down, right down to the line of the stem. The patient recovers and is released. There is a photograph of this young woman riding home on her bicycle from the laboratory where she is conducting experiments voluntarily. Here is an example of the kind of experiment performed. When the Californian housewife is asked to use her left hand to feel objects behind a screen and to name them, she cannot do so. Her right brain is mute. But the right brain-left hand combination is more skilled than its partner in recognizing objects by feeling them—provided it can identify them otherwise than by words.

I cite now a book written by a very able science writer, Nigel Calder. It is not a proper report, simply his summary of the experience he had visiting Sperry and Gazzaniga. He says:

Let me introduce Dexter and Lefty, two brothers as unlike as Jacob and Esau. Dexter, dwelling in the left of the head and controlling the right hand, is the smart one who does all the talking. Conversing with the split-brain patient, it's Dexter you meet. Lefty stands by quietly, listening politely to what is said, but saying nothing. Nor does he do anything much, unless directly asked to carry out a task. No one can say what induced Lefty to give up his birthright of language, because all the evidence suggests that at the age of three he had equal competence to his brother. Dexter can read and write fluently and does the difficult arithmetic. Lefty, though quiet, is not entirely illiterate. He can read common words, do simple arithmetic, understand most of what is said to him. For example, with the left hand groping among objects in a bag out of sight, a split-brain patient can respond to the instruction "retrieve the fruit that monkeys like best" by pulling out the plastic banana. And in many tests the patient is asked to say what the left hand is feeling. Dexter simply has no idea and makes a guess, whereupon Lefty frowns and shakes their common head. "I guess I'm wrong," says Dexter. Lefty certainly seems less bright than Dexter in the adult, but he is far from stupid. In tests of perception, comprehension and memory where he can make nonverbal responses he performs very well indeed. In double tasks of a simple kind, such as searching for different objects, reacting to different signals with the two hands, Lefty and Dexter go their ways independently. The patient can complete such tasks more rapidly than a normal person. Lefty has the keener sense of shape, form and texture. He can, for example, copy a geometrical picture much more accurately than Dexter can. The lines are clumsier, but essential features are not omitted, as they are apt to be when Dexter tries his hand. Faced with evidence of such drawings by Lefty, Dexter will say, "Well, I must have done it unconsciously." Lefty's education has, perhaps, been grievously neglected. Here is fully half of the human cortex which is relegated to the status of a second-class citizen. He may represent a great waste of mental capacity left poorly nurtured. As Sperry urges upon anyone willing to listen, the educational systems of many nations are immensely biased in their heavy reliance on verbal inputs and verbal outputs.

The Geschwind paper, with an excellent bibliography, is contained in the November 27, 1970, issue of *Science*. It would be the first reading for anyone who wishes to go further, very interesting in itself. The other book, which is an overall popular survey of contemporary neuropsychological and physiological experiments, is called *The Mind of Man,* by Nigel Calder, first published by the BBC in 1970.

The epithet "the open classroom" is now powerfully presented to us all. We hear on every hand the proposals, and even see the actual situation all the way from North Dakota to New York City and up and down the land. It has reached the status, so to say, of a sub-fad or sub-success. It isn't quite home yet. There is opposition in many places. But many people understand that it is an important and valuable direction in the schoolroom. Education, like all large-scale activities, acquires a jargon, specialized remarks of its own, specialized names. With that goes the defect of social behavior we all know: that of beginning to substitute the name for the fact, paying lip service to the proposition. The progressive school has a similar history. It is because of my concern that something may be lost in the understanding of what is meant by "the open classroom" that I propose a longer phrase: the *full* and open classroom.

It has been a good many years now since I first heard of what is now called *the open classroom,* from the people who organized the infant schools in Leicester County—not within the city—in the midlands of England. It is by no means a new idea, as I'll try to demonstrate. This is no place to describe, even in a compact way, what the term means, but I will try to give at least a caricature to indicate what I'm referring to. I may say that I am entirely enthusiastic about the proposition of such an educational reform; wherever I've seen it working well, it is superb. It is probably a *sine qua non* of satisfactory elementary schools, especially for young children; I am not at all speaking against it. I am only trying to define it a little more richly. The principal point of the open classroom is, of course, that the ideal of the unison of the class is essentially broken. In its place, the class is considered as a set of individuals or of small groups, self-paced in their instruction. But no one is ignored, rather all are nurtured with each other's assistance, with the teacher's assistance, with materials and schemes provided by the environment, all of which help the class work out its tasks. Very often physical changes are made in the sense that students may not be required to stay within the four classroom walls—special equipment is placed in the corridors or in the schoolyard, in a much more open and free situation. What you must find in the open classroom is that the conventional crystalline lattice of chairs has melted; the people are all over the room, whatever the room may be. These are the superficial characteristics of the open classroom.

From what I know of education I will present three rationalizations, or three structures of thought, which I think can independently be the basis for this open style of organization. The oldest one, and in some ways the most powerful one today, especially in the United States (much less in Britain), is what I would call the philosophical-political justification. It rests on philosophical values, not about simple things such as expediency, or effective operations or things of that sort, but much more in the quality of an ethical prescription for the development of individuals, in terms of their self-awareness, their self-images, their integrity, dignity, and so on. The oldest book that I know of that presented this very strongly is the work as educational experimentalist of the writer, Count Leon Tolstoy. Coming out of an explicit background of philosophical anarchism of the highest ethical sort, Tolstoy operated a school for the small children of the local peasantry in the village where he was a member of the gentry. For a half-dozen years or more he tried to prevent the establishment of the role of the authoritarian teacher leading a class through exercises in unison, stuffing knowledge into bags, as Professor Zacharias would say. Tolstoy was no less modern or up-to-date than George Dennison in New York, just to mention one of the many people now raising the banner of opposition to the envisioned dehumanizing aspects of an industrial society. These people present powerful discourse, placing the schoolroom with this kind of freedom in opposition to the impositions of the class-based social structure. It's impossible to argue with these schemes on any less exalted level than to discuss the nature of values. Indeed, as a critique of the social pressures which to some degree do lie behind the unison, fixed-chair classroom, there is little doubt that these people have put their finger upon a very important genetic component of the origin of the system. Very likely it is not all of the story, I might say, because the system of the verbal memorization of the Chinese classics or the *Koran,* or even the *Talmud,* go back a couple of thousand years, into communities where these class attributes were not very important.

The second attack on the fixed classroom has come in a very interesting way from what I like to call learning-theoretical experiments. I am referring to the wholly inconclusive but fascinating and stimulating experiments carried on largely by the School of Genetic Psychology of the University in Geneva, by Professor Jean Piaget and a variety of able co-workers. Although their experiments deal with subtle stuff, difficult for the layman to understand, one important message that comes through from this body of literature is that there is within humans an internal maturation of systems of concept and operation which does not allow arbitrary replacement of logical structures, arbitrary order, and so on. This has provided very strong support for the self-pacing and the independent treatment which, on very different grounds, mainly based on empathy, on ethics, on political and philosophical arguments, have been carried out in many schools for a long time. (Comenius himself was not very far from this centuries back.)

Finally, there is a kind of rationalization, not worse or better but on an entirely different level, which is coming from the scalpel and the electrode amplifiers of our present neuroscience. It is an extraordinary revelation of the degree to which the several functions of the brain, of the mind, are not to be thought of as whole things, but rather as the deeply complicated interaction of many distinct specific items, whose separation we rarely see because they are normally very closely and tightly coupled.

One gets the impression, as from the split-brain operations, that there are regions of any brain which operate very differently. If we see this aspect of human behavior vary widely over the normal population, then education must become responsible for finding means to facilitate, to help to mature and to develop all these functions in any given student. The brain regions demand a certain degree of opportunity, in the classic democratic tradition although perhaps not necessarily for each individual. To do less seems to be, as Sperry is beginning to feel from his direct participation in neurological work, as if we are neglecting to school a large portion of the population in any way whatever, allowing them to pick up whatever they know from the accidents of daily life—which, of course, is not at all a bad school, though even Dr. Illich tries to do better.

All these three different levels of argument provide support for what I like to call the full and open classroom. The ethical, value-setting, social rationalization—with which I largely agree—seems to be by no means complete. We should also look at a very powerful symbol, the Broca and Wernicke areas, and the isolation from the rest of the cortex. Let us think of such a symbol in terms of the unhappy patient who was, it seems to me, reduced very much to the level of a tape recorder: let us recall how many school examinations such a damaged brain might succeed in. Quite a few; I venture to say that in most of the structure of compulsory education of the western world for the past 200 years such a patient would have been able to pass most of the examinations. (Given, for the purpose of this symbolic argument, that she had a degree of mobility which she did not in fact have.) She would not have done brilliantly, but she would certainly have passed, because the ability to find verbal forms and to complete them—to give back what is put in—is widely taken to signify that there is some kind of productive understanding behind this repetition. But, as we now see, this is only an assumption; it is an assumption not strongly reinforced by the efforts to test it, or to develop it by the curricular schemes or the practice of the schools. I am not at all arguing against memorization, for I know it has its virtues. I wish I had been assiduous in learning some of the poems I had to learn in school; I am happy to have the few scraps of them that have stuck with me. I envy people who play with all their hearts in good plays, and for the rest of their lives can remember the sides they learned, and can cap very prettily a speech out of Shaw or Shakespeare. (As Francis Bacon said, "A ready man is always worthwhile," or did he?)

On the other hand, the fact that such repetition does not necessarily require or to any high degree statistically imply the *productive* manipulation of that same knowledge in new combinations is now clear. The neurological work tells us of unknown additional steps which we must make sure are not lost in the operation. The history is plain everywhere in educational systems, especially in those of industrial societies. The ancient history of mankind, going back to the hunting-gathering cultures, has always included some element of the school—there has always been play and imitation, and hence education. But it's quite plain that the chief experiences were very rich and diverse: to hunt or gather, you had to be alert with nonverbal abilities, recognizing, walking, peering, looking around a corner and remembering where you were, how you got there, and quite a number of other mental and physical skills. There is much evidence for this in contemporary studies of the few hunter folk we have left.

A more familiar symbol of such diverse educational experiences can be seen in the village, most particularly in the characteristic pre-industrial or early-industrial stage of west Europe, the United States and also in Japan. Although the village is no longer really autonomous because it has acquired trade flow and usually allegiance to a state system, still the characteristic life of the New England farmer in his farm village serves for most of us as the symbol of what we think of our national past. It's quite characteristic that every child there, boys and girls alike, by the time he was ten or twelve had acquired a rich, nonverbal experience of such diverse things, such utilitarian things as being able to see by the light of the full moon, understanding how wheat was ground into flour, or possibly even using a crowbar as lever to lift some heavy load around the farm. Their experiences included not only a thousand different chores productive in their nature, but also a rich acquaintance with the affective side of life— with complex personal relationships, with birth, and death—all matters much closer at hand in reality then than they are in our time, when indeed they are fully present, but outside the home, even mainly in symbol. Our present-day experience derives mostly from the newspaper, from the television set, from the leisure time of parents, not their working life. It is far reduced to a symbol flood which comes to you, a severely dependent social creature, for the most part nonparticipating and specialized away from all these experiences. This tells me that the task of the school cannot primarily be the symbolic task.

We must try to undertake the replacement of these diverse experiences of early productive life by supplying in school context what is not available in everyday living. The city kid now, of course, has much less contact with those utilitarian chores but yet he can acquire a very much richer understanding both of the world of manufacture and of social relationships in the city; although this is a complex experience, it is always more vicarious; it doesn't have the same immediate quality as the diverse rural life we were talking about. Therefore we need, not just an open classroom, by itself it will fail, because it will lead to that terrible affliction of simple situations where compulsion is lacking: boredom from having nothing to do that seems challenging. This is something we need to guard against: once a self-paced, open style becomes systematized and accepted, it must be nourished; or it will lead to doing nothing on your own in the classroom which is not for long a good feeling. A class must have much material, both physical and symbolic, but certainly largely physical, in the grades particularly. It needs a wealth of relationships with its material; not only must the children consume it, but they must make it, destroy it, work on it with several people, work on it with one person, oppose it, support it—all kinds of relationships on the affective side. We haven't much neurological information on that, but I dare say a psychiatrist could show us considerable need to reinforce the affective side of childhood with a variety of relationships to others and to material tasks.

It is important that school not be given over entirely to the high banner of individual development, to the exclusion of social and cooperative efforts in the classroom. One cannot, in my view, hold so naive a critique of our social system as to say that we have one which is not built upon cooperation, even though one may dislike the degree of cooperation. There's little question that industrial corporations, and industrial society in general demand a higher degree of socialized cooperation from its members than any other kind of economy. We are mutually interdependent never more than we are now—nowhere more than in America—this is something which has to reflect in the classroom. The competition of market and career are contradictory elements. I think therefore that shared tasks are very important; they should be tasks which are shared benignly, amiably, cheerfully, at least for the most part, entered into voluntarily, shifting associations in the classroom and with other resources in the community. Here I would tend to agree—although I do not always agree—with the critiques of Dr. Illich who has written so heavily against the schools as institutions in themselves. He has made a sound critique of the political-philosophical rationalization for new schools: it is not possible to reform social relationships and political economy at large by using primarily the schoolroom.

Finally, I think the full and open classroom has to be thoroughly open, depending of course on circumstances—open in the sense that it cannot remain within its own four walls all of the time or even most of the time, but it must allow very easy passage through the membrane of the door to the outside world, to the rest of the school, to other classrooms, to younger children, to older children, to professionals, out into the city, a whole variety of destinations.

I am impressed too by the existence of a true school community, an extremely difficult thing to build up. I don't mean the adult community in which a school is embedded, although this is an important background for every successful school; I am thinking mainly of the few private schools, dominated largely by the founder himself, whose personal presence or whose influence on succeeding generations has given the school a sense of special community, a special purpose. I know of one elementary-junior high school, for example, where for a long time now every graduating class has produced a full-tilt production of a Shakespeare play, with the whole school (up to ninth grade) involved. Thus a child who comes as a five year old to the school, who barely understands what the theater is, can march across the stage holding a branch. Then seven years later the child is Titania or Puck, or at least can stand knowingly in the wings criticizing, reciting those fat speeches to cue his small part. A very remarkable community grows out of children's *Tempests* and *Dreams* alternated year by year now for twenty years and more.

Many such possibilities exist, of course; the world is very diverse. I have the opportunity to show the sort of thing I mean, merely examples drawn from a single context of the elementary school, a mixed science-and-art activity. I happen to know very well the elementary-school teachers and developers of these plans—my wife and her colleagues. They have very kindly supplied me with a few of what could be called flow charts. (They're, of course, not entirely novel in form; such charts are in Nuffield science, too.) They begin with a single topic more or less of laboratory or classroom nature; then they surround it with a net of connections, summarized by a kind of map. This is by no means a curriculum, however, not at all a race course. Using this map many roads may be taken; of course, the map is far from complete, many wonderful places still remain to be mapped on it. This kind of map shows that you can begin somewhere, at any attractive, real, somewhere, and if the classroom is open, so that different tastes, different styles, different affective behaviors, different channels for symbols and many material things are available, and if all honestly different directions are encouraged, an extraordinary connectivity can develop out of what might seem to be a restricted and minor inquiry. I display here four such charts: Light and Shadow, Printing, Making a Collection of Seeds, and Play Games.

It is characteristic of the knowledge of scholars, artists, craftsmen, scientists that it has exactly this sort of connective structure, particularly knowledge which arises from the natural sciences. But it is by no means exclusive to science. There lie very close to it the problems of visual arts, and at the next stage the problems of myth, which is in many attributes the counterpart of science, which nourished it maternally, so that it is very close to the heart of the scientist. Indeed anthropology has been all three: a social science, a study of literature, and a natural science. This mixture of the verbal and the nonverbal, of the abstract and the concrete, of the commonplace and the hard-to-search-out, of the library, the slide projector, and the piano, is an extremely interesting presentation on a level meant for the teachers in the first three grades of school. In many circumstances this is a far better map of the mind than a single, connected line of development, unless a very special, carefully chosen one were presented.

Finally, I want above all to say that the open classroom, if you take it literally, is not a panacea. The principal concern we must have in social interactions like education is for strategems which allow different inputs for different persons. That means a mixed strategy, with a choice of some sort, either elicited or free, some sort of mixture of experiences.

What this world teaches, to paraphrase a well-known German philosopher, is that its unity is material in nature. We share that unity together with other combinations of atoms. But the world's diversity, highly important to all our activities, is founded upon that remarkable arithmetic property, the combinatoric richness of the world, symbolized by this kind of a symbol: $n!$ Once you have some appreciation of these two elements of reality, you can probably design reasonably well the beginnings of some educational experience.

References

Nigel Calder, *The Mind of Man,* The Viking Press, Inc., New York, 1971, p. 249.

Norman Geschwind, *Science,* **170,** 940 (1970).

The flow charts are the unpublished work of Phylis Morrison, Mary Eisenberg, or Jinx Bohstedt, as the initials indicate.

Primary Science:
Symbol or Substance?

I have been by devotion and vocation for many, many years a theoretical physicist and something of a bookish writer. Like many professors, I'm an awful lot of a talker. Therefore, for me to say anything that could denigrate the social importance of symbols would be insincere! You can't talk all the time and say there is no case for symbols. It's obvious that there is, obvious that language is the principal tool of humankind, that which distinguishes us from very clever hominids that came before our species. To deny it would be absurd. The schools of course do not deny it. They're founded on the three symbol systems which are most characteristic of the cultures of today: language, mathematics, and now, of course, the ubiquitous two-dimensional images fleeting across the video screen, the most important, the most frequently-used communication in our society. Put them in stereotyped language: reading, writing, and arithmetic. We'll add the images on the screen to bring them up-to-date. Of course we can't do without these things and of course that's what learning in large measure consists of—especially when we educate institutionally, at second-hand.

If you go into much less complex and large-scale cultures than our own, you will find that the learning of systems of special skill, understanding and insight is by and large a genuine apprenticeship: apprenticeship to being a shepherd, potter, preparer of daily foods, builder of the hut. There are skills so undivided in the society that everybody has them. Others are highly directed towards vocations. In many societies the potter is a professional. Not everyone pots, but all use pots. Now the potters don't go to school and read books about potting. Separation between the task and the symbolic description of the task is characteristic of literate and complex societies. This makes for schools that emphasize the symbolic.

Phylis Morrison reminded me that one of the best schools in traditional society that we know of is the famous Navigators' School, a preliterate school on the Caroline Islands, in Palau, where they teach how to get to a distant island. (An anthropologist actually graduated from this school and he can almost do it, but I felt when I read his book that he didn't quite understand what was going on. The trouble was that before he went he studied educational psychology. He should have studied astronomy.) There you learn rote rhymes until you are letter-perfect in them—word-perfect I should say—because those rhymes embody the succession of skymarks you have to follow on your way between some Caroline point and some remote place across the horizon. With no almanac, the data have to be implanted in the students so they don't forget it at sea when they're hungry and cold and seasick and the waves are breaking over the mast. The navigator still has to say "steer for that star" and not get the wrong one. It's a matter of life and death. They understand that there's a role for rote learning. I'm not against that where it has a purpose. But when the purpose is secondary, to take an examination, to see how much you learned, then a certain feedback loop appears which the computer would reject as endless.

Symbols are not very much if they cannot draw upon the real world. The aim of symbolic behavior is the manipulation of the real world. For the scientist, that is largely the natural, the physical and the biological world. For human beings in general, it includes also the social world, the world of other human beings, and their responses, their psychology and aspirations, in some connection with your own.

We know many people for whom it is fair to say that schools present and past do not do very well. The school that does best for them is the direct apprenticeship, and perhaps this is true for all people in some respect. Many are people whose degree of articulateness, whose manipulation of symbols, even of the spoken word, is not as congenial, not as comfortable, no as productive as changing small portions of the world by hand. They are artists, designers, potters, fishermen, mechanics—a lot of whom manage their complex tasks because they hold a complex, insightful understanding of at least a small portion of the world. But they have not learned it from symbols, and do not easily reduce it to symbols. It is another task to map the world symbolically, quite a different task from understanding it in reality. This distinction we lose at our peril.

Speak not about language or mathematics or even mapping, but about science. It is none of these things, though it lies at the edge of and behind them all. It employs and demands some understanding of all of them. Yet you can't talk about science and remain solely in the domain of symbolic discourse. You require some contact with that substance of which science is a symbolic representation. Science is only a halting symbolic representation of connections. It is a partial effort to penetrate the innumerable connections between the real world of which we are a part, in which we are all immersed.

People who live in a city, very far from the farm, live everyday by technical interchange, recognized as characteristic of our world. Our schools arose to teach the handling of symbols, not language, the mother tongue, which is learned much better at home than in schools, but the written word. Most literacy was the impartation of a few special texts: the Bible, the Koran, the Analects.

Phylis Morrison: Coming to deal with the symbol systems of society has been the central thing that schools have done. I asked myself for how long: since the Massachusetts Bay Colony? Certainly so. Since the Renaissance? Certainly so. Since Rome? Certainly. Yes, surely, since the beginning of towns. It has to be that way. The schools have been very successful, more and more at the core of society. Society cannot continue to exist without handing of symbolic skills for the next generation.

What is different now? It is the lives of people, of children as well as adults, outside of school. As short a time ago as a hundred years, most people spent most of their time dealing with nonsymbolic problems that presented themselves to them as physical problems. The easy example is the child growing up on the farm. That child had experience with heating the house. The child knew perfectly well how much firewood had to come into the house to deal with one sort of weather at one time of the year, or the problems of growing things, of mechanisms you used to make growing those things easier. They were all part of everyday, regular life of the child. The child grew into an adult who knew that there were many kinds of problems that one solved on the level of action, out of judgment, out of knowledge slowly built up out of experience with real, physical things. Learning that never went through the stage of being symbolic learning, even though something happened in the mind of the child that is probably symbolic.

Today out-of-school experience with physical reality is no longer true for most people. New requirements are laid on schools, where you must make sure that all the pieces are brought together that are necessary for the next generation. We see that the schools have come upon a new problem that they have not brought forward or faced squarely— the problem of enlarging the task of education in school beyond the teaching of symbolic material.

If you seek to learn the Bible or the Koran, there is no way that experience can bring it to you. There is no way you can see it in the sky or in the ground or in the well pump. You have to be given the Word. "In the beginning there is the Word," says the Book of James. That is characteristic. The book learning of nearly all cultures centers around a Grand Book, usually some epical essence of the culture's views.

People who learn the Book learn first by memorizing the wisdom and the miscellany that pass in the name of Scripture in their cultures. The enormous gain was that it formed a community, while behind it there was for the most part the life of the subsistence farmer.

There needs to be an experiential and challenging connection between individuals today and the natural world. Phylis Morrison has argued that this task must be taken up by the schools as institutions, as well as by less formal institutions like museums, clubs, and so on. Schools should give children what children need, which is different from what the farmer's children needed when Abraham Lincoln walked long miles from his farm to school.

It's our indictment of the schools that they don't see this as a central problem. Images and words flow out electronically now at such a rate, so free of central direction, that the schools can hardly match them.

Phylis Morrison: But as the schools come to science, they typically reach for science as part of the symbol system. We see again and again that the symbol is used to replace the substance, the experience, of the reality of things.

We visited the marvellous country of Yucatan (Mexico). There were no cattle in Yucatan until post-Columbian times, and there was not much use of milk except for mother's milk. There were not suitable animals for milking. Milk for adults there is a rather exotic substance. (Even in Spain, which is certainly European enough, I was taken a little bit aback by the same fact.) Of course in India, where the cow is everything, society revolves around dairy products. The Hindi word for what we could call "butter" is even simpler. It is only a one-syllable word—"ghee"! (Ghee is clarified butter, but that's what they use.) In our culture, "butter" is a good old clearly Anglo-Saxon word. Margarine is not. It is so new a product that you can hardly understand that word at all. I hardly know how to pronounce it; I certainly don't know what it really means. In Spanish, you have to ask for a simple thing like ghee or butter by asking for "mantequilla." That is a society not quite at home with dairy products!

So we came to consider a couple of symbols—symbols from common parlance, from the common language, not mathematical symbols—and to unpack them in the style of science. These two words are "milk" and "water." In the schools they are very close to each other. You might learn them in the same lesson. Now nobody will deny that these are very well-known symbols. Everybody with the slightest claim to literacy or to speaking the language knows what milk is and what water is, and will not fail to distinguish them, and will not confuse them. They're really rock bottom. But are they in fact in any way parallel categories of reality? I don't think so. From the point of view of natural science, they're utterly distinct references, utterly distinct meanings.

The concept of water, which goes with an adjective, "pure water," is well-defined in everybody's mind. There's a clear, ineffable substance—you may not have seen it, but you can approximate it—pure water. And then, of course, there is brackish water, salty water, muddy water, Coca-Cola, etc. They are a lot to do with a thing they tell you has a chemical formula (it's really a mixture of associated molecules, nearly all of which are H_2O clusters, with an occasional DHO in there). But we can pretty well identify pure water. Ordinary water is an approximation of pure water.

But now if you ask yourself what do you mean by pure milk, you've set yourself into a philosophical quandry. To most of us, pure milk would mean milk that somehow came from a cow, but was handled with care by the chain of those who brought it to you. But otherwise, what is pure milk? There's a wonderful thick book by a certain Mr. Scott, an English expert, called *Introduction to Cheese-Making Practice*. Once you have thought of cheese-making, you begin to appreciate that milk is not just a simple, single thing. The first important sentence of the book says, "Milk is not, as some people imagine, a defined substance—M-I-L-K—to be injected into chemical formulae to make cheese." That will never lead you to cheddar, brie, or anything else you can sell at a good price. You can be sure that when you put milk into a recipe for cheese, it has to be a specific milk. The milk from one corner of one department of France doesn't make the same cheese as that from the corner of the next department. There are many reasons for this that you might want to elaborate, but here I make only the point that the grammatical category is based on superficial experience. In order to come to grips with the substance of the world you need to unpack the symbols for the world, to find in them the reference to the richer real world. Then you see that a single noun that we use for social convenience has a complex reference to the natural world.

Begin with milk. I would ask anybody to name some attribute of milk besides purity. Yes, whiteness is bound to be one. "Whiteness" and "purity." Atoms would be way up at the top, but an atomic description of milk is still a long way beyond us. It takes a thick book even to begin. Only in the last four or five decades, with the enormous growth of biochemistry and microbiology and microscopy, have we come to understand a little bit about milk. As to the scientific account of cheese-making, until recently the way you learned to make cheddar or brie was to go and apprentice yourself to the best brie-making farmer in Brie. There wasn't any way to go to Cornell and find out. You could go there and make cheese all right, but by God it was sure not brie.

Phylis Morrison: When, in upper New York State, the Kodak people were making their first industrial batches of photographic film, they made gelatin from the bones of local cattle. They realized that some batches of film were faster. Using this film, you got your image in the photographic emulsion without having to open the shutter of the camera for as long. This was very troubling because Kodak really wanted to be able to tell its customers how long they would need to leave the lens open. So Kodak paid attention to everything going into that film and discovered that the film which was made with gelatin from the cows of *certain* farmers was the faster film. What was different about this set of cows? It was yellow mustard wild in the pasture. The little yellow field plant was eaten with the grass. Mustard somehow changed the property, the delicate molecular quality of the light-sensitivity of the film made with gelatin binder.

It turned out it was sulphhydryl radicals—molecules of compounds that come in mustard and that give it bite. These make impurity centers in the crystal grains. That's where photo development begins. Now they don't have to find cows that eat mustard. They can add it as they wish. This is a typical situation.

Our vision is not, of course, a property of the microscopic world, just as cameras are not of the macroscopic world. Retina and film work on light. Light is, if I may use the word broadly, atomic—the photons that do the job are not things you can localize and package and carry around like beads. They're physical but microscopic. Just as the actions within cells are molecular, in the same way whenever you deal with vision and with photography, you are dealing with the microscopic structure of the world, with the fundamental particles where ordinary experience doesn't suffice. You have to start dealing with symbols. You confront a world of manipulation which is more difficult than you can understand easily, like the mustardy gelatin, like the truth that even if you start with two milks that look alike and might even taste the same (though taste is a little closer to microscopic analysis), you don't get the same cheese. When you process milk, depending on what you do to it, you bring out different properties. Milk is not a single thing. It has properties evinced by what you use to make a different kind of cheese.

There's one simple process everybody knows because it happens spontaneously in containers of fresh milk before it spoils. Something that happens in hours, not in days—separation. I'm afraid we've forgotten because the milk we get doesn't separate (how they do that is a good trick, by the way), but separation leads to creamy top milk. That is, of course, the simplest sign that milk is not a substance. It's more an ecology of many interacting populations in different amounts. You can't even tell the way you used to: when the bottle used to come there was a line of yellow stuff on top, though our experience of milk is mostly an experience of whiteness.

Milk is not a substance. The word "milky" is the recognition of the visual experience, a recognition of whiteness. That is shown in *the* language. There is soybean milk, milk of magnesia, and milkweed; there are milky fogs and clouds, and milk-white skies. In a certain social context that whiteness correlates strongly with a thing you call milk. It comes out of a cow and you can drink it and it's good for you. By the time it enters the kitchen, our experience is very thin. You've got to trust a long chain of intervenors if the white fluid you're going to get out of that bottle is to be the thing you want.

As soon as we think of the *function* of milk, there's another thing that the word immediately connotes. It has a more specific function for us than water. Water is indispensable, but milk is nurturing. All human beings, with the rarest biochemical exceptions, have experienced this nurturing property, since we are all mammals. Nurturing, mother's milk, leaves the sense that milk is a precious and marvellous thing. Here that is not so strong, for milk is such a commodity. But in India it has salience. There is a saying that you dare not refuse the offer of milk because that represents an affront to Lakshmi, the Goddess of Benevolence and Good Fortune. Who would want to offend that powerful and coy deity? So you have to accept milk whenever anyone offers it.

By the way, it is fair to say that in India, from ancient experience, the use of raw milk is scarcely known. It's always cooked, minimally. Of course, cooking plain water does something for you, in fact very much the same thing. The boiling of water for safety, like the boiling of milk, suggests at once that water, which is not a pure substance, has something unstable in it. Milk has that *par excellence*.

What is milk? It is 85 percent water. That's not the essential part of it. In fact, when you take the water out, you have powdered milk. Powdered milk is milk less water, usually also less fat. (You can make whole milk powdered but that doesn't keep so well.)

What is it that makes the whiteness of milk? Of soybean milk? Of clouds? And allows you to have separation? And fatty powdered milk? And nonfat dairy milk? And all such things? The fundamental physical fact about milk is that milk is not a stuff like water, a substance on the atomic-molecular level. Milk has large-scale little balls in it. They are not large on our scale, but large on the scale of atoms: little balls, little marbles, a few tens of thousandths of millimeters in size but lots of atoms, a trillion atoms in each drop. These spherical drops sitting there in the milk scatter the light randomly. They don't absorb it because they're perfectly transparent. So the light comes out, as much as went in, but now in all directions. There is no color change. What we see we call "milky white." The yellow is the second kind of tiny globule, butter fat, that scatters yellowed light.

There's more than one way to skin a cat or to separate globules from solvent. This works not only for milk, but also for other milk-like things. Milkweed juice and latex paints and fogs are all the same thing, little transparent globules floating in another transparent stuff—water or air—so the result is that the light doesn't go anywhere. It bounces off and comes back to you. That's what we call white: when it comes back to you. If only some parts of it come back to you, it might be a color like blue. Once you take out all the larger globules and leave only a certain number of protein globules that are smaller and less numerous, they scatter blue better. Skimmed milk has this bluish cast; its yellow fat is missing.

Casein is the name you give to the little globules of milk. When you separate out cream from milk what you have is whey. The skimmed fluid contains mainly casein and water. Take away the water, pack the globules in a box, and sell it—casein. Dry it and it's something that can make wonderful glues. You can feed it to animals or people and you can do a lot of fine things with it, like make milk powder.

Phylis Morrison: Old barns were painted with milk, all those old red barns.

Yes, milk is a very good vehicle for paint. You have to add red lead to it, so it gets red, but it reflects the light. Its reflectivity and the pigment make a very good paint. The protein forms a hard film, like Elmer's glue, which is in fact made from the same thing.

Phylis Morrison: In many parts of the world, they don't make cheese, but they do take soybeans and grind them and mix them up with water and that makes a white substance that looks an awful lot like milk.

Which you almost always call "soybean milk." I think there's a single word for it in Chinese, but in English we translate it as soybean milk. Soybeans are not mammals, but they have something like milk.

Phylis Morrison: It's a complicated mixture. It's got a carbohydrate part and a protein part and if you want to make soybean curd, what you want is to throw away everything except the valuable protein fraction. How do you separate it? Well, it's a protein-chemistry magic, like cheese. You take a little bit of plaster of Paris and you sprinkle it in—not much, a pinch to a quart. It's very good for you, too; it has calcium. You do this and, magically, a white stuff settles to the bottom. Above there is a whey and at the bottom curds of coagulated globules, only they're soybean curd, instead of milk-protein curd. If you can't afford to feed the grass to the cow so the cow can give you back 10 percent as milk, then you want to get that 90 percent yourself, and you just take the bean out of the field and that's how it's done.

You take that same plaster of Paris and put in the milk and the same chemical magic occurs. You get a beautiful separation into curds and whey (remember Miss Muffet?). So soymilk is a lot like cow's milk.

Yes, and soybean curd is a lot like cottage cheese. It has a somewhat different origin. The amino acids, the chemical spelling of the "words" of the long protein chain, are different. Soybean is nutritious and gives you a higher yield per acre. That's the way China is going. You've got to like it though.

Phylis Morrison: In those soybean cultures there is a range of soybean products as different from each other as cheddar and pot cheese.

There's an indigenous process, especially in Java and Southeast Asia, of fermenting the curd to go from soybean farmer's cheese to soybean cheddar, brie, Limburger, etc. You grow molds in the bean curd and treat it, press it, cut it, age it, and you have a dozen different foods all made from soybean curd.

In the watery whey there is still in solution about 5 percent of lactose. Now the strange part is that all mammals start with the ability to digest lactose, which is milk sugar. They have enzymes built into the cells. They have to, because all mammallian milk contains lactose and they need to live on milk for some time. Then their chemistry changes. Nearly all mammals lose the ability to easily digest lactose at the time of weaning. Most adult human beings cannot, in fact, easily digest fresh milk and won't drink it. It makes them uncomfortable. But we of the cow world— that means Americans, Europeans, and Indians especially, and some Africans but not many people of East Asia—have the lactose-digesting ability. When we grow beyond weaning, we keep on making this enzyme and we still can digest lactose.

That's related somehow to historical truth, for people from Central Asia, people of the Indo-European tongues, associated themselves with cattle in the very early period in human history. Probably selection produced a large number of lactose-digesters in adult life among our population. The biochemical tests show that 85–90 percent of the U.S. population can digest lactose as adults. They like ice cream, etc. But some can't. In cheese, the bugs, bacteria and molds, eat up the lactose.

That's why cheese isn't sweet. It has that typically acid or strong flavor that comes from getting rid of the lactose. Only the rare cheese doesn't have that quality.

Interestingly enough, for example in West Africa, where most people cannot manage lactose and don't use fresh milk, milk-cattle-rearing groups, particularly one call the Fulani, are very widespread. They sell fermented products to the others, who are much more numerous. But they don't sell them untreated milk. They sell them cheeses and yogurts and stuff like that, from which the lactose is gone.

I think you can see that once you define milk in words as a white substance, the secretion used for nutrition, you've missed a great deal: all that comes from studying natural science and the nature of milk, its origins, its biological nature, its functions, and, of course, its place in history and the myths and metaphors it has induced.

Milk is nowhere in schools essentially, except in a lunch box, and maybe in the nutritional admonition to drink a pint a day (which is not good for all children, in fact).

I think we should leave milk and water. Don't we get to say something about computers? I think everyone knows those crazes of American life, where one thing becomes incredibly important for three years, and at the end of three years you never hear of it anymore. Computers are more worthwhile than that. They will persist because, for me, they are manipulators of symbols. The image most people have of computers is that they compute—they have something to do with mathematics, with arithmetic, with some more elaborate forms of mathematics.

I brought a home computer and what I've learned from it is that, in fact, the computer is a symbol manipulator, *par excellence.* It deals with language. Everybody knows that in an office, because there the computer processes words. It deals with words in an extremely interesting and flexible way. It has no knowledge of the meaning of words, but it will rearrange, it will tie these symbols together, if you show it how. It will sort them by any attribute it can measure in the word itself.

It can even signal you when you overuse the same words, and things of that sort. It's a tireless comparator, list-maker, and associator. Since I began by saying that the schools deal with language, reading, writing, and arithematic, I'm convinced that schools are made for computers and computers are made for schools. But not for science! In the real world, the number of computer screens that show mainly words is much larger than the number of screens that deal with graphs or mathematics. The chief mathematical use is showing the number of the flight, the number of hours it takes to go from here to there, and then maybe the same sort of thing for the payroll and the stock list for the businessman.

Of course, the computer does a heavy computing job on the weather, on aircraft design, etc. But these are not its principal uses. Nowadays, the department computer is crowded with letters and papers being composed and recomposed by various people who have to write, to express in word symbols, what they have found out by study involving maybe some other computer, but not that particular one. I'm enthusiastic about computers, but I don't think they represent very much in science, unless you take them apart. There's a lot of science inside a computer. But that's one thing you'd better never, never do—take it apart. It's really not rewarding, unless you have a microscope.

Phylis Morrison: I took a chip apart and put it under a microscope. Part of the assembly language that I really want to understand is what that thing is doing when I say "run" it.

How would a computer be used to unpack a concept like milk? You'd have to put everything into it that's associated with the word "milk." But then milk becomes a vapid thing. It doesn't have any of the evidence or the development in it. It's useful to have some of the associations. After all, we did outline it, and that could be recorded. But that's not the way science is done. Science is how we happen to know all these things. Of course, you can't know every one of these things by direct experience, but you have to have some way to begin. You have to point out at least some of the steps which lead to this broad introduction to a world of diverse experience.

Learning to use computers may be very valuable for setting kids on the right path into the symbol-using world, a very important part of the world socially. They may work in that world. That's important and perhaps the computer will be a key. But it is not a key to science. It is not going to bring anyone—rich or poor, city or country, suburban or inner city—into contact with the structures of the mind and the work of the hand which cumulatively developed science and the technology out of which computers grew. Science isn't just deftly putting together chips that look like little bugs with 32 legs into tiny little sockets just the right way—that's what most people who work in the computer world without skill or training do and it's low-grade assembly work, requiring a certain seriousness and deftness of hand. I don't mean to denigrate it, but it is not open-ended. It doesn't lead you to understanding. It leads you to working carefully, to a routine with your hands. That's not a route into power in our society, which really involves the manipulation of symbols to instruct, to inform, to direct.

I don't object to the use of computers in schools. They'll have to be there. They'll enter schools more and more. But are they helping children to read? Are they helping *the children* to write and spell? How are they helping children to figure and learn arithematic? Are they replacing spelling drills with computers? That's a good question to ask. They do not belong in science primarily. You may find a way to use them in science classes, but students better do more than just punch keyboards, because keyboards are not science, except inside and maybe in the software.

Knowing Where You Are

Before the Kalahari Desert was divided by a high wire fence and patrolled by armed jeeps, the San people dwelt there, as some still do, following their unending seasonal rounds under the sun and the stars. All that they materially possessed they carried with them. That inventory was small indeed, but their minds were full. Wanderers, though never aimless, they depended upon and enjoyed an intimate knowledge of every feature of their wide and lonely range. Every knoll, every rock, every sparse patch of growing things, bore its proper name and its own bundle of memories. As the band made its way, the very landscape steadily evoked a shifting and lively conversation. They could explain their untiring interest in the world, too: "We always like to know exactly where we are," they said.

Enlightenment philosophers like John Locke held the view that the human mind is given as a clean slate upon which a suitable scheme of education may write whatever it wills. That view is wrong, wrong for desert gatherers, wrong for schoolchildren, wrong in particular for student teachers and for their more experienced colleagues. Certainly by the time the individual has mastered any natural language, the mind's slate is well-formed and full. Plenty more can be learned, of course, but the learning process is not at all like writing on a blank piece of paper. Rather it involves annotating, amplifying, enriching with example, supplying new procedures for entry and confirmation, adding links and illuminating comparisons between portions earlier seen as distinct. Sometimes, to be sure, the blue pencil is needed, and cutting and editing improve what is there.

No one comes to a college class without a large body of quite workable theory. For not only wanderers but most of us in modern society also seek to know exactly where we are. Not knowing is uncomfortable, often dangerous. Language, even old saws, custom, direct or half-remembered advice, everyday experience with a multitude of images, locutions, daily tasks, and interaction with a variety of people, intimates and strangers, provide everyone with a conceptual and generally factual framework for coping with the world. We call it common sense.

Quite a good deal of that treasury is built-in to the circuitry of the brain, and more is selected out of experience. Consider, for example, the perceptual hypotheses that underlie most of the visual judgments we all make to cross a busy street or to catch a ball or even to reach out for a pencil. Robot designers have a way to go before they can emulate the structure of kinematic and topographical theory held by an active five-year-old, most of it of course not at all open to easy expression, but constantly seen in action and often recoverable by study.

Common Sense and Its Correlations

The critical barriers that David Hawkins describes are a working part of the well-furnished mind. Some of them at least are remarkably unvarying. He has explicitly shown that some critical barriers met in the Boulder classes led by David and Frances Hawkins express the very views of Aristotle and the schoolmen, the cause of difficulties for beginners in science that are often more grave—and certainly more interesting—than the merely "pedagogenic disorders" also reported in plenty.

There is little mystery about this persistent structure. The social anthropologists, the linguists, and the psychologists find it as grist to their many mills. For one who does not engage in producing or managing the technical bases of life in either epoch, daily tasks differ little between old Athens and today's suburb. Routine physical chores, reading and writing, the complex web of relationships with family, friends, even pupils in the classroom: all those depend upon many skills of hand and eye and on habits of mind and speech, skills that are predictive, effective, and widely shared. But they do not differ so much over the millenia. There are more images today; the TV set is prolific as the fresco painter never was, but to one who has only to regard an image, and takes no more part in its electronic distribution than to buy the set and push the *on* button, only the nature of the surfeit makes much difference. The screen is otherwise like a decorated page.

Both the necessary perceptions and the deft handling of the essential artifacts of daily life do not much differ whether the vessel of cool water is formed of clay or of the latest petrochemical plastic. Using the telephone and driving a car are unique to our time, but the swift hints of the conversation itself and the quick reflexes so neatly trained do not add much new to the demands of life as it always was for those so placed in society that they did not daily make some specialized tool-using intervention into the physical world. Human relationships carried through stance, word, gesture, and deed, differ even less. So arises common sense.

What sort of content does common sense hold? The air is invisible but always ambient; when one needs it freshened, the window is opened. Yet a glass ready on the shelf is, and always was, regarded as a matter of course to be empty, never as filled with air. A dropped ball comes to rest somewhere nearby, and cannot bounce along forever. By day, light is present to make the space useable; when daylight is lacking, one turns on the lamp, a notably shining supply of needed light. Workable, often complicated and subtle judgments about homely detail tell us whether things will fit, can be moved, will hold water or keep out the cold. Little needs to be measured or drawn; at most a length of cloth is suited to the width of a table. The stick of butter and the cup of sugar are the most used metrics; the gallon of gas and the passing time are digits well displayed. Images dwell on the page or the screen or on the mirror surface; what we most often see in a mirror is placed directly before us. Trees grow taller over the years, and by their nature stay somehow in proportion. One senses too hot a room, or too cold, and makes appropriate changes in the circumstances. Perhaps a thermometer aids that judgment. Road signs and speedometers work roughly for anyone able to read. Other instruments are rare: neither a shadow marker nor a measure of wheat are as common as perhaps they were in everyday Athens.

The conclusion can be summed up in the remark that a complicated set of judgments based on experience and reason has always gotten people through many days. The matters at hand are ordinary enough so that a subtle correlation of size, nature, motion, energy flow, or whatever it may be, is almost always present. The appropriate presence is assured by built-in design, mainly due to others. Perhaps it is based instead on the distillation of long experience, enabling common sense actions to work with good probability. They seldom need explicit calculation, nor is there any desire to pose sharp logical tests of the comfortable and usually adequate presuppositions for action.

These everyday encounters—and the theories, explicit or unvoiced, that alike arise from and motivate them—are rarely analytic. More often they are syntheses, built of several mutually weighed discriminations. They are diverse, not often seen as unified by logical inference. They come out of practice, and nowhere is much effort made to fit them into a simpler whole. In general that issue does not even arise, for what is involved are rough conclusions about a wealth of distinct details, rarely step-by-step paths to long-pursued ends. They have more to do with the comprehension of whole events than steps of understanding. The cup drops on the tiles and breaks at once, or else it remains intact after a fall to the carpeted floor. There is no need for, or gain from, a sequential analysis during the microseconds when the crack is spreading around the glass. The light goes on, or the engine starts, once the right switch is activated. That chain of familiar but unseen events is not analyzed, but taken as the effect of one initial cause, one willed action of the hand. Often small accompaniments are needed, found by precept or experience, that improve the result. A key is to be held a certain way, or a lid pressed a little to one side. That is often the closest the actor comes to reflective analysis.

But the complicated everyday world works, mostly, and therefore the point of view we take of it is naturally satisfactory. Neither generalization nor objectivity nor precision are very important to the common-sense frame of mind. Given some willingness from time to time to improvise and modify, usually by trial and error, once faced with a new artifact or condition, some novelty can be incorporated. That new learning is soon naturalized by repetition within the old domain of common sense. Anomalies, all the way from transient minor puzzles to ghosts and miracles, are rare; even when they occur, they are mere happenstance in a complicated environment, to be weighed against a lifelong chronicle of the successfully ordinary. Let a real mirage or a deft magician bring in something genuinely unexpected; even then we are prepared by experience to discount the unusual event. After all, it is unimportant in the long run.

Such is the theory of knowledge we all share under the rubric of common sense. It is the expected general correlation among varied experiences that validates our views: the light is almost always there in the room without delay once we snap the switch or kindle the candle. Whence and how light moves is not even asked and would indeed be hard to answer through our commonplace perceptions.

Beyond the Range of Common Sense

In context, there is plainly nothing wrong with the view of the natural world that seems embodied in what I have called common sense. It is a broadly workable first approximation to a limited, but extremely important, set of phenomena; what you could call the everyday. It is, moreover, strongly reinforced conceptually by the common language, for what are pretty surely deeply evolutionary reasons.[1]

But an understanding of even the beginnings of natural science requires one to transcend this range of experience. Is it the sky? Distance and gravity and inertia rule its perpetual motions, all outside of usual sensory reach. Is it the perception of light? Time is, so to speak, absent for vision, yet light propagates, as the physicist knows but common sense never needs to admit. Is it air? Invisible, it nevertheless has weight and major chemical effect—life-giving, in fact. Breathing is hardly comprehensible on common sense alone, a strange exercise of in and out, changing nothing, yet somehow intuitive and essential. Is it weight? Fundamental for anyone who changes the form of matter, every recipe touches upon it. The craftsperson who works at larger scale or with any but the most customary and routine materials depends on it pretty strongly. Its remarkably simple linear behavior violates all the rules of more complex and common-sense quantities; it never saturates, it never hits diminishing returns, it never is deceived by screens or walls. (The buoyancy of air does affect it at a delicate level.) Heat? Classroom studies make plain how complex a perception that becomes, once experience tries to deal with anything beyond the tried and true arrangements of clothing and space heating. The camper in winter, the firemaker, the careful cook, already need more of a guide to heat transfer than the feel on the hand.

Bumping Against the Barriers

It seems useful to cite examples of collisions with critical barriers. In a way here is the very heart of the matter, for the recognition of such collisions is usually the chief basis of their identification. If what I see is borne out, these collisions do not represent so much the inadequacy of logical categories and operations—something akin to the stages made famous by Piaget—but rather the systematic, if less than conscious, application of workable and even tested common-sense theories within domains where their utility fails. At least one major cause of such misapplication is the lack of experience in the novel range of phenomena. A second factor appears to be the powerful effect of ordinary language, whose statistically reliable connotations often mislead a user extrapolating to a new domain.

Consider a few cases of this curious, yet entirely explicable, narrow-

---

[1] I am referring not at all to biological but solely to linguistic change.

ness of everyday theory. One student felt that the shadow could not be described as mere absence of light. It made so strong a mark, she felt, that it must be an active presence. Yet striking graphic results are produced every day by simple tricks that merely keep out the marking substance in order to produce a figure-ground contrast. The cave painters made pigment patterns around hands placed on the wall! Painting with stencils and resist dying and a dozen such tricks would suggest that contrast is a means to a recognizable mark made by nothing at all. Experience with contrast can extend the common-sense generality that strong effects need a physical carrier: the perceptual difference is physical, but the figure itself may be a sign of absence.

The travel of light is not rigorously demonstrable without explicit instrumentation, no doubt. But it is made pretty suggestive by the use of a small hand mirror to direct sunbeam and lamplight into dark spaces, or the view of darkness outwards to the lighter space. A cloud of chalk dust makes the idea of a ray more real; talk of direction of view might strengthen the insights of those who felt, quite as do the expert mineralogical and chemical guides, that the milky appearance we call white and the clear passage through substances we call water-white are two distinct colors. The *white* light of the simple text accounts of color vision does not refer to milk-white at all; it is a jargon name for light that resembles the unbiased color balance of sunlight, called white in a second sense, that it is not the deliberately filtered green or red light of color experiments. Here it is language that enters to make marginal experience ambiguous.

As for scale, it enters everywhere, below simple perception and above it, too. Moving heavy weights or painting barns or building kits or using a crystal radio or seeking to explain breathing and fire all bring home one way or another the domain of scale and its dimensional attributes, indispensable to science down to the atom and out to the stars, and really prior to both.

Widening the World of Experience

There can hardly be a better way to lift the common-sense barriers than to extend experience beyond the everyday. Almost any direction that extension takes is full of gain. To paint the floor, to build a kite, to study arithmetic and geometry, not with symbols alone but with rods and string and paper, to cook for a dozen at table, safely to move a piano or a sand grain, to make a blueprint image in the sun . . . in short, to open yourself to new experiences, not carefully worked out by others beforehand, allowing time and concerted attention enough to permit explorative trial and error. The world to search is the world of almost every craft and sport pursued for utilitarian reasons or for simple enjoyment; but it must be entered across some easy threshold of action and not by images and symbols alone.

Did someone once refer to messing about? It is messing about—not turned to a single clear purpose, but focused on some real physical system at hand that offers the best crossing of barriers, to my mind. For messing about with boats, for example, can and will come to include knots and waves and reflections in still water and cold hands and the spawning of frogs, and even adzes and map projections, if given time to grow.

If the prescription sounds as much like art and craft and sport as it does science, all the better. There is need as well for the more academic "language arts" (if the jargon persists). What is called for there is practice in the communication of concept and action and values by a wider variety of means than words alone. Diagrams, maps, sketches, teaching of some manual skill hand on hand: those belong to human communication as much as do reading and writing (not to forget written recipes and protocols), and are perhaps even more relevant to the beginnings of science.

Finally, we scientists and especially authors of textbooks need to lend a hand in this process. We too often use time-honored words that connote metaphors not to be grasped, like *heat flow*. We freely employ diagrams and sketches without serious effort to develop a common visual grammar. We grow impatient with approximate description and partial understanding even on the way to a working grasp. In arithmetic instruction, it is notorious that the precision of algorithms good enough for accounting or number theory, but employed instead to overspecify rougher statements about the world, has induced a widespread and chronic pedagogic disease of right answers.

All that can be changed. Change requires an understanding of real learning in the classroom tells. There is a good wind astir, perhaps only Force 3 right now: "leaves and thin branches move constantly, a flag flutters." But the wind will strengthen. It must bring more than symbol, more even than images on the screen, however nimble and interactive they are made. We need to place hand, eye, tongue, and mind all together to work upon the real world. We need to invoke the shimmering variety of experiences that border upon and can extend the complex but well-worn patch of daily life for which the student is so well prepared by common sense.

People indeed like to know where they are; the trick may be to lead them to many agreeable places, until they recognize that they too can begin to know and to feel at home in almost any domain where other human beings have dwelt in pleasure.

WAR AND PEACE IN THE AGE OF URANIUM

W hat had come to chemists with the poison gas of 1915 came to physics with uranium fission in 1939. I recall vividly the year of 1938, when we stayed up till dawn in Berkeley to hear Hitler's hoarse threats at the noontime rally in Nuremberg, eight time zones away. Then came the spring months of 1939, a wider war patently close. We graduate students began to draw and redraw caricatures of The Bomb on the blackboard in LeConte Hall, We knew precious little about neutron diffusion, but our fears were well-based: explosive power would soon be increased by six orders of magnitude.

I followed the Manhattan Project in person almost all its days, from the week of Fermi's first chain reaction in Chicago, on to the desert test of the bomb, to the air base at Tinian, and beyond that to the rust-red rubble of Hiroshima. At war's end most physicists knew that a global disaster lay in the nuclear arms race, a risk our leaders harbored for four decades. Through a kind of knight's move in the chess game of unfolding history, the threat of large-scale nuclear war has now faded at least to implausibility. A vast relief is widespread but somehow still tacit. The glorious news is not fully celebrated, perhaps both because the outcome is not securely at hand, and because our tiring effort of denial over so many years has left us still fearful of much smaller threats.

These half-dozen articles sample what one witness had to say during some years of anticipation of nuclear war, one week of its embryonic reality, and four decades under ominously thickening clouds.

If the Bomb Gets Out of Hand

W e sat in a small open wooden hut, like a booth at a church fair, listening to the Japanese General Staff major from Tokyo. Around us the ground was blackened. The trees were strangely bare for September beside the Inland Sea. The advance party of the American Army mission to study the effects of the atom bomb had come to Hiroshima. In the rubble of the castle grounds, the old headquarters of the Fifth Division, the local authorities had prepared for us a meeting with the men who had lived through the disaster of the first atomic bomb. The major was very young and very grave. He spoke slowly and carefully, like a man who wants to be properly translated and clearly understood. The story he told is worth hearing. It is the story of the first impact of the atomic bomb on the structure of a nation.

About a quarter-past seven on Monday morning, August 6, the Japanese early-warning radar net had detected the approach of some enemy aircraft headed for the southern part of Honshu, and doubtless for the ports of the Inland Sea. The alert was given, and radio broadcasting stopped in many cities, among them Hiroshima. The raiders approached the coast at very high altitude. At nearly eight o'clock the radar operators determined that the number of planes coming in was very small—probably not more than three—and the air raid alert was lifted. The normal broadcast warning was given to the population that it might be advisable to go to shelter if B-29's were actually sighted, but that no raid was expected beyond some sort of reconnaissance. At 8:16 the Tokyo control operator of the Japan Broadcasting Corporation noticed that the Hiroshima station had gone off the air. He tried to use another telephone line to re-establish his program, but it too had failed. About twenty minutes later the Tokyo railroad telegraph center realized that the main line telegraph had stopped working just north of Hiroshima. And from some small railway stops within ten miles of that city there had come unofficial and rather confused reports of a terrible explosion in Hiroshima. All these events were then reported to the air-raid defense headquarters of the General Staff. The military called again and again the Army wireless station at the castle in Hiroshima. There was no answer. Something had happened in Hiroshima. The men at headquarters were puzzled. They knew that no large enemy raid could have occurred; they knew that no sizeable store of explosives was in Hiroshima at that time.

The young major of the General Staff was ordered in. He was instructed to fly immediately by army plane to Hiroshima; to land, to survey the damage, and to return to Tokyo with reliable information for the staff. It was generally felt in the air-raid defense headquarters that nothing serious had taken place, that the nervous days of August, 1945, in Japan had fanned up a terrible rumor from a few sparks of truth. The major went to the airport and took off for the southwest. After flying for about three hours, still nearly one hundred miles from Hiroshima, he and his pilot saw a great cloud of smoke from the south. In the bright afternoon Hiroshima was burning. The major's plane reached the city. They circled in disbelief. A great scar, still burning, was all that was left of the center of a busy city. They flew over the military landing strip to land, but the installations below them were smashed. The field was deserted.

About thirty miles south of the wrecked city is the large naval base of Kure, already battered by carrier strikes from the American fleet. The major landed at the Kure airfield. He was welcomed by the naval officers there as the first official representative of aid from Tokyo. They had seen the explosion at Hiroshima. Truckloads of sailors had been sent up to help the city in this strange disaster, but terrible fires had blocked the roads, and the men had turned back. A few refugees had straggled out of the northern part of the town, their clothes and skin burned, to tell near-hysterical stories of incredible violence. Great winds blew in the streets, they said. Debris and the dead were everywhere. The great explosion had been for each survivor a bomb hitting directly on his house. The staff major, thrown into the grimmest of responsibilities, organized some two thousand sailors into parties, which reached the city about dusk. They were the first group of rescue workers to enter Hiroshima.

The major took charge for several days. The rail line was repaired, and trainloads of survivors were shipped north. The trains came first from Onomichi, where, about forty miles north, there was a large naval hospital. Soon the hospital was filled, and its movable supplies exhausted. Then the trains bore the injured still farther north, until there too the medical facilities were completely used up. Some sufferers were shipped twenty-four hours by train before they came to a place where they might be treated. Hospital units were mobilized by Tokyo to come from hundreds of miles to set up dressing stations in Hiroshima. One bomb and one plane had reduced a city of four hundred thousand inhabitants to a singular position in the war economy of Japan: Hiroshima consumed bandages and doctors, while it produced only trainloads of the burned and the broken. Its story brought terror to all the cities of the islands.

The experts in the science of the killing of cities have developed a concept which well describes the disaster of Hiroshima, the disaster which will come to any city which feels the atomic bomb. That is the idea of saturation. Its meaning is simple: if you strike at a man or a city, your victim defends himself. He hits you, he throws up flak, he fights the fires, he cares for the wounded, he rebuilds the houses, he throws tarpaulins over the shelterless machinery. The harder you strike, the greater his efforts to defend himself. But if you strike all at once with overwhelming force, he cannot defend himself. He is stunned. The city's flak batteries are all shooting as fast as they can; the firemen are all at work on the flames of their homes. Then your strike may grow larger with impunity. He is doing his utmost, he can no longer respond to greater damage by greater effort in defense. The defenses are saturated.

The atomic bomb is pre-eminently the weapon of saturation. It destroys so large an area so completely and so suddenly that the defense is overwhelmed. In Hiroshima there were thirty-three modern fire stations; twenty-seven were made useless by the bombing. Three-quarters of the firefighting personnel were killed or severely injured. At the same instant, hundreds, perhaps thousands, of fires broke out in the wrecked area. How could these fires be brought under control? There were some quarter of a million people injured in a single minute. The medical officer in charge of the public health organization was buried under his house. His assistant was killed, and so was *his* assistant. The commanding officer of the military was killed, and his aide, and his aide's aide, and in fact every member of his staff. Of 298 registered physicians, only thirty were able to care for the survivors. Of nearly twenty-four hundred nurses and orderlies, only six hundred were ready for work after the blast. How could the injured be treated or evacuation properly organized? The power substation which served the center of the city was destroyed, the railroad was cut, and the rail station smashed and burned. The telephone and telegraph exchange was wrecked. Every hospital but one in the city was badly damaged; not one was able to shelter its patients from the rain—even if its shell of concrete still stood—without roof, partitions, or casements. There were whole sections of the outer city undamaged, but the people there were unable to give effective aid, lacking leadership, organization, supplies, and shelter. The Japanese defenses had already been proved inadequate under the terrible fire raids of the B-29's, which had desolated so many of Japan's cities. But under the atomic bomb their strained defenses came to complete saturation. At Nagasaki, the target of the second atomic bomb, the organization of relief was even poorer. The people had given up.

A Hiroshima official waved his hand over his wrecked city and said: "All this from one bomb; it is unendurable." We knew what he meant. Week after week the great flights of B-29's from the Marianas had laid flame to the cities of all Japan. But at least there was a warning. You knew when the government announced a great raid in progress that, though Osaka people would face an infernal night, you in Nagoya could sleep. For the raids of a thousand bombers could not be hidden, and the fire raid had formed a pattern. But every day over any city of the chain there was a chance for a few American planes to come. These inquiring planes had been photographers or weather forecasters or even occasionally nuisance raiders; never before had a single plane destroyed a city: Now, all this was changed. From any plane casually flying almost beyond the range of flak there could come death and flame for an entire city. The alert would have to be sounded now night and day in every city. If the raiders were over Sapporo, the people of Shimonoseki, a thousand miles away, must still fear even one airplane. This is unendurable.

If war comes again, atomic war, there will not even be the chance for alerts. A single bomb can saturate a city the size of Indianapolis, or a whole district of a great city, like Lower Manhattan, or Telegraph Hill and the Marina, or Hyde Park and the South Shore. The bombs can come by plane or rocket in thousands, and all at once. What measures of defense can there be? To destroy the bombs in flight many measures will be attempted, but they cannot be a hundred per cent effective. It is not easy to picture what even one single bomb will do. We saw the test shot in the New Mexico desert, and we pored over and calculated the damage that a city would suffer. But on the ground at Hiroshima and Nagasaki there lies the first convincing evidence of the damage done by the present atomic bomb.

The streets and the buildings of Hiroshima are unfamiliar to Americans. Even from pictures of the damage realization is abstract and remote. A clearer and truer understanding can be gained from thinking of the bomb as falling on a city, among buildings and people, which Americans know well. The diversity of awful experience which I saw at Hiroshima, and which I was told about by its citizens, I shall project on an American target. Please do not believe that there is exaggeration here; this story will be conservative, it will allow for no increase in the effectiveness of the bomb. It will tell of only one where, if there is atomic war, twenty will fall. Your city, too, is a good target.

The microwave early-warning radar towers on the Jersey coast and up past Riverside had recorded the approach of the missile. It was 12:07 when they noted the end of the signal, and the operators wondered what the thing had been. When the telephone circuits failed and the teletype stopped, they grew worried. When they listened to the shaky and disturbed news report from WABC a few minutes later, they knew what had made the mark on the screen. One of the men walked outside with his camera and looked north in the bright noon sun to see the great pillar of cloud he knew would come. The wind had been from the northwest all day, and it is interesting to note that the radioactive cloud passed over the same radar installation which had first remarked the missile. The recording radiation meter at the station showed a harmless quantity of gamma radiation, but the photographic film was badly fogged.

The device detonated about half a mile in the air, just above the corner of Third Avenue and East 20th Street, near Gramercy Park. Evidently there had been no special target chosen, just Manhattan and its people. The flash startled every New Yorker out of doors from Coney Island to Van Cortlandt Park, and in the minute it took the sound to travel over the whole great city, millions understood dimly what had happened.

The district near the center of the explosion was incredible. From the river west to Seventh Avenue, and from south of Union Square to the middle thirties, the streets were filled with the dead and dying. The old men sitting on the park benches in the square never knew what had happened. They were chiefly charred black on the side toward the bomb. Everywhere in this whole district were men with burning clothing, women with terrible red and blackened burns, and dead children caught while hurrying home to lunch. The thousands of brick and brownstone walk-ups, huddled closely to the elevated and packed thickly between the rivers, were badly shaken in a few seconds. The parapets and the porches tumbled into the streets, the glass of the windows blew sometimes out and sometimes in, depending on the complex geometry of the old buildings. The plaster fell on the heads of the tenants, old floors and stairs collapsed under the terrible wind of the blast, and only the heavy walls stood to mark the homes. Closer to the center nothing much was left. Many of the narrow streets passing between the old five-floor brick or stone tenements were choked with rubble, until it was difficult to walk down the street. Here and there collapsed buildings had piled a great heap of pitiful debris, all the wares and effects of living, into a useless and smoldering jumble. Everywhere there were fires, usually licking at already useless wreckage, but making heartbreakingly difficult the escape of the injured and the slow work of the half-stunned rescue parties.

The elevated structure stood up comparatively well. All the elevated stations from Fourteenth almost to midtown were wrecks. The steps were gone, the flimsy flooring and the old baroque railings lay in the street below. Only the clean steel frames were for the most part intact. In the blocks near Twenty-third even the main frame had gone, and the twisted vertical columns remained above the nightmare of steel below. The loss of life was very large from this alone. A train had been pushed off going at full speed north on Second Avenue near Twentieth, and the flames which burned the whole of the district seemed to begin from the wreckage. A few concrete garages and warehouses stood up over the gaunt frames of the elevated tracks, but the whirlwind which tore through them left the interiors wrecked. Fire usually finished the job.

The great buildings were not destroyed; none had been very close to the blast. But they were not unharmed. The high Metropolitan Tower was the worst damaged. The steelwork stood unharmed nearly to the top, though it was badly twisted where a whole ten-story wall section had come down into the street. The interior partitions from the sixteenth floor and up were completely gone, and even some floors had failed, leaving a kind of half-filled honeycomb of a building above the twentieth floor. More than fifty people were later said to have managed to clamber down from the wreck. It is known that eighteen of the radiation deaths recorded in the St. Louis hospitals later were of people who had been in the higher floors of this building when the bomb struck. The people below the tenth floor were not fatally injured for the most part. Fractures and lacerations from glass were the principal cause of injury. A good many hundreds of people from the south side of the building died two or three weeks after the blast from radiation. Among them was a well-known aeronautical engineer who had managed to remain uninjured by the flash burn or the blast, standing as he was behind a steel beam column on the south side of the first floor, near the windows. He bravely worked the whole day as one of the rescue party for the Tower. The bad nauseous symptoms which he underwent at six o'clock caused him to seek hospitalization at Philadelphia, where he died in twelve days, while working on a report for the Air Forces on the extent of the damage to steel structures.

The Empire State building nearly a mile away was strikingly little damaged. The radio structures and the external ornament of the high spire were swept clean. The windows of course were shattered and much damage was done to the light partitions and even to the glassy exterior walls on the higher floors. Elevator machinery was badly damaged by a freakish falling beam and many were trapped in immobile cars. The flash scorched papers and window screens and set fires going in all the offices on the side facing the blast. These fires were brought under control in a day or so. For months after the blast the high tower seemed to stand defiantly at the upper edge of the vanished district, but the building was useless except in the very lowest floors. The tenants of the building had not fared so well as its steel and concrete frame; the great dressing station established in the corridors and rooms of the first five floors handled many of them, and sent many of them to the Police Department's common graves.

The underground world of the city had been relatively safe. When the power failed in the whole lower eastern Manhattan district because of the destruction of the transformer sub-stations, the subway power alone was restorable. The Lexington gratings collapsed, and near the blast one or two large street cave-ins had stopped traffic on the IRT and flooded part of the tubes from broken mains. But the greater number of subway passengers and crews escaped. A few hundred were trampled in a bad panic at the Thirty-fourth Street entrance, and one train piled into the wreckage below ground very near the aiming point. Some people walked north underground all the way to the Bronx, not believing it safe to come up any closer to the bomb. Men in the sub-basements of the great buildings were horrified when they came up to see why the lights had failed; they had known nothing of the great blast but a ground tremor and the dust of falling plaster.

The nearness of Bellevue Hospital to the blast—about half a mile— was tragic. The long brick walls collapsed. Only a few patients here and there survived. The doctors and nurses had no time to salvage even the carefully prepared emergency supplies. Fire attacked the ruin, and the scenes which followed are indescribable. The knocking out of Bellevue was a hard blow to the rescue organization of the city and delayed for some time the proper organization of relief.

There were many stories of unbelievable good fortune and magnificent heroism. One man, a glassblower apprentice, was walking along Lexington south to Twenty-fourth. He described the great flash, but he was protected from a direct view by the corner of a building. The blast knocked him down along the broad street, but no heavy object hit him, and he escaped without serious injury. All day and night he worked, leading the badly injured north and pulling many people from the wreckage. Though he was only a few hundred yards from the point below the point of impact, he suffered no symptoms of radiation injury. He was the only person on the streets of the city within a ten-block radius who is known to have survived without serious injury, and not more than a thousand of the hospitalized but recovering victims were as close as he.

The most tragic of all the stories of the disaster is that of the radiation casualties. They included people from as far away as the Public Library or the neighborhood of Police Headquarters downtown, but most of them came from the streets between the river and Fifth Avenue, from Tenth or Twelfth to the early Thirties. They were all lucky people. Most of them had had remarkable escapes from fire, from flash burns, from falling buildings. The people around them had never gotten away, but they had crawled, injured but alive, from the wreckage of homes or shops, from the elevated platforms, or from cellar stairways. Some had seen the great flash, felt the floor collapse, and picked themselves up ten minutes later from the rubble of their homes. Others had gotten free of the bus or auto they were in when it was thrown into a wall and had pulled out after them the dead and dying who had been their fellow passengers. They were all lucky, as they said. Some were dramatically uninjured, like the aeronautical engineer. But they all died. They died in the hospitals of Philadelphia, Pittsburgh, Rochester, and St. Louis in the three weeks following the bombing. They died of unstoppable internal hemorrhages, of wildfire infections, of slow oozing of the blood into the flesh. Nothing seemed to help them much, and the end was neither slow nor very fast, but sure. They were relatively few in number—the doctors quarreled for months about their census, but it was certainly twenty thousand, and it may have been many more.

The people far away who lived through the days of the aftermath suffered too. Homes and offices were systematically and badly damaged as far away as Fifty-seventh Street and Fulton Market, and across both rivers. Every block had its collapsed brick structures and its many walls carried away, and its dozen dead. There were not many windows intact on Manhattan Island, and there were many thousands wearing the face dressings that marked the target of glass splinters. But their lives went on, the damage was slowly repaired, and those who had no job to do there stayed far away from the scar that had been the Twenties. The rerouting of traffic and the repair of the telephone and the electrical and the water systems had its effect on the economic life of the whole city. The damage was felt in many ways as a drain on the recuperative power of the whole of New York City, and the loss of one-tenth of the people and the property of the city was enough to lessen the work of the city by half. People moved away and tried to forget.

The statistics were never very accurate. About three hundred thousand were killed, all agreed. At least two hundred thousand had been buried and cremated by the crews of volunteer police and of the Army division sent in. The others were still in the ruins, or burned to vapor and ash. As many again were seriously injured. They clogged the hospitals of the East and turned many a Long Island and New Jersey resort that summer into a hospital town.

There was no one of the eight million who had not his story to tell. The man who saw the blast through the netting of the monkey cage in Central Park, and bore for days on the unnatural ruddy tan of his face the white imprint of the shadow of the netting, was famous. The amateurs who collected radioactive souvenirs from the strong patch of radioactivity which sickened Greenwich Villagers for weeks were matched by those who found scorched shadow patterns in the wallpaper and plasterboard of a thousand wrecked homes.

New York City had thus suffered under one bomb, and the story is unreal in only one way: The bombs will never again, as in Japan, come in ones or twos. They will come in hundreds, even in thousands. Even if, by means as yet unknown, we are able to stop as many as 90 per cent of these missiles, their number will still be large. If the bomb gets out of hand, if we do not learn to live together so that science will be our help and not our hurt, there is only one sure future. The cities of men on earth will perish.

Physics of the Bomb

Nuclear Energy Controlled: The Pile

The Hiroshima explosion signalled to the world that the large-scale release of nuclear energy had been achieved. The spectacular and terrible detonation drew all attention to the catastrophic release of energy which in the bomb had been made possible. Even now that the facts of the matter are known, a belief remains that the "control" of nuclear energy is still beyond our means. Nothing could be further from the truth. The nuclear chain reaction has operated literally every day from December 2, 1942, until now. It has worked in one of the several existing forms of piles, or chain reactors, whether the uncooled pile of graphite blocks and UO_2 lumps which was man's first nuclear reactor, or the huge plutonium-producing units at Hanford which quietly warm the Columbia River. This is operated with a small reproduction factor so that the number of neutrons produced in a generation of the chain by a single neutron reaches unity *only* after the delayed neutron emission from the fission fragments has occurred. By this means the chain reactor is made into a smooth and trouble-free device. Probably no other machine can be made to operate successfully over such a great range of intensity, and stabilized at any operating power level allowed by the engineering design by the simple motion of a single control element. No complicated moving mechanism or delicate vacuum system forms part of a chain reactor. The steadiness and smoothness are noteworthy; the system is inherently stable. And why not, for the only moving parts are neutrons! At both the Argonne and the Clinton piles, one minor problem has been to provide a routine of meter-reading and recording exacting enough to ensure that the man who is operating the pile does not find it too easy to fall asleep. The normal demands of keeping the pile intensity constant are so easily met that the operator has very little to do. Starting and stopping the reactor is simply a matter of displacing the control rods by a predetermined amount.

The problem before the whole Manhattan Project, given the production of quantities of fissionable material, was to make the chain reaction uncontrolled and explosive. The Los Alamos laboratory was given this problem, and its practical solution was in almost every detail conceived and executed there. While the scale of the operations at Los Alamos is far smaller than that at the great production plants, the diversity and the difficulty of the research and production there carried out make the site a unique and incredible one. Military security regulations still prohibit the discussion of many of the interesting problems there solved, but following the Smyth report and making obvious extensions of previous knowledge very much of the physics of the bomb can be discussed.

The Meaning of a Microsecond

It was clear from the beginning that the great piles of uranium and moderator could not make satisfactory military weapons. In the first place, they were too large for reasonable means of transport. Even more so, while they are capable of running at any power, and can certainly be so set that the intensity will melt any structure, they cannot violently explode. The rapid rate of energy release is what makes an explosion. In the pile, the balance of neutrons made in the chain against those lost by non-fission capture is too narrow. The gain in number of fissions per generation is small, so that many generations are needed to produce a doubling of the intensity. Even more important, the chain operates on thermal neutrons. Such neutrons move at rifle-bullet speeds, and in travelling the many feet that are required to escape from the pile, they make a large number of collisions and eventually spend a lifetime of a thousandth of a second or so in each generation. Since many generations are needed, the time which would be required for the explosion of a pile may be measured in hundredths of a second. A block-buster explodes in the time it takes for a detonation wave, travelling some five thousand metres per second, to cross the mass. This time is perhaps one hundred microseconds.[1] The deliberate explosion of a pile would have the character not of a bomb, but perhaps of a steam boiler bursting.

All the criteria for a weapon are satisfied by a very different kind of

---

[1] Throughout this essay we shall find it convenient to measure time in millionths of a second, or microseconds. A high-velocity rifle bullet could cross the letter "o" in one microsecond.

chain reactor, the kind from which an atomic bomb is actually made. The reactor is not a large and carefully arranged lattice of moderator and natural uranium, but a small compact mass of nearly pure fissionable material. Instead of the tons of even the best pile, the critical mass of a sphere of pure plutonium or highly-enriched U235 is measured in kilograms. The loss of neutrons to the chain by non-fission capture is nearly zero, so that the reproduction factor is large, much greater than the near-unity value of the pile. Perhaps most important, the chain no longer operates on thermal neutrons. Even the primary fission neutrons have a good chance of producing daughters for the chain. Only a few collisions make up the whole life-history of even an unusually long-lived neutron, and the fast neutrons move not at rifle-bullet velocities, but at velocities many thousands of times greater. A typical fast neutron would travel from Los Alamos to London in a second. A generation in the small chain reactor which is a bomb, is measured by the time that it takes a fast neutron to travel a few inches. The time is perhaps a hundredth of a microsecond. And the intensity increases markedly in a single generation if we use a mass well above the critical mass.

How many generations are needed for an explosion? We know that the energy released from the fission of a pound of fissionable material is about that resulting from the explosion of eight thousand tons of TNT. The present atomic bomb, which yields the energy of some twenty thousand tons of TNT, must therefore consume a couple of pounds of material. If to every fissioned atom we assign one neutron—the one that did the job—we must have released two grams of neutrons at least in the explosion. But in two grams of neutrons there are about 10^{24} particles. All of these descend from a single neutron which initiates the chain. If we needed, for the sake of example, only one generation to double the intensity, this would mean about seventy-five generations of the complete chain. The whole process would take only a microsecond, and the violence of the true explosion would clearly be present. This is the very opposite of the controlled reaction of the pile. It is worth while noting that the delayed neutrons, which are not even emitted for tenths of seconds after fission, play no part whatever in the explosion of the bomb, though their role is determining in pile operation. The two applications of the nuclear chain are as different as the burning of coal and the detonation of TNT. It is not the least remarkable property of the fission chain that the same mechanism can act in two situations physically so different.

The Chain Starts

From all this it is perfectly clear how to make a bomb. You must bring together enough fissionable material to have more than a critical mass. The more you assemble together, the less the proportionate leakage of neutrons out of the mass, the greater the increase of neutron intensity per generation, and the more rapid the whole process. The trick, of course, is to make the initial assembly. For the geometrical conditions and the nature of the material alone determine the rate of change of the neutron or fission intensity, and hence the energy release. There is no switch to turn the chain reaction on or off. You must do this by modifying the shape. If you have, say, two pieces of material each weighing nine-tenths of the critical mass there will be no self-sustaining chain until you begin to bring the two pieces together. As they draw close the neutrons which once escaped out of the neighbouring surfaces of the two pieces now have some chance to take part in the chain, because they are captured and induce fissions in the other piece of material. Long before you had caused the two pieces to make contact, the total configuration would have become critical. As the pieces approached even closer the reproduction factor would steadily increase, and the neutron intensity, had a chain started, would grow at an ever-increasing rate. If your actions were slow, the energy released would have melted the pieces long before they had approached contact. Under these circumstances the energy released would not be that of thousands of tons of TNT, but simply that sufficient to melt or otherwise destroy the assembling mechanism. Obviously this will not do. The solution is clear: bring the pieces the together rapidly. Assemble them by making one piece the projectile of a small cannon, the other its target, placed directly at the mouth of the cannon. But recall the time scale of the whole phenomenon. Only microseconds are required for a tremendous energy release. Even special artillery will not move our pieces more than a fraction of an inch in this time. It is clear that the movement needed must be of the order of the size of the masses assembled, certainly several inches. There is only one way out of this dilemma. Nothing will happen to our assembling device if the chain does not start until the pieces are assembled.

Let us try to follow in words the incredibly rapid events which make up the nuclear explosion, beginning with the role of chance and ending with the catastrophe that follows.

The Toss of a Coin

All the spectacular effects of the atomic detonation are started by an event as purely chance in its nature as the toss of a coin or the flip of the ball in roulette: the presence of a single neutron. If the neutron appears, by sheer chance, before the assembly has reached the degree of supercriticality which its designers had hoped for, the bomb will fail, though all of its components function exactly as planned. For the first time, perhaps, a single atomic event, a necessarily uncontrolled circumstance, can measurably affect the world of men. Again, if no neutron appeared, though some were expected, the projectile might pass through the target, and the bomb parts fly apart again without detonation. Of course, some ingenious though simple device can be introduced to prevent the fiasco of failing to get started on the chain, the fiasco of post-detonation. Any neutron source—such as a beam of radium alpha-particles arranged to strike a beryllium foil when the position of target and projectile is the best possible one—which can be turned on at the right time, will make it very sure that a neutron is present, by emitting many of them in the time the projectile moves only a short distance. But nothing can save the bomb from *pre*-detonation except chance.

However he may be intrigued by the philosophic implications of the predetonation problem, the designer of the atomic "gun" must face it realistically, which is to say quantitatively. While he cannot escape pure chance, he may estimate how likely is his failure. He will not worry if after he has finished his design he is willing to wager odds of 100 to 1 that his gun will explode. No military venture has a chance of success much greater than that, neutrons or not. Is this high reliability possible? Let us make a guess at the numbers, using only pre-war information, and not the precise data now available to the workers of the Manhattan District. You will remember that the critical mass of U235 is measured in kilograms. Such a mass—even if we are pessimistic and take Dr. Smyth's upper limit of one hundred kilograms—will be only about six inches in diameter. The projectile and the target will each be of this size in a possible design. Now when these two active pieces are within say six inches of each other, the neutrons from each piece will find their way with considerable probability to the other. The assembly will be supercritical. If the chain begins, it will build up to a high enough number to destroy the whole device in a few microseconds, as we saw above. The stray neutrons must be kept down. How well must this be done? The projectile in a light naval gun moves six inches in a fraction of a thousandth of a second (a long time in our nuclear scale, some three hundred microseconds). During this time, we should have no neutrons, or at least none to start the chain. It is clear that not all neutrons will start the chain. Some will be absorbed in the steel structure of our gun, some may even produce a fission, but the neutrons from the fission fail to find the active material again to continue the chain. About the average neutron we can make the sure prediction that it will begin a divergent chain reaction in a supercritical mass; but about a single neutron we must quote only probabilities. But we will leave this out of account, so as to plan for the most difficult case. We shall assume that any neutron appearing in this crucial third of a millisecond can make the bomb fizzle out. We must then require that during that time the wager is a hundred to one against any neutron appearing. Since the neutrons may come at any time taken at random, this is about equivalent to saying that in thirty milliseconds on the average not more than one or two neutrons appear. Where will neutrons come from? The Smyth report lists the sources: (i) cosmic rays, (ii) nuclear reactions in light impurities in the heavy metal, in which the weak alpha rays of uranium may induce neutron emission (analogous to the familiar radium-beryllium neutron-producing reaction), (iii) the release of neutrons by the spontaneous fission of the uranium itself. Let us estimate these stray neutrons. Cosmic rays produce a neutron at the rate of a dozen or so per square foot per second. The impurities, it can be shown by simple calculation, cannot produce even this many if the metal is quite reasonably pure. The spontaneous fission rate is publicly known for normal uranium through work published in 1940 in the *Physical Review* (the chief American physics journal) by two Russian workers, who first observed this queer phenomenon and there described it. Just as uranium spontaneously disintegrates by alpha-particle emission, it spontaneously disintegrates, though very rarely, by fission, without any incoming neutron to cause the event. The rate of such disintegrations is very slow. For every uranium atom which dies by spontaneous fission, tens of millions die by the emission of an alpha particle. The Russian work of 1940 indicates that between five and fifty neutrons will be given off spontaneously by a kilogram of normal uranium in one second. This rare event—so slow and improbable that it would take uranium from ten to a hundred million times the age of the earth to decay by this process alone—is the controlling factor in our problem. We do not know how this rate varies from isotope to isotope; let us assume it is the same for U235 as for U238. Then spontaneous fission will produce some five thousand neutrons per second in our bomb, and is by far the chief source of stray neutrons. Our design cannot get the reliability we had hoped. Even if we take some advantage of the fact that every neutron will not start a chain, the chance of pre-detonation by a neutron appearing in the crucial three hundred microseconds is something like even, and not a hundred to one. We have been too pessimistic, of course. For some of the six inches of motion is near the fully assembled position and here the predetonation will make very little difference to the final energy release. The problem is after all one of continuous gradation from dud to great explosion, and we have spoken as though it were all or none. But our numbers, guesswork as they are, still show that the controlling factor in bomb design may very well be predetonation.

Against this danger there are few measures the designer can take. He can reduce the cosmic ray and impurity neutrons by shielding and by painstaking purification. But their importance is not great. He cannot affect the spontaneous fission rate. To reduce the probability of predetonation he may reduce the amount of material in the bomb. When he does this, of course, he reduces the potential energy release. He must use at least a reasonably supercritical total mass. The spontaneous fission rate of the particular isotope which is his active material will thus help determine the size of his bomb: it cannot be too small, or it will not be sufficiently supercritical, nor can it be too large, or the spontaneous fission may cause pre-detonation. Only one thing is at his choice. As the Smyth report says, he may "reduce the time of assembly to a minimum". He may shoot the two parts together at great speed. In this way he can cut the chance of a fizzle just in proportion to the amount by which he reduces the time of assembly. The atomic bomb is simple, but its design is not easy!

The Chain Is Under Way

Let us trace the explosion still further. By chance and design, let us get projectile fully meshed with target. A neutron initiates the chain reaction. For illustration (we do not know the real numbers) let us assume as we did before that a single neutron may be absorbed, and the resulting fission yields two to take its place. These two each give rise to two, and the chain is under way. In some seventy generations, one neutron has multiplied to a thousand million billion (10^{21}) and enough energy has been released to turn the materials of our bomb into vapour. This is still no nuclear explosion. Only an energy equivalent to that of some ten tons of TNT has been liberated. Everything now depends on the next generations.

If the reaction stops now, the atomic bomb is nothing but a costly and unreliable block buster. But if after seventy successful generations, only ten more can be sustained, the chain will yield the twenty-thousand-ton explosion which devastated Hiroshima. Time is indeed of the essence. As we have said before, the fast neutrons which carry the chain in the bomb require very little time to multiply. The whole of the first seventy generations can take place in about one microsecond, during which a bullet would travel hardly a tenth of an inch, or an explosion in TNT proceed a quarter of an inch further into the mass. And the next tenth of a microsecond will make all the difference between an atomic bomb and an ordinary one.

The Chain Reaction Stops

The nuclear properties of matter are not affected even under the conditions now existing within the bomb. The chain proceeds as usual. It will stop only by exhausting all the fissionable material—in a perfectly efficient bomb, this is the mechanism of the end of the reaction—or by some geometrical change which will alter the supercriticality of the assembly. If the now vaporized bomb expands, its surface will increase, and the chance for a neutron to leak out of the less dense active material into the outside world will grow. When sufficient expansion has taken place, when the chance for neutron leakage has so increased that the chain reaction no longer multiplies the number of neutrons present, the release of nuclear energy has stopped. The bomb reaction is ended. It is clear that if the bomb we have described expanded in all directions a foot or so, its density would have very much reduced. The chance that a neutron could escape from the mass without colliding with a fissionable atom on the way would have increased very much, and the excess of one neutron would probably no longer be enough to sustain the chain. To expand a foot in a tenth of a microsecond implies a motion at a speed of a few thousand miles per second, and an accelerating force corresponding to suddenly created pressures of millions of tons per square inch. It is as though the weight of ten battleships were suddenly brought to bear on every portion of a surface as big as a sixpence.

It is clear that some expansion will occur. This is in fact what limits the reaction. The details of the physics of this spectacular event will have to wait for later publication. Obviously the strength of solid materials plays no part in this phenomenon. All that matters is the behaviour of the intensely heated gases, their inertia, and the way in which the great blast wave can eat its way through the material of the bomb and the matter surrounding it, pushing out with these great pressures, until the whole device has expanded and the energy release has stopped. The release of energy in the form of the energy of motion of fission fragments, its transformation to heat, at temperatures beyond those of the centres of stars, and the expansion wave of the hot gas are all subjects of importance to the designer of bombs. For the key to the efficiency of a bomb is the speed at which this expansion takes place, compared to the speed at which the last few generations of the chain cause the energy release to multiply. The last tenth of a microsecond is the crisis of the end of the reaction, just as the first three hundred microseconds were the critical time for its initiation. The measure of the difference between nuclear explosive and ordinary cordite is found in this vast change in time scale. Nuclear explosions are fast.

The role of the so-called tamper, a heavy wall of matter placed around the active component of the bomb, is now clear. Such a wall will, of course, act as a cloudy neutron mirror: it will reflect back to the fissionable material some of the neutrons which might otherwise have been lost into space. This will make a given mass of material more supercritical than without the reflector. This is obviously a sought-for result. Even on the controlled piles such neutron reflectors are used. In the bomb they are less advantageous than at first glance appears, because the time that a neutron takes to wander into the tamper and come back to the bomb again will find it arriving long after its birth. By that time the neutrons present in the bomb have multiplied considerably in number, and the inheritance of reflected neutrons from an earlier and neutron-poorer epoch is no longer so important. On the other hand, a simple mechanical property of the tamper is now of great importance. The heavy tamper must be pushed out by the expansion of the heated active material. The blast wave must move into the tamper before expansion can end the reaction. The tamper really does "tamp" the nuclear explosion, almost as the wooden tamping rod of the miner tamps the dynamite into the drill hole. It increases the pressure required for expansion, delays the final expansion, lengthens the time of the reaction, and hence increases the energy release.

The Expansion: The Ball of Fire and the Mushroom

In a fraction of a microsecond, then, the energy of the bomb has been released. From heat, part of it has become the mechanical energy of the outrushing material of the tamper. Some fraction of the total is in the form of "penetrating" radiation—neutrons and gamma-rays—which must be considered later. But just now the interest centres on the hot ionized star-stuff of the exploded bomb. Out the edge moves, until in a few score microseconds the whole mass occupies a sphere something like fifty feet in diameter. When the heavy uranium and steel bomb has been spread as thin as this, it no longer has the density of metal, but that of ordinary air. From this time on the expansion is slower, and the hot material mixes with the air around, by now also heated to an ionizing, beyond-white, heat. The air and the vapour of the bomb continue to mix and the heated mass eats further and further out into the undisturbed air, displacing and heating the layer beyond the edge of the hot sphere. When this hot gas has expanded until the pressure within is not greater than the pressure of the atmosphere it stops growing. Perhaps a millisecond has elapsed. The gas has formed the hot "ball of fire" which measures many hundreds of feet in diameter and glows with a white heat for as much as a second or so. It is this hot ball sitting on the desert sand which turned it into the iron-containing friable green glass that carpeted the desert floor in the test explosion called Trinity.

The ball of fire then cools a little, by radiating away its heat. In a short time it begins to rise, like the hot air balloon that it is, and the cool currents rush in around it. It rises, leaving beneath it as it goes a trailing column of dust and cloud. The great column rises, carrying within it the active fission fragments left by the burned material of the bomb. It is incredibly radioactive. For the twenty-thousand-ton-equivalent bomb has within it about a kilogram of fission fragments. The disintegration of this radioactive debris in the first few seconds or minute of its rise represents the activity of a million tons of radium! The fission fragments, of course, are principally short-lived, and in a few hours or a day the far-spread activity has dropped by a very large factor. We who watched at Trinity could see the violet glow of heavily ionized air around the rising column as the material irradiated the upper air.

Everyone has seen the fantastic grandeur of the rising column of heated gases over the ruined cities. Two phenomena of special interest are seen in those fine photographs. The column itself is mainly a cloud, just like any thunderhead, except for the dust and vapour it holds, radioactive and generating heat. The cooler air of the upper air causes the water vapour brought up in the hot column from the air near the ground to condense and the cloud forms. As the column rises like the smoke from a chimney it may come to what is called an inversion by the meteorologists. This is a layer of air warmer, instead of cooler, than the air below it. When such a layer is reached, the gas of the column will spread out and rise no longer. This is the formation of the mushroom. But some of the gas of the column is still being warmed by radioactivity. This gas is warm enough to break through the inversion layer, and the column sends another stem from the first mushroom cap. This too mushrooms, now very high in the air, perhaps six to eight miles. All of these grand and ironically beautiful phenomena can be seen in the moving pictures of the attacks on Hiroshima and Nagasaki.

The Shock Wave: How a City Is Flattened

We have left the ground for an account of the relatively lasting spectacle of the mushroom. But a grimmer and more important phenomenon has already taken place below. A large fraction of the energy of the explosion went into pushing aside with extraordinary speed a large mass of air as the ball of fire formed. The pressure wave which pushed aside the tamper as the bomb expanded continues through the ball of fire, and, leaving the ball of fire when the gases stop expanding rapidly, hammers against the undisturbed air with terrible force. This wave is called a shock wave. The physics of the shock wave, which differs only quantitatively in this atomic explosion from the similar wave produced by any explosion, is a subject once overlooked by physics in the main, but forged to a real completeness by the requirements of the war.

A shock wave is simply a sharp sound wave of very great intensity. In a sound wave the pressure increase, followed rapidly by an equivalent decrease, is small in magnitude. But the great compression which the push of the atomic explosion gives to the air not only increases its pressure by a very large factor, but also heats it, as a tire is heated by rapid pumping. The sound wave so started travels faster than sound in the cool air still unreached by the shock. The shock front penetrates further and further. As it moves it represents a sharp boundary between two regions of air: one region ahead of the shock, yet untouched, still cool and at normal pressure; a second region, hot, compressed, just behind the front, where disturbances travel at high speed in the hot dense air. As the shock front moves it loses some energy by fulfilling its purpose, that is by pushing over anything in its path, but more and more its energy leaks away as heat, by conduction and radiation. The air in the very front of the atomic bomb shock is at first actually red-hot. The shock becomes increasingly less sharp; as it spreads in a great sphere the pressure excess and the temperature rapidly fall, and far away, the disturbance has faded out into an ordinary sound wave, and a rumble is the only evidence of its passage. All the phenomena of sound are associated with the shock wave: it may be reflected, it may leave shadows, it may diffract around obstacles. Such phenomena are responsible for the not quite predictable behaviour of a shock wave passing through the complex pattern of a city. All who have seen the results of an air raid incident will fully appreciate what I mean.

The wave itself is simply a transmission of pressure through the air, successive layers in turn feeling the compression. Behind the compression there follows generally a rarefaction, representing the wake of the original displacement. After the entire wave, both the increase and the decrease of pressure, there comes a real current of air, a wind, which may in such great explosions reach hurricane velocities near the centre, especially in constricted places like buildings.

How is damage done? The high pressures of the shock hammer on the surfaces of buildings. The air inside—ahead of the shock—is, of course, still at ordinary pressures. The difference in pressure represents a force which can break the strongest walls. Through every opening the pressure difference will force a current of air as well, a wind which devastates the interior of most structures lucky enough to withstand the main force of the shock. Here a few figures will make very clear what happens.

At about a thousand feet from the atomic explosion the excess pressure in the shock is as much as ten or fifteen times the normal pressure. This is the pressure needed to cause the death of a man from blast alone, from crushing the walls of the chest and smashing his ribs. At about twelve or fifteen hundred yards, the over-pressure has dropped very much indeed. The excess is here only about a third of the normal pressure already present. This seems small, but it is enough to demolish most houses, brick, wood or stone. On an ordinary wall such a pressure difference means a thrust of fifty tons, the weight of a locomotive. Still further out the over-pressure drops more and more. At two miles, it is only a mere fifteen percent of the normal pressure. But this is as much as the effect of a tornado, and will damage structures of any ordinary kind, and demolish the weaker ones. At three miles, the effect is down to that of a strong gale, and only light roof tiles, signs, or such weak objects will suffer damage. In Hiroshima we found the first missing roof tiles and broken window panes at eight miles from the blast.

British readers will be impressed by comparable figures for the V-1 explosion: the killing over-pressure is there found only at fifty feet; the demolition pressure at seventy-five yards, and the tornado effects at two hundred yards. The explosions of the atomic bomb are by far the greatest man-made blasts ever observed.

One difference between the blast phenomena in the atomic explosion and those of a block-buster is the duration of the over-pressure, that is, the depth of the strong shock front. The shock from a block-buster passes in a hundredth of a second or so: that from the atomic bomb will generally last many times longer, exerting its force against the structure walls for a longer time, and usually causing more severe damage.

The shock moves at first much faster than sound, but in a rather short distance comes down nearly to the speed of sound in normal air, about a thousand feet per second, or seven hundred and fifty miles an hour. Thus the two or three miles of Hiroshima which saw destruction from blast were flattened in the time it took the shock to cross that distance, perhaps ten or fifteen seconds.

The Heat of the Bomb

We lay on the ground nine miles from the tower in the desert where the first bomb was ready to detonate. The night was cold; the thin air of the high desert just before dawn made us shiver even though it was mid-July. Then came the blinding blue-white flash of the bomb and simultaneously the heat of the noonday sun full on the face. That great heat out of the cold dawn at ten miles distance is for all of us who were at Base Camp the most deeply remembered experience. The bomb radiates a not negligible fraction of its total energy in infra-red, visible, and ultra-violet light. Its average temperature in the first second or so, when the ball of fire has opened up until there is a sizeable surface to radiate, is like that of the sun, or perhaps somewhat hotter. The effect at ten miles is that of many seconds of solar radiation delivered in a much shorter time. At one mile distance, the radiation intensity is one hundred times greater than that we experienced. This terrible flash heat deeply burned most of the men and women in the street within about a mile from the explosion at Hiroshima. Many light cloth and wooden objects—curtains, mats—were set on fire. The flash was so short that no one could move during its main brilliance. As a result the shadow of the nose was sunburned into the skin of many present at the explosion in Hiroshima. We saw a horse with the shadow of a rail fence unburned on his flank, which was elsewhere raw and hairless. This radiant heat is after the blast the most important source of damage. Tens of thousands of dead and injured, and scores of secondary fires can be ascribed to the radiation of the bomb.

Nuclear Effects: Gamma Rays, Neutrons, and the Fission Fragments

For every uranium atom fissioned in the bomb, two fission fragments are produced. By the time the ball of fire has formed, these fragments are no longer screened by the dense matter of the bomb, but only by the dilute vapour of the bomb's debris. Thus the delayed neutrons they emit, and the gamma rays associated with the beta decay of the typical fission fragment, can freely irradiate the outside world.

About one neutron is produced by the fission fragments for every hundred fissions in the bomb. These neutrons spread out in all directions. They are however rather rapidly slowed down and finally absorbed by the nitrogen nuclei of the air, and do not reach any sizeable distance in great number. At Hiroshima, Professor Nishina and his staff collected samples of every kind from the region below the bomb. Artificial radioactivity had been induced by neutron capture in many elements. In samples of phosphorus which they prepared from the bones of the many dead whose bodies were recovered within a kilometer of the blast the fifteen-day activity of phosphorus was especially strong. From such measurements they were able to estimate the number of fissions which had taken place. But the amount of neutrons reaching the ground from the height at which the bomb detonated was too small to induce any activity of more than laboratory significance. At the desert test, sodium activity in the salt of the desert floor was strongly excited. It is likely that few persons were injured by the neutrons emerging at Hiroshima or Nagasaki.

This is not true for the gamma rays. They, too, are emitted from the fission fragments into the air. We normally think of gamma rays as being very penetrating. Compared to the neutrons, this is true. But compared to visible light, gamma rays cannot pass through very thick layers of air. The energy of gammas is so much larger than the binding energies of electrons in atoms that every atom looks much like any other atom to a gamma ray. Ordinary light, on the other hand, finds air transparent, metal totally opaque. But to gamma rays air or earth offer the same barrier, weight for weight. Now, a thousand yards in air corresponds in mass to about one yard in water, or two feet in earth. Two feet of earth represent a rather heavy shield for gamma rays, and a few thousand yards, if you remember the added effect of the inverse square law as the distance increases, is a formidable shield, better than that installed at Hanford. Let us make a small computation. Suppose that two or three gamma rays are emitted for every one hundred fissions (as with delayed neutrons) in the seconds that it takes to produce the ball of fire. This represents more than 10^{23} gamma rays. At one kilometer, neglecting the effect of the absorption of the air, about 10^{12} gamma rays strike every square centimetre. The air shielding is small at this distance—perhaps as much as a factor of ten, but not more—and the actual irradiation would be about that which physicians estimate as a minimum lethal dose. Many persons at even larger distances in Hiroshima and Nagasaki died from gamma radiation. Three thousand yards further, however, the dose received would be down by a factor of nine from the inverse square effect of distance alone, and by about a thousand from the six feet of earth equivalent which the additional air represents. So at two miles distance, there can be no noticeable effect of the gamma rays from the explosion.

Some of the fission fragments may be blown into the ground. At the test shot, Trinity, the bomb exploded only a hundred or so feet above the surface on a steel tower. The ball of fire containing the vaporized fission fragments rested on the desert sands. A portion of these active atoms became fixed in the glassy slag which was the imprint of the ball of fire. This material remained on the ground, and did not rise when the ball of fire drifted into the air. The region of the crater thus was made very radioactive. Early entry into this area was possible only in the lead-lined Sherman tanks which rumbled in the first two hours into the centre of the crater. The short-lived fission products gradually decayed. About a month after the explosion, the activity was down by a factor of hundreds from the activity observed when the tanks made their initial survey. A party of reporters toured the central region. But even then the ground was active, and no one would have liked to make camp at that blasted spot. The collecting of active samples of the green glass, which has become tacitly named "trinitite," is still possible. The piece on my desk will still drive the counting dial of a Geiger-Muller detector set at staccato speed.

At Hiroshima and Nagasaki, the bomb was detonated high in the air. This was done both to enhance the energy of the shock wave, by taking advantage of a reflected shock from the surface of the ground, and to lessen the radioactive contamination of the earth. The ball of fire did not touch the ground. Very little of the fission fragments reached the ground, and no dangerous activity was observed at any time. No casualties were caused in these cities by early entry into the region of the explosion.

As the cloud and the mushroom drift across the countryside, the fission fragments they hold will in part condense out on to dust or water droplets, and the activity will gradually shower down on to the surface of the earth far below. The details of this phenomenon are strongly dependent on the weather: a hard rain through the active cloud in the first few hours would deluge the surface with dangerously active water. As time passes, the activity decreases and the cloud spreads; the danger from the contamination lessens. The direct gamma radiation is well shielded by the few miles of air between the surface and the cloud high above the inversion layers. Only if the atoms of the fragments are brought closer to the earth, as by actually falling upon it, is there any sizeable irradiation of the surface. About twenty or thirty miles from Trinity there is a mesa a couple of miles wide and ten miles long on to which a rather considerable activity fell. The New Mexico range cattle which were grazing there developed very noticeable loss of hair, greyness, and even ulcerated sores on their backs. It is dry in that country; the cattle received the dust from the cloud, and it remained for weeks on their backs, close to the skin, representing a serious though not fatal local dosage of radiation. The beta rays from the active material in contact with the hide of the animals much increased the effect.

A thousand miles away from Trinity, in Illinois, the cornfields lay in the hot sun and strong rains of the Midwestern summer. The corn straw from certain of these fields is used for manufacturing the paper in which some sensitive photographic film is wrapped for sale. The batches of paper prepared from corn harvested a week or two after the test shot proved unacceptable to the film manufacturers; it contained enough radioactive contamination to fog some of the film. A physicist testing some Geiger counters in Maryland, two thousand miles from the test shot, noted an inexplicable increase in the background activity of his instruments on July 19, which went away a few days later. Only when the news of the bomb over Japan was released did he see a reason, which he transmitted to the *Physical Review*. Local rains may have brought down some material from the dilute cloud passing over Maryland in the stratosphere winds. While there is no definite proof that either of these remote events were direct consequences of the first bomb explosion, their timing is most suggestive. It must be said that even more remote effects are not absurd. There are some 10^{24} active atoms of fission fragments produced in the explosion. Weeks afterwards, there are still 10^{18} atoms decaying per second. If these atoms had been uniformly mixed by the winds over a belt a hundred miles long, thousands of miles wide, and the full height of the atmosphere—which certainly overestimates the dilution—the resulting activity would still be conspicuous on an ordinary Geiger counter.

The Death of a City

In all that has gone before we have described the physics of the atomic bomb. But the bomb did not seem like an experiment in physics when we walked through the long half-ruined shed of the rail station at Hiroshima and saw the litters of the blackened wounded and dying. The bomb is a weapon: the most deadly and terrible weapon yet devised. Against any city in the world from New York and London to the hundreds of large towns like Hiroshima and Nagasaki, the bomb is a threat. In any of man's cities a strike from a single atomic bomb will claim some hundred thousand deaths and some square miles of blackened ruin. The shock wave will wreck brick as it does wood, the gamma rays will kill men with white skin as it does those who have a yellow pigment, the fire will burn the cotton clothing of Manchester manufacture as well as it did that of Kyoto. And the bombs, if they come again, will not come in ones or twos, but in hundreds or thousands. Their coming will wreck not cities, but whole nations.

Physicists have learned to say more than physics these days. I hope the reader will see the propriety of ending this partial account of the work of the great laboratory at Los Alamos with a few sentences from a speech of its wartime director, Robert Oppenheimer:

". . . If atomic bombs are to be added as new weapons to the arsenals of a warring world, or to the arsenals of nations preparing for war, then the time will come when mankind will curse the names of Los Alamos and of Hiroshima.

"The peoples of this world must unite, or they will perish. This war, that has ravaged so much of the earth, has written these words. The atomic bomb has spelled them out for all men to understand. Other men have spoken them, in other times, of other wars, of other weapons. They have not prevailed. There are some, misled by a false sense of human history, who hold that they will not prevail today. It is not for us to believe that. By our works we are committed, committed to a world united, before this common peril, in law, and in humanity."

Accidents with Atomic Weapons

Igh explosive has claimed a toll of life and buildings by accidental explosions ever since it was first used. A whole shipload of high explosive went off in the convoy port of Halifax, on its way to France in 1917, and killed nearly two thousand people. Nowadays aircraft are daily in flight in numbers, carrying in their bomb bays explosive loads so potent that should a single bomber crash and have its load go off on impact, brick apartment buildings would collapse for a distance of five or six miles in all directions, and frame houses for eight miles around would for the most part lie in ruins. A few thousand square miles would be contaminated. Since aircraft are never wholly safe from disaster, this fact has caused wide concern.

One can estimate the hazard only by knowing something of the design of atomic weapons. It is all but certain that every atomic bomb, whether it is a fission bomb, a fusion bomb or some of the popular high-yield mixtures of these devices, contains, besides its nuclear explosive charge, a quantity of ordinary chemical explosive. Only with such explosive can the rapid motions needed for initial assembly or compression of the nuclear material beyond its critical dimensions be produced. The Hiroshima bomb was the simplest possible one; the bomb was simply a small cannon, with a projectile of fissionable material and a short barrel ending in a fissionable sleeve. When the projectile was set in motion by an explosion of the propellant charge, it moved down the barrel a few feet like a common shell. When it reached the interior of the sleeve, the nuclear explosion began. The Nagasaki bomb was far more sophisticated. It held, not a mere charge of a few pounds of propellant, but some tons of high explosive. The high explosive charge was a sphere surrounding the nuclear explosive. Here the high explosive had to be detonated simultaneously at many points on the sphere surface. The spherical shock wave so initiated con-

verged on the central charge, and squeezed the nuclear material into su-
per-critical density.

No information is published on the design of any later nuclear weapons. It seems very probable that they still reflect the general ignition principle, though not the geometrical configuration, of the Nagasaki bomb.
Insofar as this is true, every such bomb contains a fair amount of high explosive. This was confirmed in the recent impact of a nuclear weapon in
South Carolina, which caused a ground crater rather like that of a high-explosive bomb. Now, to produce a nuclear explosion, it is essential to cause
rapid relative motion of parts of the fissionable material. Heating alone
will not release any nuclear explosion, nor can a shock, friction or spontaneous internal changes set it off. In this respect, the nuclear material,
whose explosive properties derive from the very core of the atom, is far
more stable than chemical explosive. For many chemicals of explosive
type are safe enough against accidental detonation when pure, but subject
to triggering by friction or fire or slight impact because some small portion of the charge has been improperly prepared, or has spontaneously
changed into a less stable form which can set off the whole mass. The
greater sensitivity to outside forces is characteristic of chemical explosive; but nuclear explosive cannot be detonated without rapid and precise
motion of the sort its designers foresaw. An airplane crash, or accidental
dropping, will usually set off high explosive, almost always if the more
sensitive explosives used to detonate the high explosive are in place.

But it does not at all follow that a nuclear weapon will be exploded
when its chemical charge goes off. This would be true in the old form of
the Hiroshima bomb, and indeed the projectile would hardly be carried inside the barrel in any training or patrol flight, if only to protect the airbase. "Arming" such a weapon means placing the projectile within the
barrel, and perhaps withdrawing some metal barrier placed athwart the
path. An armed Hiroshima-bomb is a real hazard, and is very likely to go
off in any sizable crash. But such bombs are probably obsolete. The more
modern sort require not simply a single chunk of chemical explosive to
detonate, like the charge in a gun, but rather a dozen or so must go off all
at once. If they go off one after the other, as by a shock-induced firing of
one piece, whose detonation itself sets off the others, the small time delays between the firing will spoil the symmetry of the resultant wave,
which will not squeeze the nuclear components properly together, but instead act merely to scatter them abroad in one or more pieces. Under such
treatment the nuclear explosion cannot occur. The more and the larger the
explosives which must be fired, the less the risk of accident.

The ignition of the nuclear explosive demands a simultaneous firing of its many pieces of high explosive, simultaneous to a matter of ten microseconds or so. This is normally brought about by an electrical signal, sent from a central source to a number of small sparking wires, each embedded in its priming charge sunk in the various blocks of high explosive. "Arming" such a weapon very likely means closing all the circuits which enable such a signal to travel. After that action, a strong central electrical pulse detonates the nuclear bomb. It seems sure that a crash *cannot* succeed in detonating an "unarmed" weapon, but only in causing a badly timed explosion of the chemical charge, and a scattering of the nuclear material. Once the bomb is armed, an act which cannot take very many seconds, and is probably done only near the target run of the aircraft, a crash might possibly set off the charge, if somehow a strong signal pulse is caused to appear in the electric circuit, as by a shorting of the wires with a lead from a functioning generator at the moment of impact. Safety tests carried out by the U.S. Atomic Energy Commission in Nevada, held since 1955 suggest, however, that the most modern weapons, especially igniters for fusion bombs, are intermediate in safety between the two old types. These tests have not shown any hazard, it seems, but that they are considered to be useful is itself some source of worry.

From this account, it seems plain that the risk of accidental explosion of a nuclear weapon dropped before arming, or involved in an aircraft crash or collision, is from very slight to negligible. It is true that other safeguards—insertable parts or multiple switches of various sorts—can be used and probably are provided. These added safeguards are extremely worthwhile, and public attention to the hazard is very desirable, to ensure that the weapons designers continue to pay close attention to their grim responsibilities in this rather paranoiac enterprise. But impact and fire, alone or together, will *not* set off most post-Hiroshima nuclear weapons.

Nuclear weapons factories and magazines are all but safe from accidental nuclear detonations. Sympathetic explosions of nuclear weapons can hardly occur, though the high explosive may well go off from a nearby explosion. Plants making or storing fissionable material itself can, however, rather easily suffer radiation bursts very dangerous to their own personnel, but accompanied by little explosive force, and by not very widespread radioactivity. This can follow the simple misplacement of nuclear explosive, which is proof against shock and impurity effects, but may become dangerous if placed too close to other samples of the same material. Nuclear power plants can also explode, releasing large amounts of radioactivity, dangerous to whole city districts. But the explosion can hardly be severe, more like a boiler than a bomb, and the long-range fallout hazard is almost nil. Here the care of designers must however be extreme. As far as we now know, it is, and plants like the British Calder Hall seem quite safe. (This is despite the fuel element fires which contaminated the neighbourhood of Windscale plant not long ago; "safe" in this context merely means "safe against accidents that are large compared with normal hazards.")

All in all, the greatest dangers are not accidental, but deliberate ones. The high-yield tests produce much fallout, close and far. The acts of an irresponsible bomber or missile commander, or simply a pirate, are more to be feared than crashes. It seems probable that before too many years of this hair-trigger era have passed, a nuclear explosion will have taken place by human error or mad design. It is to be hoped that the military-political controls within the weapons-delivering establishments are delicate enough to enable immediate distinction between a single bomb and a planned surprise attack.

Whoever does not trust the sobriety and judiciousness of a bomber general that far, must, like the rest of us, seek a degree of nuclear disarmament. Unsupportable nuclear attacks are now always as close as the flight time of the vehicles that deliver them. With rockets, that will be measured in tens of minutes. It is hard to see how the states can live forever on such a mined brink.

Caught Between Asymptotes

The threat of nuclear war, so gross that it numbs the intellect, must be illuminated from every angle if it is to be comprehended, if it is to sink into our collective consciousness, there to become the root of action.

My title is inspired by remarks of John von Neumann, published (in *Fortune* magazine) shortly before his death. Von Neumann, a brilliant mathematician and father of the digital computer, was also a systematic promoter of large-scale modern weaponry. He was, at the same time, a man of clearsighted vision who recognized the terrible hazard posed by escalating weaponry. Our problem in the second half of the twentieth century, he said (I am paraphrasing), is that our weapons grow—in numbers, in accuracy, in destructive capacity—more or less monotone. They don't go backwards. But the area of the surface of the earth and the volume of its atmosphere remain fixed, gaining not an acre nor a cubic kilometer as time goes on. An extrapolation is painfully clear. We are *caught between asymptotes,* with ever diminishing room to maneuver. The distance between von Neumann's asymptotes has been halved, and halved again, since he offered the metaphor in the early 1950s.

World War III vs. World War II

Physicists understand numbers. But not even physicists can readily grasp the significance of the nuclear numbers. A "frame of reference" is lacking. Perhaps it is of some value, therefore, to compare "World War III" (defined as the use of a significant fraction of current weapons) with World War II. It is hard enough even to comprehend the awful death and destruction, the six-year cumulative total, of World War II.

What are the nuclear numbers? A few years hence, if present plans mature, there will be, in the combined arsenals of East and West, between 20,000 and 25,000 warheads capable of intercontinental distances. These will be complemented by about 30,000 warheads capable of shorter range, carried on artillery shells, torpedoes, mines, anti-aircraft rockets, and so on. In total, a panoply of weaponry, 50,000 or more nuclear-tipped devices of every size and type that man can build.

How can one assess the "size" of this pile of weapons as a destructive force? It has become conventional to translate actual megatons to "equivalent megatons" as a way to describe the total area of damage (which is not necessarily the most relevant measure over such a wide range of sizes and purposes). The world's arsenals add to some 10 to 15 gigatons equivalent. A prodigious number, easier to write down than to understand.

Going back to the late 1940s, I find an excellent effort by P. M. S. Blackett to forecast and assess the situation with nuclear weapons. Blackett's whole bent was against strategic warfare, which at that time meant the mass bombing of civilian targets. (Since then, it claims to include the "pinpoint" nuclear bombing of hardened military targets, notably missile silos. In the event it will probably revert to area destruction.) Throughout World War II, he fought against the mass bombings of Germany and Japan, and was most determined to try to show that strategic bombing was indecisive. I mention Blackett's bias because it serves to set his estimate of the military effectiveness of nuclear weapons at the most conservative end of the scale. To the atomic bombs dropped on Japan, he assigned an equivalent of only two kilotons of TNT each—strikingly low, but he had technical reasons for his estimate. (The actual energy yield of the Hiroshima bomb was about 12 kilotons, and of the Nagasaki bomb, about 20 kilotons. In terms of human casualties, the TNT-equivalent of these bombs would be greater, not less, than their actual yields.)

Using Blackett's equivalence, we find that it would have taken some 100 to 150 megatons of nuclear explosives to equal in overall effectiveness the destruction wrought by the actual high explosives and incendiaries rained down in World War II from the B29s, B17s, Lancasters, and so on (I hope that your attention to old movies makes these terms real). This World War II bombardment inflicted about 1 to 1.5 million casualties—deaths and serious injuries at the two ends of the Axis.

Now (working still with equivalent megatons), we and the Soviets have about 100 times that total destructive capacity poised for war. It seems to me impossible to deny that the consequence of using any substantial fraction of these weapons would be—at least—50 to 100 million serious casualties. These casualties, let it be noted, would be more disruptive to the social fabric because they would be inflicted in a matter of hours, or at most weeks, not years. Most computer simulations of World War III give larger, probably more realistic, numbers, but I wanted to take the most conservative view.

How are we to comprehend a World War III, its destruction magnified over World War II by two orders of magnitude, its time scale compressed from six years to perhaps six days? World War II decimated a generation; its political ends are, in an important way, unaccomplished; its military lessons have not been learned; and, in ways we are scarcely conscious of, it still dominates our daily behavior. If material recovery from World War II took determined and well-assisted lands from eight to ten years (psychic recovery may still lie ahead), I would estimate that it is very unlikely that material recovery from World War III could occur in less than eighty years. Adding the unknown effects on morale, and the incalculable environmental effects on soil, air, and water, I would be surprised if it would not be a century or two before the northern hemisphere found itself in anything like its present state, if indeed it ever did. Would the population have the spirit to build again the structure that had served it so poorly?

Technological society is built on productivity—output per unit input of capital or labor. So long as productivity can increase, there is no lid on production. The same rule applies to weapons. It is worth recalling that the incredible arms buildup of recent decades has been made possible by lower unit costs of destruction, by greater military "productivity." Nuclear destruction is cheaper than conventional destruction. The capital cost is lower, and so is the labor cost. In World War II, strategic bombing, at its peak, required the labor of 1.4 million men and women in the armed forces of the allies. Today, it is unlikely that as many as one-third that number are needed to manage the strategic "deterrent" of the Western alliance, a hundred times more powerful.

The Convergence of Two Trends

The present state of mutual terror—the mutual hostage relationship of the superpowers—is not unrelated to two centuries-old historical trends, trends which seem to be converging as surely as von Neumann's asymptotes.

First is the increasing size of unified units of population. True, small nations have proliferated since World War II. True also, some large nations existed long ago, brought about through occasional bursts such as that of Ghengis Khan out of central Asia, who made a kind of nation, united for warfare, that stretched halfway around the world from Vienna to Peking. But, by and large, the trend over centuries has been from smaller to larger units of population, communities bound by language, customs, behaviors, and beliefs, and perhaps mustered for common purposes by common fear of a perceived enemy. Where Los Alamos, New Mexico, now stands was once the battleground of several tribal "nations," each with its own language and customs, each holding dominion over what now might be a county. Hundreds of European rulers once divided among themselves what is now called a Common Market.

In the twenty-first century, this trend may hold hope, hope for one world. But in what is left of the twentieth century, it poses danger. The communities of several hundred million people who now face each other in fear command vast resources to apply to war.

The second trend is the exponential growth of science, and of its companion, technology. To physicists, this is a trend of special poignancy, for as we reach farther and deeper into physical reality, as we expand the boundaries of human knowledge, so too do we make possible the application of science to ever more horrible engines of war.

Unfortunately, the second trend is overtaking the first. The ability of mankind to wipe out populations is growing faster than the unified units of population are growing. Although the powerful nations, and allied groups of nations, are larger and perhaps more resilient than they once were, they are at greater risk than ever before. The escalating technology of warfare has increased the *percentage* of a major nation's population that can be, or is likely to be, destroyed. Another pair of asymptotes squeezes mankind.

The Instrument of War as a Thing of Beauty

Can the horror of war be dulled by the esthetic appeal of the instruments of war? In a recent magazine advertisement (two pages, in color), a jet fighter is shown hugging the earth at high speed as it "flies contours" to elude radar. In the background, dawn lights cirrus clouds over jagged mountains. It is beautiful.

It was Ruskin, I believe, who in the early nineteenth century, rhapsodized on the elegance and beauty of the ship of the line, its compact, graceful volume filled with the work of every kind of skilled artisan, its ropes the strongest, its design, its workmanship, its navigational equipment the best that money could buy. It was, in modern terminology, a superbly organized weapons system, which could capture the imagination of the poet as well as the admiral.

We can scarcely avoid looking at a ballistic missile in the same way. It lies at the leading edge of technology, a realization of nuclear physics and applied astrophysics, of hydrodynamics and thermodynamics and metallurgy, of computer science and solid-state physics, a package guided with incredible gyroscopic accuracy to within a few hundred meters of a target thousands of miles distant. Small wonder that we have our modern rhapsodists[1] or that the average taxpayer sees it as money well spent.

We can reflect, too, that nearly every weapons system has a nonmilitary sibling that is itself a technological marvel. A luxury liner, a supersonic transport, or a rocket to bear men to the moon commands its own esthetic response. All are devices that we build as much for the thrill of achievement as for their practical utility. The same may be true of the endless evolution of weapons.

Beyond Nuclear Weapons

From the seeds of physics have flowered war technologies of vast proportions, which, along with physics itself, have been enormously developed and systematically applied. One cannot see clearly what other technologies may lie latent in our growing knowledge of the world. I think they are there. But I cannot be sure. Perhaps there is nothing else in science that is so demanding, so evidently applicable to war, as the sudden release of nuclear energy that we have in our hands.

Yet it is worth glancing at other great advances, and asking: Do they carry the seeds of military application? Microbiology is much talked about. Its relevance to crops and to human life is plain. Geology is not irrelevant. Studies toward understanding, and therefore perhaps controlling, earthquakes have been made, enough to give a sense of foreboding. Knowledge of the atmosphere, especially the chemical kinetics of the upper atmosphere, is growing. Will man tinker with the breath of life itself? I even mention neurochemistry, a popular basis for controlling humans in science fiction. Will new advances turn fiction to fact?

It is not my purpose to suggest new weapons systems. The ones I hint

[1] Norman Mailer, *Of a Fire on the Moon* (Little, Brown, 1970)

at above may well never eventuate into anything. But the next generation promises more and more fundamental insights in areas outside of physics. Will every possible military application of new knowledge serve the ends of the state? Will other weapons comparable in terror to nuclear weapons be devised? And if devised, will they be deployed?

Warfare in the Third Dimension; the Erosion of Defense

1914: Darts dropped by hand from the canvas and wood aircraft of the day. 1915: Hydrogen-filled Zeppelins dropping kilopounds of bombs on the coastal cities of England. 1945: Tokyo and Hiroshima. 1982: Thermonuclear ICBMs, SLBMs, and cruise missiles by the thousands. Such has been the evolution of air war. It was Sir Hugh Trenchard, who, in World War I, suggested that erasing the boundary, flying over the coastline and over the national frontier in a third dimension, topologically distinct, was the preferred and irrestible mode of future warfare. We know very well how much misery has resulted from warfare in the third dimension, and how little military certainty.

From coats of mail and walled cities to heat-seeking anti-aircraft missiles, war has been a game in which each offensive move has been matched by a defensive countermove. But now there is a striking change. Those who wage war from the third dimension or propose to do so no longer claim the existence of a defense. Instead of defense, we have retaliatory measures, planned or anticipated, and the associated psychological stress; we have Mutual Assured Destruction. This is the dilemma of defenselessness that further shrinks the distance between asymptotes.

Is There Reason for Hope?

In April 1947, the number of ready nuclear weapons in the world was zero. It is now, or soon will be, about 50,000. (That is four new bombs added to the world's arsenals each day for thirty-five years.) The most conspicuous feature of this buildup is, of course, that no use has been made of the weapons since the count was zero. This is at least as important as the other side of the story. The fact of no-use does not offer a naively automatic hope, but it does offer a sense of opportunity. There is indeed a profound wish and need for survival that has inhibited the use of these weapons, without yet being able to inhibit their steady growth and their ever-present metastability as threat.

The signs of our times suggest that the decade of the 1980s is a decade of realization, a decade in which peoples everywhere are coming to understand that they face the kind of asymptote closure that I discussed at the beginning. This heightened consciousness may turn out to be no promise and bring no fulfillment, but I would prefer to be hopeful. We do know very well that the first time we saw these signs, just after World War II, the hopes were dashed.

Let me invoke another physicist, J. Robert Oppenheimer, a man no longer here, who has become, in a way, a mythical figure. Like many lives that have been transformed to myths, his life has seized the imagination of the informed world by his suffering, by his having gone so high and having been brought so low in service to the state.

When Oppenheimer spoke about the implacable fate of the nuclear-armed world, beginning in the fall of 1945, what he said was that the atomic bomb never offered a new argument for making a lasting peace but rather a new *opportunity* to do so. I think that opportunity, at that time, was lost. Perhaps it was illusory. Perhaps the notion of using the opportunity to exploit the new energy for economic good, to share weaponry among recent allies with deep distinctions between them, was illusion. I think, myself, that it was largely illusion. Be that as it may, enough was done so that we are still here.

But now there is a new opportunity, an opportunity to unwind the store of metastable energy and to curb self-centered nationalism. Nationalism has so much that is noble within it—I recognize that fully—but the decisive point, if we are to live, is that we have, as Oppenheimer said, "a common bond with other human beings everywhere."

I cannot end more fittingly than to read from a letter written, very probably, by Robert Oppenheimer, reporting for the Scientific Advisory Panel to the Interim Committee of the Secretary of War only two days after the end of the Japanese war, a month after Trinity. The letter ends with these words:

We believe that the safety of this nation, as opposed to its ability to inflict damage on an enemy power, cannot lie wholly, or even primarily, in its scientific or technical prowess. It can be based only on making future wars impossible. It is our urgent and unanimous recommendation that all steps be taken to this one end.

Very Sincerely,
J. Robert Oppenheimer, for the Advisory Panel.

The Spiral of Peril

Xerxes, they say, led three million armed men toward Thermopylae. The walled cities of the Old World offer substantial evidence that ancient warfare could hold as target a whole population. But the growth of the state, of cities themselves, and of human control of the physical world have brought swift war on intercontinental scale, new in our time.

From roots deep in the cruelest past, strategic war made against distant enemy homelands, the source of troops and of arms, the seat of the national will, first grew to gain strong adherents in what we now call World War I. Reflective military thought was then appalled by the stinking mire of the trenches, that bloody gash across Europe to whose brink young men in millions were brought in fruitless sacrifice. For years the shells mixed blood and dirt, yet no line moved. Against that frustration the war theorists found grim hope in the air, the third dimension, newly available for human combat. In 1918 the Associated Press carried dispatches like this: "This navy of the air is to be expanded until no part of Germany is safe from the rain of bombs. It is a thing apart from the squadrons attached to the various army corps. The work of the independent air force is bombing munitions works, factories, cities, and other important centers far beyond the German lines."

But the power of the aeronautical industries grew too slowly. These first air raids with wood-and-canvas planes and hydrogen-filled Zeppelins dropped bomb loads measured in hundreds of kilograms. Even by the end of World War I, bombers of wood and wire lagged far behind the well-made guns for weight of explosive. It was World War II, a generation later, which foreshadowed our present peril. Most of the weapons systems that peril the world today barely began during World War II: the jet engine, the cruise missile known then as V-1, the orbital ballistic mis-

sile called V-2, the atomic bomb over two cities in Japan. Then too, fully
developed after 1918 in Great Britain and in the United States, there was
the air war doctrine of Sir Hugh Trenchard and of General Billy Mitchell,
that "the real objectives are the vital centers." Guernica, Rotterdam,
Coventry, Leipzig and Dresden, Tokyo and Hiroshima, all bear rubbled
witness. Bomb loads had grown to maturity.

It is scale rather than kind of destruction which chiefly marks today's
war potential: A city no doubt burned under the rain of firebombs—just as
it would under thermonuclear attack of a few warheads today—to rust-red
ashes. While of course gross energy expenditure is not the whole story, it
offers a simple common measure as rough guide. Its unit is the equivalent
weight in TNT, the commonest military explosive of two world wars. One
ton of TNT is the nominal load of one big aerial bomb or of the tip of a
V11 flying bomb over 1945 London. Hundreds of heavy bombers, each
with a belly full of bombs, thus sum up to thousands of tons of TNT.

When explosive nuclear energy came into use, it was scaled within
military experience by measuring it in terms of tons of TNT. Counting
only single ton by ton, the comparison would be futile, but with the scientific Greek prefixes of kiloton—1,000 tons, megaton—1 million tons, and
now gigatons—1 billion tons, the event could be at least abstractly described, if not easily grasped by the mind. One megaton of TNT as cargo
of a long freight train of boxcars would rumble past not for a couple of
minutes, but for six or eight hours: a train 250 miles long. A gigaton begins to engage the whole transportation system of the United States, rail
and road; it implies TNT more voluminous than a national wheat harvest,
or a year's stock of coal. Nuclear explosive transcends thereby the entire
past of warfare.

In World War I, aircraft were in their infancy, but cannon had achieved
modern scale. Along a carefully-prepared front like the 100 kilometers or
so along the River Somme in 1916, the cannon stood ranked hub to hub,
and the sweating gunners hauled shells steadily day and night. Such an artillery attack might sum up to tens of kilotons per day of attack. The
whole of World War I might be measured in a hundred such attacks over
the four years, a few megatons of TNT. But those were delivered mainly at
battlefield range, not flying far beyond the barbed wire of Flanders to
London, Paris or Berlin. In World War II, the new machines and the
massed forces led in five years to about the same overall explosive release. But by 1945 war could come by heavy bomber to any city, to Tokyo
as to Dresden. The volume of attack rose at most to some tens of kilotons
per day, like the Somme. Even the first two atomic bombs remained at
that scale; some 10 or 20 kilotons per attack. The whole war again

summed to multi-megaton scale, in all like that of a few hundred atomic bombs. Destructive war had reached the city, but its overall scale remained that of the World War I.

Since 1945, a heedless steady investment in the cheap destructiveness of nuclear war has entirely transformed the scale of potential ruin.

The size of single weapons has not yet passed beyond the more familiar scale of such natural disasters as volcanic eruptions. The world remains large compared to the reach of their immediate effects, although both for the natural and the man-made eruptions grave long-term effects subtly modify the whole environment. Still it is true that, just as Krakatoa was seen in Europe only as colorful sunsets, so the biggest nuclear weapons tests, which are truly volcano-like close up, could be noticed across continents only by instruments or by some of the uncertain if fearful delayed effects of airborne dusts or of unseen radioactivity. Immediate effect dwindles with distance from the explosion. Therefore strategic nuclear war demands two coordinate technologies: the energy release from nuclear weapons; and some means of swiftly, surely and irresistibly bringing the explosion near its intended target. In the neutral vocabulary of the analysts this task is alluded to as *delivery*. The twin growths of explosion capability and of distant delivery are the technical substance of that nuclear arms race which has gripped the world since the late 1940s.

Megaton War

The first atomic explosion took place at the site called Trinity in the desert of New Mexico in July 1945. It released about 20 kilotons of energy. Just three-and-one-half years earlier, the first controlled chain reaction with uranium had been produced quietly in Chicago under Enrico Fermi's direction. The first chain reaction outside North America took place on Christmas of 1946, in a suburban site along the Moscow River, under the direction of Igor Kurchatov. That was four years after Fermi's entry into "the New World." The Soviets were in pace with the American-United Kingdom effort of World War II, but with a four-year lag.

Such bombs must have a delivery system. Up to 1960, only bombers using gravity bombs were of much long-range operational use. The United States took its first steps toward atomic war readiness by sending some World War II prop-driven B-29s to British bases, closer than Maine or Omaha to Soviet targets. Two groups of these planes were deployed to Britain in the crisis summer of 1948, the season of the Soviet land blockade of Berlin. They were probably not yet modified—nor all crews trained—to deliver nuclear weapons; true atomic bombers may have reached Britain only in 1949.

From that time until about 1953, nuclear forces threatened war at the megaton scale. U.S. atomic bombers were numbered at many hundreds. Some were able to fly from U.S. bases without refueling; more were of shorter range but based farther forward. Before the mid-1950s all were slowish, propeller-driven planes, with bomb loads measured at a few atomic bombs apiece. The strike could reach a scale of tens of megatons—the same scale as that of all of World War II, but now attained against cities and other strategic targets. That attack would take only days or weeks rather than years, but its dimension, if not its rate, was within experience.

The Soviets, of course, had a condign response, although their forces came later and remained smaller. They had in the late 1940s built the Tupolev-4s, copies of the B-29s. These could not easily reach U.S. targets, though some might perhaps get through. The Soviets had no overseas bases to match those of the U.S. Strategic Air Command. It was in the 1950s that the Soviet air defense matured: its swarms of high-flying fighter-interceptors, short-range supersonic jets, countered the big, slow intruding bombers of SAC. But megaton nuclear war against the enemy homeland was a U.S. near-monopoly. The ultimate Soviet counter was more likely to be warfare by air and land against the great cities of NATO Europe. It was in this period that the critics, especially the British physicist-analyst P.M.S. Blackett, made their case that atomic war was after all still only megaton war, not yet overshadowing by its scale the terrible experiences of World War II.

Gigaton War

In January 1950, after brief but intense debate, mostly in secret, President Truman declared it a priority U.S. task to go beyond the atomic bombs of the day to the thermonuclear weapon, the so-called super. That possibility had been latent in the wartime success of the atomic bomb; it is still necessary to trigger every thermonuclear weapon with an atomic one. The atomic, or fission, bomb uses the neutron chain reaction which serves to split each nucleus in a macroscopic mass of the heaviest of atoms, uranium or the transuranic plutonium. But the transient high temperatures so made can, with sufficient skill, be used to ignite the fusion reactions, which are then self-maintained by high temperature in the manner of the innermost core of the stars. Such devices unite, or fuse, in pairs and fours the nuclei of the very lightest element, hydrogen.

The two uncommon hydrogen isotopes, that of mass two, deuterium, found in small amounts in all natural waters, and that of mass three, the unstable tritium, which must be made afresh in nuclear reactors, are the chief reagents of a thermonuclear explosion. (A lithium isotope not difficult to separate from the natural minerals of that rare element also plays a key part.) Fusion design is more complex than fission, and in any case its use implies the technical ability to make an effective fission bomb, to use as trigger.

Fusion yields vary between about 50 and 150 kilotons TNT equivalent for each kilogram of nuclear fusion fuel expended (that is, deuterium, tritium and lithium-6); for the fission reaction, one gets about 17 kilotons of TNT per kilogram of uranium or plutonium consumed. The fusion reactions can be used in two distinct ways: Fusion fuels can be simply enclosed within a fission weapon to enhance, or "boost," its yield, very cheaply, though only by a rather modest factor. Or they can be adroitly disposed so as to allow the propagation of the fusion explosion within the explosive fuel more or less without limit, once ignited. A third step, fission *after* fusion, is also in use. The largest explosion yet made was a 60 megaton device detonated by the Soviets in 1986, which used mainly fusion.

The first demonstrated step to the eventual super bomb was a U.S. test at Eniwetok Atoll in 1951, which tested the booster principle. The first fully self-propagating thermonuclear explosion was also at Eniwetok, on November 1, 1952. The yield was nearly a thousand times that of the Hiroshima bomb, the first entrant ever into the megaton range. By late 1953 the Soviet experts too had fired a boosted fission bomb; and by early 1954 the U.S. had tested its first practical thermonuclear weapon, again in the multimegaton range. A year later the Soviets followed suit, using the design of Academician Andrei Sakharov as the United States had earlier used the related scheme of Edward Teller and Stan Ulam. The H-bomb was here. Robert Oppenheimer displayed "lack of enthusiasm," according to E.K. Lindley, with the remark: "This thing is the plague of Thebes."

Large fission bombs are the more expensive, especially beyond a couple of hundred kilotons yield; fusion is cheaper and more or less limitless in yield. The fission megatons of the bomber attack became thermonuclear gigatons as the 1950s passed. By about 1960, the U.S. nuclear stockpile amounted to some 30 gigatons of TNT; the Soviets held a comparable, if somewhat smaller, amount.

Delivery kept up. The Strategic Air Command built fleets of jet bombers, developed ways to refuel them in flight, spread its bases for them in

Europe, North Africa, the Far East and the Pacific. By the end of 1959 SAC held some 1,400 B-47 jets, plus 500 B-52 jets of still longer range, and more than a thousand jet tankers to refuel its bombers in flight.

The Soviet counterpart forces were less powerful, though formidable still. They had between 1,200 and 1,500 intermediate-range jet bombers, the Tu-16s, not unlike the B-47s but without that chain of overseas bases. Their bigger bombers—Tu-95s—were a third as numerous as B-52s, and were not jets but slower turboprop planes. Theirs too, however, was a gigaton air force; to this day the world overall nuclear weapons yield has never been greater than it was in the early 1960s. Not much of the Soviet air-strike force could count on entering U.S. air space, but the threat even of that fraction was nevertheless terrible, and U.S. allies were easily in range. Gigatons are hard to ignore, even when flown in turboprop bombers; even reckoned only over distant bases and allied capitals.

Gigaton War from Orbit

V-2 was the first ballistic missile, with an explosive yield around one ton. A missile's supersonic rise into airless space and back again is swift, free of weather impediment, almost impossible to intercept. At the close of World War II the superpowers began a long quest toward a force of ballistic missiles with nuclear warheads for warfare on megaton scale at intercontinental range. The rocket seemed then, as it in fact became, the most daunting of strategic weapons. (I was approached in 1946 to join the nascent RAND Corporation to study "the waging of intercontinental nuclear warfare." By the spring of 1947 the Soviets began a long-term development as well.) By 1953 it became clear to the United States that the Soviet Union was fully engaged in an effort toward such rocket weapons, an effort which had priority over Soviet development of bombers. By 1954 both sides were hard at work on ICBMs, drawing on the German experience, even enlisting some of the leading Peenemünde personnel.

The first successful long-range missile test was carried out by the Soviet team in August 1957 across Siberia. The use of the same technology to launch the first artificial Earth satellite, Sputnik, in October of that year was a brilliant public exploitation of that success. The United States seemed bested, but that was largely in appearance. Within three years there was indeed a "missile gap," but it had opened the other way. For a variety of reasons, the Soviets did not succeed in deploying a strategically significant number of missiles with nuclear warheads until years after a strong U.S. force was well in being. The numbers of missiles on each side became about equal only around 1969. In 1983, there was still crude parity, but at an inordinately high level.

The nuclear-powered submarine was a remarkable U.S. innovation. In the present missile submarines the energy wealth of the nucleus is used in two quite distinct ways: First, the submarine is itself propelled by a nuclear power reactor. That gives it true submersible performance, the dream of Jules Verne. Free of the need for air or fuel for a long time, a nuclear submarine can—since the launching of USS *Nautilus* in 1955—voyage for round-the-world distances underwater at a sustained speed greater than that of any big surface warship. Give such ships precision orbital nuclear missiles, and the subtle means of navigation which enable them to strike very close to a target on any continent from an underwater hiding place at sea, and you have an unprecedented weapon. About 100 of these exist now on the two sides, with a considerable edge in sea-keeping and silence at present in favor of the United States. One such big submarine disposes of a 10-megaton war like World War II by itself from its unknown and shifting position in the depths. The point of course is that a missile submarine—unlike a land-based underground launcher which is targetable by a well-aimed nuclear warhead—is hidden and all but safe from pre-emptive attack. Efforts to find means to locate submarines at sea are prodigious, but neither side can yet reliably depend on finding the hidden boats within the foreseeable future, almost surely not within the century.

In the 1960s, the practice was to fit one nuclear warhead, with a yield measured in megatons in one explosion, into each missile, like the one or two big bombs in each early jet bomber. But beginning with the U.S. innovation of MIRV, the multiple, independently-targeted re-entry vehicle, first tested in 1968, the scene changed. Now each missile launcher could strike accurately with its dozen heads not at one, but at several different widely-spaced targets overseas. If those targets were enemy missile silos, then a few missiles could be used to destroy many more. There would be an implied imbalance; some advantage would accrue to the first to fire. Each side could have more deliverable warheads than the other side had vulnerable missiles, all at the same time. This clearly worrying stance was implied by U.S. Air Force innovators from 1970.

By the end of the 1970s, the turn of the spiral had been completed; U.S. MIRVed missiles faced many MIRVed Soviet missiles, and U.S. leaders explicitly expressed doubts about the stability of the situation into which the United States had been plunged by its own efforts. No clearer example is to be found of the self-frustrating nature of this generation-long race for nuclear dominance.

The next turn of the gyre lies ahead: more missiles in the silos, or shifting about onland, or packed tightly to allow some hope of anti-missile defense, or . . . ? An independent observer will not miss the hints that these steady elaborations of weaponry have roots extending beyond the military rationale offered for them, often domestic rather than international, in both political perceptions and interservice rivalries. It is necessary to recall that the theoretical advantage of a first strike is gravely diminished by complex reality. The complicated operation is not testable; the adversary might well fire on warning alone; his submarines are hidden. The gamble is an incredibly uncertain one, and the stakes are life and death for the nations. So far prudence has kept the nuclear peace.

The Nuclear Battlefield

From the mid-1950s the gigaton strike of the big bombers fed upon a new plenitude of nuclear destruction: big new fuel plants, new mining enterprises, new designs. As a kind of counter to the uninhibited heavy strike on cities and bases, long gospel to SAC, there developed especially in the United States a plausible doctrine which sought to return nuclear war "to the battlefield." Nuclear weapons then came into being for many purposes, with yields from the fraction of a kiloton up to 1,000 times larger. Since 1956 they have become part of the arsenal assigned to every branch of the armed services. These weapons are meant not for long-range strategic delivery but for use in every kind of ordinary military engagement, by land, sea or air. Known as *tactical,* or *theatre,* nuclear weapons, their variety is striking.

Almost every military organization has found use for some nuclear weapon. The land forces can fire short-range nuclear-tipped missiles against targets even 1,500 miles off; the artillery has nuclear shells for its eight-inch cannon with ranges of 10 or 20 miles. There are man-portable nuclear demolition mines to destroy roads and junctions. In the air, the bigger fighters and attack planes have nuclear gravity bombs; the strategic bombers now mount air-to-surface stand-off missiles with nuclear warheads. The fast interceptors have nuclear missiles for air-to-air use against intruding bombers. At sea, carrier-borne aircraft have nuclear bombs, and ships mount nuclear missiles against aircraft attack at ranges up to 100 miles or so. There are nuclear depth charges to drop against submarines, while submarines have nuclear-tipped anti-submarine and anti-ship weapons of their own. That is all on the U.S. side. The Soviets have a similar panoply: The surface-to-surface rockets operated by their infantry and artillery are many; their attack submarines, both nuclear- and diesel-powered, routinely mount nuclear-tipped torpedoes targeted against NATO shipping.

The estimates count the present supply of such nuclear weapons, distinct from long-range strategic weapons mainly by the nature of delivery, at about 15,000 to 17,000 on the U.S. side. Some 6,000 of these are deployed in Europe, including the United Kingdom; the rest are stored in U.S. magazines or held on bases and on ships of the fleet around the world. The Soviet stockpile is estimated to be somewhat smaller, perhaps 10,000. Many of these weapons approach megaton yield; perhaps half are in the low-kiloton range, 10 times smaller than the Hiroshima yield. Nonetheless, any wide use in Europe would clearly amount to a fierce multi-megaton war across that populous continent, at a rate of destruction and with perils of radioactivity and fire unprecedented in the long annals of conflict along the Rhine, the Vistula or the English Channel.

The complex interaction of nuclear weapons with the pattern of alliances in Europe since World War II cannot be treated adequately in this summary account; it remains perhaps the most urgent issue of them all. One can read the whole story as an effort of the European governments to guarantee that a nuclear war in Europe would engulf the superpowers as well, while far away the United States has been anxious to avoid that very outcome, while reserving its right to initiate—perhaps better, to threaten to initiate—nuclear war on that continent. It should be remarked that West German forces do not possess nuclear weapons of their own, only some held under U.S. concurrent control. Soviet control of nuclear weapons in Eastern Europe is judged to be even tighter.

Admittedly, the superpowers' nuclear forces in Europe have long been a little contradictory in nature and purpose. Eager to avoid mutual destruction, the outside powers could target only European points. Yet both sides have over the years held intermediate-range nuclear ballistic missiles in or near Europe. Those on the Western side could reach Soviet home targets. For a decade or more, the United States had no such missiles, though it has strategic missile submarines offshore targeted on Europe and nuclear-strike aircraft of adequate range based in Europe. Soviet counterparts could strike London or Paris but not domestic U.S. targets. Such weapons seriously tangle the issue, and have since 1979 been again a focus of U.S. policy in Europe. NATO has agreed to increase such capabilities sharply with new U.S. weapons, claiming response to an impressively powerful Soviet modernization of its own long-installed intermediate-range systems. Very likely intentions on both sides are mixed:

nothing requires that NATO target Soviet territory instead of Eastern Europe, even if it can do so. Political maneuvers surely play a major role as well. The dangers of such ambiguities are plain.

The Nuclear Club Grows

France and Great Britain both possess strategic nuclear forces; that is, they hold nuclear warheads with delivery systems able to threaten the Soviet Union. Geography does not require of their weapons a fully intercontinental range. China, too, possesses a strategic nuclear force, though its missiles are of intermediate range; they might reach many eastern Soviet cities, but hardly Moscow. Taken together, France, Britain and China have about 100 land-based ballistic missiles, a couple hundred nuclear bombs or short-range air-to-surface missiles for aircraft delivery, and more than 100 submarine-based missiles. For those three states, their nuclear forces represent a genuine deterrent; they surely appear to think so. That deterrent is at a damage level 50-fold smaller than the swollen arsenals of the superpowers. It implies the sub-gigaton consequences of a nuclear World War II, not the 10 gigatons of the United States and the Soviet Union.

India has tested one fission bomb, and it has some bombers. Rumor strongly suggests that both Israel and South Africa may possess as many as some tens of untested fission weapons; both of those states have aircraft which could deliver nuclear weapons over a modest range.

Public entry into "the nuclear club" has been signalled with test explosions by the five nuclear powers—the United States, the Soviet Union, Great Britain, France and China. All five have since repeatedly tested both fission and fusion weapons.

India's single test was a plutonium fission device in 1974. South Africa seems to have prepared a test in 1977, but cancelled its plan after warnings from both Moscow and Washington, if we are to credit a series of quasi-official newspaper reports. (The 1979 detection, by a U.S. satellite, of a possible test in the South Atlantic region has not been confirmed and may well have been spurious.) That Pakistan undertook development of a nuclear bomb under a previous government seems sure; subsequent progress is not known, but there is reason to believe it may be substantial. Iraq has officially denied that its recent developments in the nuclear energy field are weapons-related; that did not prevent an air attack by Israeli bombers on a reactor under construction in Baghdad.

By the Non-Proliferation Treaty of 1968, three nuclear weapons states—the United States, the Soviet Union and the United Kingdom—have undertaken to prohibit transfer of nuclear weapons or their technology to any other states. The non-weapons signatories undertake for their part to accept international safeguards against their own diversion of nuclear materials from peaceful uses. There remain quite a few non-signatory nations. Proliferation of nuclear weapons has certainly been slower than some have feared. While there is danger in the existence of many independent national stocks of weapons, it is still the case that the enormous quantitative difference between the two superpowers and all other aspirants to the nuclear club makes superpower conflict the greatest world danger. Even if somewhere an irresponsible leader did initiate nuclear war with a small arsenal, it seems more likely that the superpowers would join in suppressing the spread of the danger than that the event would catalyze world nuclear war. That is not to deny that a city or two in ruins is tragedy enough; but today the stakes go much higher than that. It is noteworthy that no nuclear weapons state has ever given much help even to its closest allies, let alone to unreliable and distant clients. The break between the Soviet Union and the People's Republic of China arose at least in part over Soviet reluctance to assist China in becoming a nuclear weapons power. The nuclear relationship between the United States and the United Kingdom is based on joint wartime efforts, yet the subsequent story has not been one of easy sharing. The United Kingdom has U.S.-made missiles, but its own breed of nuclear warheads. Since nuclear weapons are so dangerous, their possession has been sobering at least so far.

Once nuclear weapons were rather easily taken on as a small operational addition to a force of long-range bombers. They were only new heavy bombs, if terrible ones, which fell of their own weight onto a target visible below the plane. The special arrangements needed were not major. By now, however, the nuclear weapons themselves are a still smaller portion of the remarkably complex and varied systems needed to control and to deliver them to worldwide targets.

The era of nuclear plenty in yield and variety came during the thermonuclear breakthrough of the mid-1950s; ever since, warhead types and sizes have elaborated greatly. Warhead variety is a matter now of sound engineering rather than of technological revolution. New warhead types differ in choice of yield, in weight and size, in reliability, in safety during transport and storage, and the like. There are no more large changes to be expected in nuclear warheads. The most novel new U.S. device—the so-called neutron bomb to be fired in Lance battlefield missiles—derives from a concept and design 20 years old. The neutron bomb spends its energy preferentially in extending the area of radiation lethal to human targets rather than in effects of blast and fire. The change is quantitative rather than qualitative; tactically its consequences seem marginal. In both the United States and the Soviet Union nuclear weapons technology is now mature—between them they have carried out about 1,200 tests of nuclear explosions—and no great novelty lies ahead.

All evidence points to two decisive properties of nuclear weapons: They are such copious sources of destructive energy that one nearby explosion finishes almost any target. And they are so inexpensive for their energy store that they can be launched in considerable number. Yet even if only a few arrive, their work is final.

Around these two properties have been drawn all the newest technologies of postwar history. While modern military technology and organization have extended to ends other than the service of nuclear weapons, advanced nuclear systems distinguish the armed world of today. This is true even though conventional warfare along the lines set by modern technology itself transcends the historical record we must draw upon for estimates of the future.

A quick way to grasp the transformation of warfare in the nuclear epoch is simply to list in their order of appearance a selection of modern systems for war—not only for nuclear weapons and their delivery, but also for physical control and command of their use by national leaders. At the same time the list conveys the nature of the superpower competition—the race against the clock of change.

Military nuclear systems have grown steadily more complex and powerful over more than three decades. The decline in overall energy yield by a factor of three since a peak around 1960 was in fact an outcome of measures of weapons modernization; warheads with high accuracy and swift employment could be smaller than the big bombs dropping from aircraft only after flights of many hours. But in most other respects there has been no respite; the two sides have steadily built, diversified, improved. The U.S. lead was clear from the start to the mid 1970s; a long pull toward crude parity has been the Soviet role since the start of the 1960s. Now again the plans look gloomily familiar. The United States proposes a considerable build-up, to regain its initial lead, and the prospect of determined new Soviet effort to remain a close match is expected.

Three periods of restraint can be found in the chronicle:

- The first arose out of the steady worldwide increase of radioactive fallout after the first thermonuclear tests of the early 1950s. That led in time to a world protest to which the powers bent, both by self-restraint and by the formal Partial Test Ban Treaty of Moscow in 1963.

The Treaty indeed ended the chief danger of test-produced fallout to public health, but it failed as a broader measure to curb the arms race, for underground testing continued at a faster pace than even the atmospheric tests.

The crisis of 1962 over deployment of Soviet medium-range missiles in Cuba had its influence on both leaderships; there was a slow move toward detente, which from 1972 to 1976 led to a useful if minor limitation on arms, SALT I. Even more important was the treaty of 1972 which forestalled the race for anti-ballistic missile systems. The Non-Proliferation Treaty of 1970 (signed in 1968) was a still earlier measure of value though as we have seen it bears little on the superpowers.

There are other useful specific agreements from that period, from the hot line to the exclusion of all nuclear weapons from Latin America.

- The second period was that of the vexed SALT II Treaty, signed in 1979 but never formally put into force because of the collapse of detente as a political stance.
- The latest opportunity dates from 1981. A million and more people have come into the streets across Europe and the United States. A newly-articulate public concern is expressed at many levels. This is the best chance for a generation to bring under a rule of reason the superpower obsession with gigatons of weaponry, whose use is intolerable to our species by a hundredfold.

Insecurity Through Technical Prowess

Prologue from 1945

Everyone lives in the present precarious world. But those physicists of 1945 who still live in it share a sharp sense of early recognition and a heavy responsibility. That my text is a quiet one does not mean its topics do not arouse the deepest and most powerful of feelings. Evidently they signify nothing less than life and death, set before us all to choose in wisdom. When the physicists shifted their science from its peaceable study of the ultimate structure of matter to the desolation of Hiroshima, they may have known sin. So Robert Oppenheimer put it long ago. Nonetheless, from the visible tragedy of Hiroshima and Nagasaki and from the future they could infer, scientists soon learned what would need to be done if our culture were to survive for a long time. The statesmen have not learned the necessary lesson so quickly; it is to be conceded that their task is the much harder one.

In evidence I cite an old official letter. It was written to Secretary of War Henry Stimson on August 17, 1945, by a panel of four physicists, A. H. Compton, Enrico Fermi, Ernest Lawrence, and J. R. Oppenheimer. Here it is, cited in part:

Dear Mr. Secretary:
. . . In examining these questions we have, however, come on certain quite general conclusions, whose implications for national policy would seem to be both more immediate and more profound than those of the detailed technical recommendations to be submitted. We therefore think it appropriate to present them to you at this time.

1. We are convinced that weapons quantitatively and qualitatively far more effective than now available will result from future work on these problems . . .

2. We have been unable to devise or propose effective military countermeasures for atomic weapons. Although we realize that future work may reveal possibilities at present obscure to us, it is our firm opinion that no military countermeasures will be found which will be adequately effective in preventing the delivery of atomic weapons . . .

3. We are not only unable to outline a program that would ensure to the nation for the next decades hegemony in the field of atomic weapons; we are equally unable to insure that such hegemony, if achieved, could protect us from the most terrible destruction.

4. The development, in the years to come, of more effective atomic weapons, would appear to be a most natural element in any national policy of maintaining our military forces at great strength; nevertheless we have grave doubts that this further development can contribute essentially or permanently to the prevention of war. We believe that the safety of this nation—as opposed to its ability to inflict damage on an enemy power—cannot be wholly or even primarily in its scientific or technical prowess. It can be based only on making future wars impossible. It is our unanimous and urgent recommendation to you that, despite the present incomplete exploitation of technical possibilities in this field, all steps to be taken, all necessary international arrangements be made to this one end . . .

Very sincerely,
J. R. Oppenheimer for the Panel

Look back now at what in fact took place. War has been frequent and extensive the world around; even the United States has been caught up in two major conflicts. But so far nuclear war itself has not returned. So much is hopeful. Yet the basis of its prevention has been exactly that transient development of more effective atomic weapons about which the Panel long ago expressed grave doubts. Today, we who recall the past share those doubts, more troubling now by far than ever. The rest of these pages will look at the past and present as this author sees them, to draw once again conclusions confirming those our colleagues reached years ago.

Nothing Can Stop the Army Air Corps: Strategic War

Warfare is at least as old as national states themselves. At first it was land warfare, hand-held swords and clubs, augmented by flung spears and arrows and in time by heavier projectiles hurled at and over walls. But target and launcher were close at hand, on solid ground, where people dwell, where they reap their crops and hold their wealth. Mobility was early made part of tactics, and the rise of the chariot, the charioteer, and the special breed of fleet horses that drew the chariots of war is of Biblical antiquity. Artillery (replacing catapults) and tanks (replacing cavalry) plainly take a familiar place in this same pattern with no enormous change in the scale of times and distances: battlefields and cities are still of familiar size, though armies sometimes move much faster.

Navies came in time, as soon as the narrower seas were mastered in the interests of trade. Finally the caravels and the big junks came to be at home on the open ocean; a blossoming sea trade between continents and across oceans marked the Age of Discovery, and the power of Europe spread worldwide primarily as a power of her navies and their landing parties. Island states like Great Britain long relied upon the sea as a "moat defense," before their skills and ambitions impelled them to sea-linked empire.

For the last century, the United States has been one of the only major nations without major border threats. Since 1814, it has been a fact of political geography that no foreign power was able to mount an invasion of this country against even a modest naval defense. Our few border countries are now much less powerful than the USA, and largely at peace with us over the long term. An invading army that must fetch supplies through a long sea voyage is hardly a real threat to American security. Consider how the English Channel has been a successful moat ever since Norman times, and compare its width to the Atlantic span. The only modern war Americans have seen on our soil that compared in its toll with the centuries of warfare in Europe was the Civil War, which was organized, equipped, and manned within these States themselves.

This geographical isolation has left its mark on the American mind, just as the memories of invasion have formed the traditions and opinions of the people of Russia or France. But today the people of Omaha, say, although they live at continental distances from any coastline, know that their city is as open to large-scale destruction by contemporary weapons as was any walled town in the lowland plains of Flanders, northern Italy, or Poland.

So profound a change is one consequence of the determined application of modern technology to warfare. But realization of that outcome could follow not only technical and scientific discovery and invention, but also protracted intellectual argument, sustained political decision, and unprecedented public investment within several nations.

It is perhaps natural enough that two sea-girded nations, at low risk, enterprising and industrially powerful, should have taken the lead in the development of strategic war from the air. Of course, the leaders—Great Britain and the United States—were not alone in that. The first ballistic missiles were the German V-2 rockets; the first intercontinental ballistic missiles were Soviet. But the doctrine, the demonstrations, the large-scale preparations and use of long-range air war against industry and population, and the indispensable nuclear warheads were all achievements of the United States and Great Britain.

A look at deaths during World War II will show how important for both nations was their isolation from land battle and siege. Together Britain and the United States lost less than a million lives, including civilians at home, and soldiers, sailors, airmen, and merchant seamen abroad. But their allies and opponents alike counted deaths well more than an order of magnitude higher. The incentive for a means of warfare that would not expose so many British or American expeditionary soldiers to battle abroad was plainly high, always provided that a similar attack on the American or British homeland remained out of the question.

There were plenty of authors who foresaw strategic war from the air even in the days of the Wright linen-and-wood aircraft. The novelist H. G. Wells wrote of nuclear war from the air before World War I. But it was the 1917 report made by General Jan Smuts to the British War Cabinet that for the first time gave authoritative political support to a vision of decisive war waged in the air from behind the battle line. The provocative English historian, A. J. P. Taylor, ironically called the Smuts report "epoch-making." "There stem from it all the great achievements of our contemporary civilization: the indiscriminate destruction of cities in the second World War, the nuclear bombs dropped on Hiroshima and Nagasaki; and the present preparations for destroying mankind."

If soldier-statesman Smuts supported air war, there were professional airmen who had long worked for it. Conventional doctrine saw aircraft as close support for the other arms, as observers aloft, swiftly mobile artillery and machine guns, transport of troops, and defense from the strafing and bombing aircraft of the enemy. Those are real functions, seen even today as indispensable by land and sea. But Sir Hugh Trenchard and his American friend, the celebrated Billy Mitchell, saw an entirely different future for the air arm: their new weapon would be indifferent to trenches and walls, save as targets, and easily ignore borders and coastlines.

Figure 1. General Billy Mitchell's DH-4B bomber, c. 1922.

In 1918 the British acted. The Royal Air Force supplanted the Royal Flying Corps. It was a new service born coequal with the British Army and the Royal Navy. It was still charged with the expected support of the surface forces, but there was formed within the RAF—if with very few squadrons—a new Independent Force under Trenchard, its bombers to be used as "an offensive and not a defensive weapon," to overfly the muddy trenches, to carry the attack directly to the war factories and their workers at home. But the war ended before 1918 was out.

Brigadier General William Mitchell commanded air combat for General Pershing in France. He too was a prophet. His 200 bombers used in ground support in 1918 were described in the press in the phrases of the future: "This navy of the air is to be expanded until no part of Germany is safe from the rain of bombs. It is a thing apart from the squadrons attached to the various army corps. The work of the independent air force is bombing munitions works, factories, cities and other important centers far beyond the German lines." (See Figure 1.)

World War II fulfilled the prophecy. After the Luftwaffe had opened the door by its attacks on cities, the RAF and the US Army Air Corps over Germany and the USAAC alone over Japan carried out exactly the plan foreseen by their founders. Now the forces were armed with heavy bombers and radio navigation of performance beyond the dreams of the pioneers. Rotterdam and Coventry and London and Pearl Harbor were avenged in full measure. In Europe, where flak and the Luftwaffe fighters offered a terrible resistance, the air battle was broadly indecisive. The bombers killed half a million civilians and dislodged a dozen times more, but the losses of costly aircraft and well-trained men were heavy. Cost for cost, there was not much to choose. German war production rose steadily during the years of 1943 and 1944 under heavy air bombardment, and fell only after ground troops appeared on the Western front.

In Japan it was different. The heavy bomber was at hand at last, in numbers reaching six or eight hundred of the formidable B-29s by the end of the campaign. Their bases were islands, secure and isolated fortresses with local populations less numerous than the airmen. Japan lacked by 1945 the wherewithal for adequate defense at B-29 altitudes, either by fighters or by flak. The redoubtable Curtis LeMay, Major General, Twenty-First Bomber Command, unloaded the guns and gunners from the aircraft in order to load his planes with maximum cargoes of about six tons each. They dropped 20 million small incendiary bombs filled with jellied gasoline, to set alight about sixty crowded cities of wood and paper houses throughout Japan between January and August of 1945.

The cost in lost planes and crewmen's lives was small; the damage on the ground was about as great as that in Europe, but was wreaked in a much shorter time and within a much smaller area. Perhaps 10 million homeless and half a million dead could be counted among Japanese civilians.

It is hard to decide if that attack by fire was decisive. Terrible it was. But Japan was so badly beaten by land and sea, so much reduced by blockade, that its powers were failing everywhere.

The two atomic attacks then came as symbolic of a still worse future. Looked at from the ground, they were not so very different from the firestorms made by jellied gasoline in Tokyo and perhaps in Nagoya. The difference that nuclear explosive would one day come to mean was evident, but still potential. It can be reckoned that it took around a hundred sorties of LeMay's aircraft to destroy a square mile of city; the first crude atomic bombs knocked down and burned a few square miles each, though each was brought by a single bomber and an escort or two. Those bombs, moreover, became very cheap; one such bomb today costs much less than a single modern bomber or rocket. Nuclear damage remains cheaper by a hundred-fold.

Strategic war from on high had matured. As the war ended, it had found an unexpectedly apt and terrible new weapon, in fact, two of them—the ballistic missile foreshadowed by the German V-2, and the nuclear explosion.

The National Security Act of 1947 finally gave the United States an independent force in the air, the United States Air Force. The change was chiefly a formal one; though the big Army Air Corps of WWII was not independent in name, it was all but sovereign in fact. (This author can recall the visit made to him in 1946–47 by two leaders of the technology of war from the air. They sought to recruit nuclear weapons experts to staff the research organization they had newly founded with good will and a few million dollars from General H. H. Arnold, commander of the USAAC. The new California lab would be called the Rand Corporation, the name coming from the phrase "research and development." It had a bland public charter, but in secret, its founders said, it would put first the waging of intercontinental warfare by any and all means.)

Billy Mitchell had written it all down in 1930; nor was his the first voice. The "vital centers" of the enemy's will to resist are "the cities where the people live, areas where their food supplies are produced, and the transport lines . . . the real objectives are the vital centers. The old theory that victory meant the destruction of the hostile main army is untenable. Armies themselves can be disregarded by air power if a rapid strike is made against the opposing centers."

By 1947 we had a Strategic Air Command, its chief the same Curtis LeMay who brought wildfire to urban Japan, its weapons a US monopoly of atom bombs (once again, more of them were on paper at first than in the munitions depots). Its image of air war was based on the fiery ordeal of Japan under the B-29s, this time armed also with atomic bombs. The newspapers of the time carried proud Air Force ads. They showed a new world map, the whole globe as seen by air power. It was a ball without seas or lands, blank except for target spots, the great world cities. But among those there would one day be entered Washington, Omaha, Seattle, and all the other cities within our own borders. That fact the ads of the 1940s did not make very clear.

Year by year, ever since aircraft were new, incremental steps of technical and organizational improvement—many technical fixes, many real achievements of transient superiority, each one rationalized as an advantage—have led to an overall plan for US defense in which strategic war is fundamental. But exactly that form of war, and no other, has brought to an end the long-held advantage of the geography of the New World. By the nature of its long-range weapons, strategic war entails also the likely abandonment of almost every restraint upon the savagery of war against civil populations. That has been the record, and no change is yet at hand. Americans, like all the rest, live today in that insecure regime.

Our USAF has the greatest resources among all three American military services. It is in effect the largest and one of the most able technical and industrial organizations in the world, largely devoted to its great task of intercontinental warfare, and tirelessly innovative. But the example of its doctrine and weapons, no longer unique and unopposed, has come to threaten our United States no less than our adversaries with the unprecedented risk of destruction.

Perhaps in this narrative a lesson is embedded.

Trinity: Nuclear Weapons

In the New Mexico desert, early on a Monday morning, 16 July 1945, the sun could be judged to rise twice. No one of the men who lay there in the open air on the ground ten miles from the tower holding the first nuclear explosive device—certainly not this author—can forget that initial false dawn, when the cool of the night was ended for a few seconds in a flash of brilliant light and a wave of noonday heat that suddenly irradiated exposed hands and face.

That was the Trinity test, the first nuclear explosion. A plutonium sphere was imploded to yield the explosive energy of 20 kilotons of TNT, a couple of thousand times the energy of any earlier single bomb. That test device was the close counterpart of the bomb detonated three weeks later over the edge of the city of Nagasaki. The Hiroshima bomb was of a different and never-tested design. Its principles were so straightforward and its components so amenable to separate trial in detail that there was high confidence that the untested weapon would function.

So much has been written about the fateful decision taken by President Truman under the urging of all his advisers that no summary will be attempted here. Let it suffice to say that the two bombs were dropped more as part of the steady heightening of the strategic war against Japan, two more cities destroyed after sixty already reduced to ruin, than as the world's entry into warfare at a wholly new magnitude. No specific warning was given; time was hardly allowed between the two attacks for any response.

It is hard to look coldly at that frightful event, the more since we now know that the bombs in fact presaged a new order of warfare for which the powers are now so inordinately prepared. As weapons of 1945, barbarous and sudden as they were, laden with the cruelties of long-delayed effects as they turned out to be, they added only a final intolerable increment to the on-going strategic war over the Empire of Japan. If LeMay had dropped only a couple of dozen bombs like those instead of his hundred thousand tons of fire and blast, the people of Japan would have endured more or less the same sufferings, at least within the war years. But the atomic bombs in fact belonged to the future, now our past and our present.

The key to the unmatched danger implied by Trinity is just that: the powers are armed for strategic war at an unforeseen level of damage, far beyond what any nation has yet endured. For with nuclear explosives, the 1945 level of damage would require only a tiny fraction of the forces they can afford to build and staff. We see that in 1985 both superpowers maintain a force of a couple of thousand long-range launchers, not so very different after forty years from the nearly 1000 bombers of the Twenty-First Bomber Command. True, they are faster, bringing destruction in hours rather than months.

What has remained more or less the same is the economic effort of war preparation. The money and the men required to wage strategic war are determined by launcher costs, just as they were for the Army Air Corps of World War II. Those costs remain more or less the same in real terms, only a fraction of the total cost of all armed forces, within a much larger economy. But the expected cost of the outcome of nuclear war, the physical and human effects of actual use of all those nuclear weapons, has risen fully a hundred-fold, even when measured by human lives lost.

That is one concrete consequence of the years since Trinity. No one can be unaware that another consequence has been the avoidance of direct war between nuclear powers in those years. The stakes have grown so high that the statesmen-gamblers are cautious. War has indeed been far from absent; it has involved non-nuclear belligerents, and even superpower forces have opposed smaller non-nuclear foes. The tendency towards avoidance is clear. Yet all history suggests that the risk we face remains far too large; reason does not always control what happens between states when for years they stand with dagger drawn.

The effect of extending the air war over Japan into the atomic age was to endanger our nation and the world. Tactical victory, even a war-winning one, in the end led to an overwhelming risk few now are willing to accept as permanent. Once again achievement by innovation in weapons, "technical prowess," did not bring national security, but instead an unprecedented and growing risk.

In this story is Lesson Two.

Hydrogen: The Decision

In January 1950, President Truman publicly set a new national directive for the Atomic Energy Commission: "to continue its work on all forms of atomic weapons, including the so-called hydrogen or superbomb." The concept was not new. Even before Los Alamos, the theorists had recognized that the fierce temperatures within a fission explosion might be able to kindle the other source of large-scale nuclear energy, known to power the sun and stars: the fusion of the lightest nuclei, the isotopes of hydrogen. It is in a way complementary to fission, which is the splitting of the heaviest nuclei, the isotopes of uranium and plutonium. A fusion explosion, sometimes called thermonuclear because astronomically high temperature is the essence of the event, yields, per pound of fuel, several times as much energy as does fission. The novelty is nothing like the change from chemical explosions to nuclear, but important nonetheless for delivery of nuclear weapons in missiles. Fusion weapons can be made of unlimited yield, far exceeding the half-million tons of TNT that measure the largest fission explosion up to now. Every fusion weapon needs a fission weapon as trigger; but fusion provides a cheap way to increase by a factor of some tens the energy of the fission device, without large increase in its weight. The cost of the device does not increase even proportionately with yield; the new materials are widespread and relatively cheap.

It was not these properties that led to Truman's call, nor to the "general chord of nonpartisan congressional support," as the *New York Times* put it. It was the riveting fact that in September 1949, four years after Trinity, atmospheric samples gave conclusive evidence that the US/UK monopoly on atomic weapons had ended. An atomic weapon had been exploded within the USSR on 29 August 1949. Our reply was to attempt once more to reassert a matchless American technical superiority.

The debate over the superbomb had been brief, sharp, largely behind closed doors. The promoters stressed the necessity of regaining our lead, more in psychological terms than in military ones. The opponents argued first of all that the weapon was not clearly within reach, and second, that it raised the risk of the game without any compensating benefit to the United States. Perceptions won out easily, and about mid-1951 the first test of a booster, a small fusion contribution that enhances the efficiency of a fission explosion, was held. It is a kind of halfway stop on the road toward a true hydrogen fusion device. A remarkably original idea of Edward Teller and Stanislaw Ulam that would lead to the first fully thermonuclear explosion came at about the same time.

There was a moderate course of action; that shrewd Yankee, Vannevar Bush, put one forward, it is said. Study, develop, even prepare for tests of any promising weapons design. But do not in fact test it, except after effort at an agreement to set aside any US test unless and until someone else tested. The Bush scheme was, in effect, a plan for no first test. A no-first-use pledge was more widely urged.

Instead, the first full thermonuclear test was made by the United States in November 1952. It erased the islet of Elugelab and for a while burst the bounds of secrecy. Teller and Ulam had indeed had a workable idea. The first Soviet booster test followed only a year and a half later. The United States test-dropped a high-yield, practical H-bomb in spring 1954, and the USSR tested its first counterpart of the Teller-Ulam scheme, a device invented by Andrei Sakharov, late in 1955. The time of hydrogen had come, to both sides, and by now to Great Britain, France, and China as well.

This can be read as Lesson Three.

How the Arsenals Grew

The hydrogen bomb meant an era of nuclear explosive plenty. Delivery would rise in step with the new view of numbers and yields (see Figure 2). The Strategic Air Command's piston bombers, mostly of medium range, could carry by 1949 some 300 A-bombs from peripheral overseas bases into Soviet territory. There were no Soviet A-bombs yet, but Soviet manpower began to rise somewhat, and of course its air defense grew speedily. By 1952, SAC began to fly its first jet bomber, the B-47, and the weapons stockpile started to grow in the era of hydrogen. The numbers of SAC bombers and strategic bombs grew steadily upward to a broad maximum reached in the early sixties: as many as 2300 to 2500 jet bombers with 13,000 strategic bombs. The B-47s were based on foreign soil, from Greenland and the United Kingdom across North Africa to Turkey and the Persian Gulf, and facing the Pacific coast of the USSR and China as well from Taiwan, Okinawa, and the Philippines. The longer-legged B-52s could fly direct from US bases, and they numbered some 650. In-flight refueling was widely available. The yield of the thermonuclear bombs deployed has never been so high as it was then.

Soviet short-range fighter-interceptors and anti-aircraft guns and rocket launchers on the ground grew in proportion to SAC. That sort of a reaction—offense, defense—is clearly visible.

But meeting a growth in offensive strength by counterpart growth in offense does not show up clearly in the record as a simple action-reaction sequence (see Figure 3). There the changes are slower; they appear to come on each side in execution of a long-term policy confirmed by interpretations of the actions of the adversary, rather than in a tit-for-tat mode. The USSR might have counted 2000 nuclear bombs by 1957. Soviet bombers were mainly of theater range throughout; their intercontinental

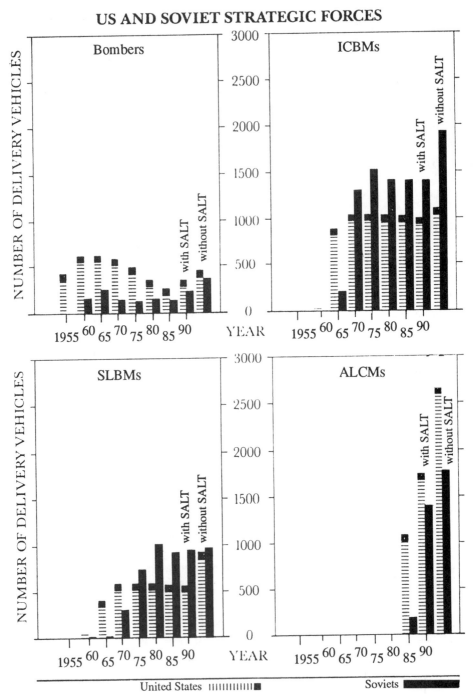

Figure 2. The United States maintained a large lead in nuclear delivery systems until the 1970s, when the Soviets achieved the rough parity that still characterizes the balance of forces today.

THE RACE

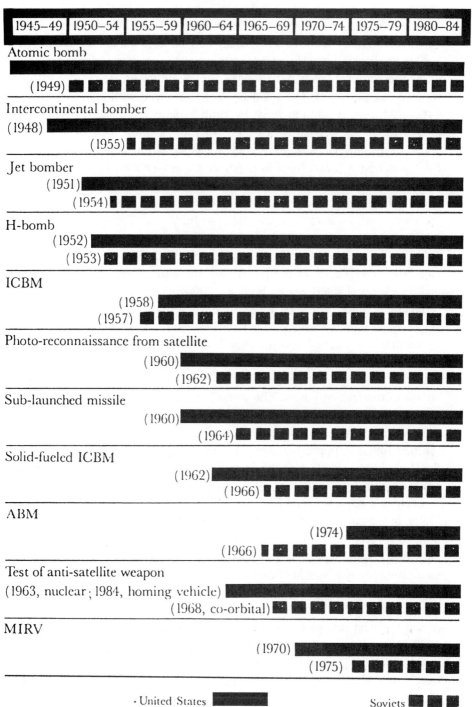

| 1945–49 | 1950–54 | 1955–59 | 1960–64 | 1965–69 | 1970–74 | 1975–79 | 1980–84 |

Atomic bomb
(1949)

Intercontinental bomber
(1948)
(1955)

Jet bomber
(1951)
(1954)

H-bomb
(1952)
(1953)

ICBM
(1958)
(1957)

Photo-reconnaissance from satellite
(1960)
(1962)

Sub-launched missile
(1960)
(1964)

Solid-fueled ICBM
(1962)
(1966)

ABM
(1974)
(1966)

Test of anti-satellite weapon
(1963, nuclear ; 1984, homing vehicle)
(1968, co-orbital)

MIRV
(1970)
(1975)

· United States   Soviets

Figure 3. For each innovation in military technology, two lines are shown to represent when each of the superpowers tested or deployed the innovation. As the figure shows, each innovation is quickly matched.

bombers reached a maximum of about one-third of SAC's B-52s in the early 1960s. That is the present state of aircraft forces as well. To be sure, stand-off missiles are now the endangered bomber's chief weapon, and counts of aircraft are no longer quite so central.

As the bombers reached their maximum deployment, the time of ballistic missiles had come. The US decision to go ahead with long-range ballistic missiles was taken in 1954, H-bomb in hand, but the first ICBM test at long range was carried out by the USSR in the summer of 1957, a year ahead of the first US test. Thereafter it was the US missile deployment—SAC was now a rocket force as much as it was an air-breathing service—which grew much more rapidly. At first intermediate-range ballistic missiles (IRBMs) were deployed, with the United States placing a hundred or so Thors and Jupiters in Great Britain, Italy, and Turkey in the early 1960s. Soviet counterpart IRBMs appeared in eastern Europe, and temporarily in Cuba. The United States deployed its first ICBMs—the liquid-fueled Atlas and Titan—in the late 1950s and early 1960s. Between 1962 and 1967, the United States built a 1000-missile force of solid-fueled Minutemen. Since then, silos, missiles, re-entry vehicles, and warheads have been upgraded more than once, but launchers have not increased in number.

Submarine-launched ballistic missiles, hiding in the wide and opaque seas, entered US Navy service between 1960 and 1966, plateauing in launch-tube numbers soon after. Treaty-limited, they are beginning to increase in number only now. The overall count of strategic delivery vehicles—by no means the only measure that matters—stood sharply in favor of the United States until the early 1970s, when the Soviet ICBM force reached roughly its present size. The long-standing disparity in favor of the United States is clear on the curves; five to one in about 1965, it fell to crude equality by about 1972, as measured by launcher numbers.

Each side has displayed periods of rapid growth: the United States initially, in the first half of the 1950s when SAC built up its thermonuclear bomber force yield, refueling, and forward bases. Between 1961 and 1965 the United States supplemented its bombers and forward bases with home-based missiles and launchers on subs, increasing its total steeply to a maximum in 1967. The B-47s—smaller loads, less penetrating against new missile defenses and more vulnerable—were rapidly removed from service during the next few years, so that our total fell to its present range about 1970. Russian delivery-vehicle totals rose steadily between 1963 and 1975, mainly in ICBM count but including sub-launched missiles as well. The matching totals and the lack of much change since 1972 or so are the most significant facts of the race. Saturation was near, at least for launch vehicles.

Our H-bomb lead did not grant security for very long. On the contrary, megaton thermonuclear weapons in the nose cones of Soviet ICBMs threaten every US city and military target. Dr. Bush's idea looks better in hindsight; even Robert Oppenheimer's alleged "lack of enthusiasm" for the H-bomb no longer seems so antagonistic to national security.

Another essay would be required to treat the spread of nuclear weapons beyond SAC and strategic war onto the plans for the battlefield, where every military organization stood ready to stake a claim on the new power. The Marines and the Army Engineers, the destroyers dropping depth bombs, the artillerymen at their howitzers and the interceptors waiting for Soviet bombers to appear, all ordered and now hold nuclear weapons. These tactical nuclear weapons first entered US forces beginning in 1952, with the Mark 9 ten-inch howitzer shell. They became diverse and numerous by the late 1960s. Soviet forces have such weapons too, as usual lagging by a few years.

There are twenty-five distinct models of nuclear weapons in use today in US forces, many in several versions, and almost as many models have run their course of utility and been discontinued, their nuclear fuels reworked into new forms. The effect has been to unite nuclear and conventional war to a degree, though a firebreak still exists as long as unit commanders in general follow orders. Some of the twelve or fourteen thousand such weapons in the US stockpile cannot physically be used without code inputs from higher authority; many others can. The consequence of this battlefield nuclear arsenal for war in Europe, where the United States maintains about 6000 nuclear weapons—both its risk and its cost—is a topic of major importance not discussed here.

SAC remains in the mode set by General Mitchell long ago, but SAC and the Air Force are not alone in might. Indeed, SAC has shared strategic targets with the Navy missile submarines in a joint plan (the Single Integrated Operational Plan) since about 1962, and the Army long-reach missiles in Europe have also been included.

Megaton yields and small yields alike are now efficient engineering options, with overall weights at a couple of megatons yield per ton of warhead weight. The variety of weapons available in small-weight packages suited for lighter ballistic missiles and for cruise missiles of long range is the outcome of control over hydrogen explosions of all sorts.

MIRV: A Premium on Striking First?

The most completely realized example of an innovation whose technical success led us to a sense of enhanced vulnerability is the rise of the multiple independently targetable re-entry vehicle, or MIRV. The idea grew somewhere in Air Force planning circles, likely in Rand, in the early 1960s. Missiles were fine weapons, but against them the anti-ballistic missile threat of Soviet technology was growing. What better idea than to send the re-entry vehicle and its warhead in to the target along with a "company of lies"? The waiting radar could be fooled by metallic chaff, by decoy balloons, by a variety of light, cheap, and clever aids to penetration. These deceptions were best aimed near the radars they were to fool, not always straight at the target. Guidance could do that, it was thought. Such schemes were carried out; they exist today.

Effective ABM systems never materialized. But MIRV came all the same, in a new and sinister form. First proposed as indispensable to the deterrent, the real deployment, thorough and expensive, turned out to be for offense, even for the intricate stratagems called "counterforce." For it was not aids to penetration, decoys of one kind or another, that the multiple RVs mainly carried; it was complete nuclear warheads. Now one missile might attack a few or even a dozen targets, and the options offered to the planners grew. They included, of course, a pre-emptive attack on any targetable missiles, with the prize a trade-off of five or ten missiles for one. Many critics observed that as soon as the other side deployed this intricate technology, the very silos in which MIRVed missiles were initially based would be made vulnerable, at least in principle. Why start a game whose next play would bring instability?

We went ahead to cash in our clear, though temporary, technical advantage. That meant turning down, it is claimed, a feeler from the Russian side to prohibit MIRV by treaty. The first full MIRV system anywhere was deployed beginning in 1970, with three warheads per missile on each of 550 USAF Minuteman III missiles by 1975 (see Figure 4).

The submarine missiles followed: in 1971 the Poseidon C-3 missile entered Navy service, each missile fitted with 10 to 14 RVs. The plateau of US missile numbers by land and sea was maintained, but warhead and hence target numbers went up steeply. In 1985, we had only about 1000 land-based and 650 submarine-launched missiles, but they can strike 8000 targets or more. Upon MIRVing, warhead yields must be subdivided too, but present guidance makes the smaller warheads—none of them so small as the A-bombs of 1945—menacing even to hardened targets. If one strikes first, up to ten warheads can be killed for each one launched.

We were duly warned of exactly this "window of vulnerability" in the election campaign of 1980. For, of course, Soviet missiles appeared with MIRV warheads operational around 1977. General Scowcroft and his colleagues on the special commission set up by President Reagan during his

US AND SOVIET STRATEGIC WARHEADS

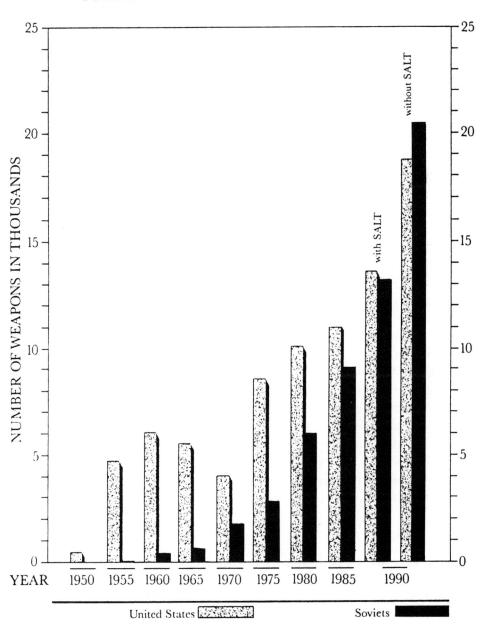

Figure 4. The number of warheads in the US and Soviet arsenals has grown steadily, but increased sharply after each began MIRVing their missiles in the 1970s. Figure shows strategic warheads on ballistic and cruise missiles, and on long-range bombers.

first term considered this "window of vulnerability." They saw the MIRV technology the United States had introduced and pushed into powerful reality as a current threat to our own safety, though not yet a fatal threat, and proposed that US policy seek to return by example and agreement to the simpler days of single-warhead missiles.

The circle of opinion closed on MIRV in under twenty years, including about ten years of deployment. That innovative plan essential to our missile forces became, as forecast, a present danger to those same forces. Not all clever new weapons confer increased security. MX is our newest MIRVed missile. It will be based in fixed Minuteman silos, with nominal concern for the "window of vulnerability," it was supposed to close as a mobile missile. Its offensive value was what remained. There is a great deal more to be said here: force scale, deterrence and counterforce, the distinction between the silos visible and targetable from orbit and the hidden subs, the gambits of launch-under-attack, the limiting role of treaties, and more. Our focus on change over time does not allow us to enlarge upon the intricate issues.

It is hard to avoid a disagreeable inference: chief motive for the adoption of novel technical schemes for nuclear war is often less that they offer a clear improvement in national security than that they enhance the scale of organizations, revenues, duties, and prestige, especially among those who decide and vote for them, or who design, build, staff, and command them.

We cannot afford the stimulus of even that natural and enthusiastic self-interest if we are to survive in the long run. We need firm and rational control over unending zealous support for dangerous growth. No other growth industry so threatens the future as the unneeded growth of nuclear weapons along with novel schemes for their delivery.

MIRV surely offers a Lesson Four.

The Present: Insecurity by Choice?

So long a chronicle is of small use unless it leads to some insight into the proposals and problems of the day. But indeed they are here in plenty to be examined, and that examination is much aided by a cool look at the sweep of technology in the past.

The first system, by now not so new, which merits attention is the long-range cruise missile. Present US plans are to produce three to five thousand nuclear warheads for use in a variety of such missiles. Around a thousand (of the first air-launched model) have been deployed by the USAF so far in ninety of our B-52Gs. Such airborne missiles serve to preserve the ability of strategic bombers to strike targets so well defended in the missile age that jet aircraft cannot approach them closely. Were this the only use, the system might find some rationale, for it does not increase the launcher numbers. The SALT II treaty limits these missiles, in number, to deployment on such strategic bombers, and even provides that strategic aircraft carrying such weapons must be recognizable, to allow weapons counts; so far the United States is complying with the proposal.

But the cruise missile is more widely used. Ground-based versions are appearing by the hundred in several European bases. Testing of the similar naval version is well under way (see Figure 5). These weapons are cheap and small by previous standards for nuclear delivery vehicles. They are V-1 successors, air-breathing and rather slow, at present sub-sonic.

Their high risk is plain to see. It is likely that a substantial deployment would mean that any naval warship down to a frigate or even a merchant ship, certainly any modern submarine, as well as a large number of relatively slow, big aircraft (including wide-bodied civil jets), could threaten nuclear strategic war. For the devices promise to fly 1000 miles or more to seek a precise target, identified by the radar pattern of the terrain. They spend hours in flight, so that before they arrive a missile retaliation might be well under way, but the potential is still menacing. A shoreline as long as our American one is no place to defend against a possible sudden launch of up to thousands of little robot jet planes, any one of which can ruin an inland airbase, headquarters, or even a city. Their low altitude of flight protects them from the orbital lasers of the imaginative future. Their widespread deployment against us would generate a large, specialized US air-defense system of aircraft and missiles, unlike any we have ever built.

The cruise missiles are simply the latest and much the most troubling of the military developments of what are called dual-capable delivery vehicles. These are subs, ships, and aircraft designed for and often armed with conventional weapons, yet able to make effective use of nuclear weapons as well. No one doubts the military utility of the cruise missile armed with high explosive against ships and similar targets. But making them able to do damage at nuclear scale and far away is raising the ante in a game no one can win. The same danger in a lesser degree has been suggested for a long time by the many fighter-bombers now prepared, on both sides, to carry afar whatever sort of bomb any particular sortie might demand. Escalation is put right around the corner.

It is the European theater that is most affected by dual-capable systems problems. There even nuclear artillery shells are deployed. A given eight-

Figure 5. Test of US submarine-launched cruise missile.

inch gun can lob a nuclear shell of 10,000 tons yield about as easily as it can hurl the twentieth of a ton of TNT that was the artilleryman's stand- ard. (Here we enter the very general reminder that the area of damage does not grow in proportion to the nuclear energy yield. Instead of two hundred thousand times more damage per shell, the nuclear artilleryman can in fact only expect to destroy three thousand times the area of high- explosive damage. But one such nuclear shell ends one European village.) That nuclear explosion takes place only some fifteen miles away from the firing cannon.

Such minimal range gives strength to the quip of "use 'em or lose 'em" to enemy forces nearby. The presence of 1000 nuclear shells in NATO forces seems more a sign of evenhanded "nuclearization" of all arms of the military than a useful part of the extended deterrent NATO maintains.

TABLE 1. Increasing accuracy of United States and Soviet missiles

Missile	Type	Year Developed	CEP (ft.)
Pershing II	IRBM	1983	145
MX	ICBM	late '80s	400
Trident II	SLBM	late '80s	400
Minuteman III	ICBM	1980	700
Soviet SS-18 (mod 5)	ICBM	1985	800
Soviet SS-19	ICBM	1979	1000
Minuteman II	ICBM	1966	1200
Soviet SS-20	IRBM	1977	1300
Poseidon	SLBM	1971	1500
Soviet SS-17	ICBM	1975	1500
Soviet SS-N-18	SLBM	1978	1900
Titan II	ICBM	1962	4200
Soviet SS-11	ICBM	1966	4500

Note: Improving warhead accuracy is indicated by the decreasing size of the "circular error probable" (CEP). Fifty percent of warheads will fall inside the circle formed by the radius indicated for each delivery system. For example, a Pershing II warhead has a 50 percent probability of falling within 145 feet of its target.

(There is a current program to replace and modernize these and other nuclear artillery in NATO.) Here the dual-capable system is seen in sharpest light. Such systems are more technical risks than they are technical fixes for the intricate problem of European defense.

The technology of guidance is one which has shown steady increase in capability (Table 1). That change rests on the remarkable improvement in packing computing power in small volume, familiar now even in the home. An intercontinental warhead can today be placed on a target overseas with an expected uncertainty in distance of only a city block. That gain in placement accuracy—there is more to come—has meant that the relatively modest yield of multiple warheads can be used to destroy small, fortified underground targets, including military and civil command centers and missile silos. That in turn is a large part of the rationale behind nuclear warfighting scenarios that contemplate such attacks in the hope that the war will not escalate to centers of population and industry. Again the assurance seems a faint hope in the context of history, in a world where people dwell near military targets, or at least downwind from them, and in a world where destruction on nuclear scale is cheap. Guidance improvement is another sweet technical triumph with a very bitter center, once two sides play at the same game.

One development under way is of momentous importance, less technically perhaps than in its possible effect on US military doctrines. The commanders of missile submarines at the moment cannot know their own shifting location as closely as do the commanders of stationary silos on land. That means submarine-launched warheads cannot seek targets with the accuracy of the land-based warheads. It is now judged, as a result, that those sea-based and hence much more secure missiles cannot be used well to attack small but hardened targets, the silos and control centers. Both the high security of the hidden submarines and the absence of scenarios developed for the dangerous stratagems of nuclear counterforce and decapitation have allowed the Department of the Navy a more prudent stance towards its strategic mission.

Land-based missiles are subject to prior destruction. Their Air Force managers view nuclear war as sudden and swift, a matter of hours, unless it includes the elaborate trading games for protracted war under the counterforce theories of duels on the razor's edge. Those commanders naturally run scared, and by intention or perhaps even without it, transmit some of their urgent sense of menace to the public.

The submarine forces are by contrast silent, private, even reflective. Will that attitude be maintained once the Navy acquires the same capabilities as the USAF land-based missiles to attack hardened targets? Once before, during the early 1950s, naval officers of the highest rank spoke out against nuclear strategic warfare. But decisive naval opinion changed as the missile subs were brought into being.

The test may come soon. Provision for guidance corrections by celestial navigation using the starry skies at mid-course, and by radio location using artificial constellations of clock-bearing satellites now being orbited, are planned for the D-5 missiles, to be deployed aboard the new Trident fleet of submarines before 1990. Those boats will remain hidden in deep bluewater, so they need not act in a rush. Yet they will newly be able to attack the opposing silos and command centers, so they hold the option to undertake the same sudden nervy games. It will be important to see how far this may modify the public views of the Department of the Navy towards the nature of strategic war.

It is worth a reminder that the counterparts of D-5, accurate submarine-borne missiles with many warheads, are likely one day to appear in the ocean off our own shores. They will then threaten, at least in theory, the simultaneous attack on airfields and silos from which we are now secure. The "window of vulnerability" will be at least arguably present in fact. Warning time might be reduced to ten minutes over so short a range; first strikes might appear still more advantageous.

New and dangerous remedies will become more attractive, too. On the one hand, we will prepare for shorter and shorter times of decision, possibly arranging to launch the missiles on first warning of attack. Even worse, the virtues of a pre-emptive attack could become clearer and clearer to our planners as well, and both sides will grimly set the triggers of destruction more delicately. The MX and its earlier Soviet counterparts, SS-18 and 19, the Pershing II in Europe and its Soviet opposite number, the SS-20, all represent steps more or less in the same direction. But the accurate and quick-to-arrive D-5—and its eventual counterpart against us—still lie ahead. If wisdom prevails, they might not appear.

The story of the sub-launched D-5 missile is a paradigm. New technology—more precise guidance by corrections made in flight—is prerequisite. But such developments do not come automatically; they must be conceived, worked out, ordered and paid for. The consequences, moreover, are generally not unique, nor automatic. What such changes mean depends greatly on the conclusions drawn by the leaders in whose charge the systems are placed. A choice remains fully possible; strategy is guided in part by technical circumstances, but finally it rests upon human intention. It is open to change.

A less time-urgent but even more tangled issue is raised by our efforts at anti-submarine warfare. In part these are in support of our surface shipping, defending naval ships and the lines of supply against submarine attack. But we seek to extend these tactics even to prepare intrusion into the Barents Sea, where the Soviet retaliatory subs might hide under the Arctic ice. A treaty limitation to anti-submarine warfare directed against such missile subs could promise more security for both sides. There is time for consideration in this regime of confrontation, since so far—to paraphrase one chief of naval operations—the more we learn of the oceans, the more opaque they seem. The quiet, swift, deep-running nuclear subs are not soon to be at risk.

A kind of warfare in time has been a by-product of accurate guidance and its counterforce premises. The missile age cut the decision time for retaliation from many hours to under one hour. The effort to devise means of attack on warning and communications satellites may reduce this further still, in the end even displacing human decision from the loop, a topic already familiar in fiction.

These remarks lead us to consider war in space, the most debated of all new technologies today. That a space defense against orbiting missiles can ever render nuclear weapons impotent is plainly not to be expected. It is doubtful whether it can make even the missiles "impotent and obsolete." For it is a commonplace that nuclear weapons can be delivered not only by ballistic orbit but by a variety of much less conspicuous or even covert means, by air, on land, or underseas; one large undersea weapon can destroy so large a coastal area by tidal wave that even close approach to shore is not required.

In fact, discussion and action under the Strategic Defense Initiative is centered on proposed systems in which important components are stationed, or perhaps suddenly popped up, into earth orbit. The technology of powerful lasers, large orbiting mirrors, and supercomputers in heavy armored satellites in space has a long way to go. Demonstrations of suggestive but in fact not genuinely important capabilities of this kind are planned over the next few years, more as public relations than as military R and D.

Consider here only one plausible early outcome of the Strategic Defense Initiative. One of the intermediate steps along the 25-year road to President Reagan's goal (or more likely, its eventual abandonment) is pretty surely a working anti-satellite system. The old US anti-satellite efforts used nuclear weapons, not the place to begin legally or popularly. The USSR space-mine system with conventional explosive has performed very unevenly. The first workable weapons system for combat in space, is the USAF anti-satellite system that uses small homing projectiles that kill by impact.

Long before there is any system out there waiting and able to kill suddenly launched missiles by the thousands as they rise to orbit, there is sure to be something of a US system good against a few satellites as they make their repetitive rounds (see Figure 6). It will be tested, and as SDI grows, various early precursors to the eventual, powerful SDI orbital stations will enter test orbits. Soviet satellites will be put at high risk. Soviet anti-satellite efforts will then become urgent as a safeguard to their command and warning systems. Their technology will improve. It is all too foreseeable that the presence of early SDI systems in orbit will provoke an end to the long regime of peaceable toleration of satellites by the states whose territories they overfly, long before the SDI goals are within reach.

We can recall the famous bringing down of the U-2 overflight, once the USSR had mounted rocketry able to attack the high-altitude jet. A rerun of that history seems likely to be one of the earlier international consequences of SDI, whose continuation for long will surely bring a measure-countermeasure arms race in the transparent new medium of space warfare, and with it a state of chronic crisis between the superpowers. That very outcome is perhaps the most realizable of all the "goals" of SDI; it is one desired by some of its most zealous proponents, but certainly not by everyone.

More prudent people share the view that SDI brings exactly what we

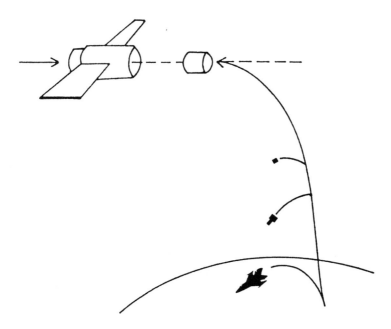

Figure 6. The US anti-satellite weapon is a direct-ascent, homing-kill vehicle that is launched from an F-15 fighter-jet.

do not need, not the promised slow-won safety from the missiles, but instead a persistent technological rivalry that will speedily touch upon or even enter the Star Wars of its popular name. If ever there was a system whose realizable technical accomplishments, not its fancied goals, would in fact lower American security—and Soviet security with it—it is the SDI.

Not So Ill a Wind: Some Technology Promotes Security

The dangers of the unique technology of nuclear weapons and of the much more complex mix of new capabilities that have grown up around its cheap destructive power have been the burden of all these pages. It would be gloomy indeed to conclude with the feeling that future technology promises nothing but greater danger.

Indeed, that would be untrue. The broad developments that have increased guidance accuracy and led to the dangerous prescriptions for counterforce and damage-limiting pre-emptive attack have also given us quite other powers. Everyone knows how valuable are the views from orbit that give firm information on the opposing forces, their growth and their readiness. It was as long ago as LBJ's time that a President praised the expenditures on reconnaissance that ensure against surprise. The same goes for various related means, like those for early warning and location of missile launch. Both sides hold similar abilities, and no doubt both gain by them. Those systems make a lasting contribution to the reasoned management of nuclear weapons, a skill of which we can hardly have too much.

There is an easy extension of the same argument. It is the obvious strength of these means, the so-called "national technical means," for verification of agreements on testing and deployment of a wide variety of weapons systems, those here and those still to come. One striking example is operating now on the North American continent, a network of wonderful, unattended robot seismographs each contained in a little fiberglass tank above the deep drill hole its underground sensor uses. Daily they report by satellite link to Albuquerque all the ground tremors they sense. They provide the best present information on earthquakes. But they were designed for installation also within the Soviet Union by treaty, to verify the absence of underground nuclear tests. Only the will to agree is lacking, more on our side than on the Soviet, whose negotiators seem years ago to have agreed to the emplacement of this ingenious and unobtrusive but reliable and hard-to-fool system.

I would like to look a little outside the confines of the world of weapons. Modern technology and applied science promise much to make this world safe against other threats to our fast-growing species. Preservation of the essential free resources of the air, the sea, the very soil, the provision of energy by means that will not bring a return of the ice nor yet an epidemic of tumors, the reliable increase in world crops and widespread freedom from microbial disease: these are goals still unwon. They mean much for wealthy but troubled lands like our own, or the Japanese; they promise even more for the majority of the world's people in the villages and the mushrooming cities.

How can we in good conscience continue to use an important fraction of the world's trained talent in the pursuits of war? Cannot the nations do better than that? They can and they must. For a scientist there is still more involved: these powers rest on gains of the human mind and human experience that ought not remain beyond the grasp of most human beings. We need education in science to sow abroad the nourishment it brings with its search for our origins, our nature, and our destiny, and its unfolding of the universe in which we dwell. We have little right to spend so much on mutual threats of conflict, and posterity should judge harshly the forty years since Trinity if we do not manage a shift of interest, resources, and energies, even should we succeed in the minimal goal of keeping the nuclear peace. There will be some healing in scientific and technical cooperation itself, once the nations can find their way to work together at some scale comparable to that at which apart they now so ingeniously and tirelessly arm.

Remedies from the Lessons of Experience

In this account of the "technical fix" and its repeated miscarriages, it has become plain enough that technology is not an autonomous agent. It usually demands strong and consistent political and organizational support before its accomplishments have much effect. It should be clear also how often it turns a double agent; what looks like a neat engineering advantage while it is one-sided is all too soon seen as a worrying challenge, once it has joined up with the other side as well.

The remedies sketched here for the chronic pathology of "modernization and improvement" are no less politically embedded. Their utility depends on political understanding and public readiness, on sensitivity to historical and social trends. No mere naming of the task by a physicist suffices to make the proposal a viable one. But a short account of a few directions that seem useful will not, I hope, be misunderstood to mean that no more subtlety is needed than the bald outline here set down. The physicists who have spent years sounding their alarm, mostly unheeded, against the steady multiplication of Hiroshimas can hardly claim to know how to influence national decisions.

The first and least contentious proposal is almost a technical one. We ought not to choose forces whose nature is to increase the instability of our stance. The claim that we need to have every possible kind of force is not in the end credible. Hydrogen, MIRV, better and better guidance—each one offered some claim to increased security; each one was something of an illusion. The decades have really seen a single political effort in this domain: to find a way to make more usable the power of nuclear weapons. One outcome is real, even hopeful. It is of course the avoidance, not of war, which has thrived on four continents, but of war direct between the nuclear states. The inhibition against war has certainly come in part from the acute recognition of the higher stakes of nuclear conflict. So far nuclear weapons, by their threat, have had that function. The task is to lower, and not to raise, the risk of intolerable escalation without losing the inhibition as a whole. Certainly the careful choice of stable forces is the first piece of wisdom. At the level of decision, it might be wise to present expert criticism of new proposals within the Department of Defense along with expert support in the case made before Congress.

One must go farther. Restraint—in the end, one hopes, a fully reciprocated restraint—is the next step. It is not necessary to keep open the path to "improvement" so zealously. It is enough to keep close watch on potential change, by modest research and development. A policy of caution in proceeding to change the mix of weapons is a workable one. Such a policy comes close to a freeze of this or that technical step, and might indeed lead to some mutual agreements in that sense. It is fairly sure that explicit treaties of arms control form a more enduring basis for the national security than the pell-mell pursuit of superior achievement.

Another corollary of this mood is the constraint over development implied by agreements that outlaw testing. Military prudence has for a long time meant that no new weapon can be made part of planning until a test has been made of its workability under some sort of field conditions. Calculations and simulations, even partial tests of components of complex systems, do not usually convince the hard-headed staff officer, who can recall how reality has dealt with many a plausible scheme. The requirement is certainly not absolute; the Hiroshima bomb was of a type never tested before its actual use. Underground nuclear weapons tests do not give as much information as those in the atmosphere, but that has not stopped the conduct of a thousand such tests since the others were banned by treaty in 1963. The users adjusted their expectations of test reliability. That might happen again, but it seems nonetheless valuable to try test limitations, both by self-restraint—like the Bush idea around the hydrogen bomb test—and by actual agreement. A verifiable comprehensive nuclear test ban is certainly a worthwhile goal.

An effort to limit tests seems especially workable in media as transparent as space and for weapons as conspicuous as ballistic missiles, whose flight can be monitored from long distances away. Tests of that sort can be checked upon with very high confidence. Verification is not primarily a question of one-by-one scorekeeping and of moral judgment, but one of practical security. The thought that an adversary might be getting away with something is disagreeable, and certainly goes to one's estimate of his attitudes, but unless an illicit test sequence long enough to make an actual difference to security can be concealed, a logical imperfection in verifiability is not a real danger.

These prudent steps are not all that one can conceive. It would be a mistake not to mention the wider hopes that we must entertain. Whatever leads to measures of control and moderation, of restraint and planned reductions and adjustments, there is more to do. Those explosives will remain an intolerable risk at least until a cut in strategic weapons by a factor of at least ten, and more likely by twenty or thirty, has been made by the superpowers and has been seen to be made. Then perhaps we can live, not yet easily, but in hope, under minimal deterrent, until one day the suspicious nations learn to bring to a close this nightmare epoch of the readiness of the unprecedented nuclear weapon.

The Conclusions Drawn in 1945 . . . again

What remains is sustained public action towards ends the panel put forward in their letter of August 1945. Their vision remains today a guide to survival. Once the public steadily insists upon a policy for nuclear weapons that rests not on the shifting mirage of technical advantage but on the reasoned growth of international agreement, decisive and saving changes will come.

The pulse of heat at Trinity's false dawn was more than a warning; it was the touch of necessity. We cannot much longer put off our response, a change deep enough to match that insistent signal from the fireball.

Nationalism, Science, and Individual Responsibility

This is unquestionably a remarkable time in history. We are now a very numerous species. We are very powerful. We have begun to make a conspicuous mark on the overall surface of the earth; this was not the case in previous generations, except most locally. We can now be detected from space because our works and our numbers are great enough.

Our technical prowess is not a new development; it is a very old one. It certainly goes back to the days when we were a tiny, furtive species in the African Rift. Those who made the most important discoveries and gained the most important insights are unknown to us. (We physicists usually take the view that the first step is more important than the slogging through to the end.) The discovery of fire, the invention of the container, and a number of other technical developments are truly fundamental breakthroughs so old that we don't know much about them. (I hope that you are not so cynical as to think that we don't report them because they probably were done by women, but that may be true, too.)

But I must pass lightly over the grand history of the neolithic and the rise of the great empires. Although I am discussing world history, in *our* time, over the past three or four centuries, the key ideas and ways of doing things, for good and for bad, come from a little peninsula to the west of Asia, often called Europe. Two of these have had and still have maximum current impact. We live now three or four centuries after the rise and export to most of the world of these two great systems of ideas with their proponents, with their material consequences, and with their uses. They are big systems: nationalism and science.

I could utter a sequence like Gilbert, Galileo, Leibnitz, Newton, and you would recognize right away the 17th century rise of modern science.

Modern science is not strictly an invention of Italy, Germany and England. It has indispensible contributions from many times and places. But two-tenths of the world has done most of it, although now it has spread and joined into a worldwide current. The European contribution is characterized by extension of the in-built senses, by telescope, microscope, voltmeter and so on, in contrast to the ancient and venerable science which is closer to craft and doesn't take much technical instrumentation. This period is a phase, but an important phase.

At much the same time, and much in the same place, was another development that also has its analogies, its roots, even its greatest exemplars in other times and other places. But it acquired its peculiar salience for our time in formulations by people whom you would recognize as well— Elizabeth I of England, Peter the Great, Voltaire, Lomonosov, Humboldt. They looked away from the ancient linguistic heritage of classical learning, and developed in the vernacular new ways of talking, thinking and organizing people. Of course, there came boundaries, anthems, passports, armies, navies, titles, wars, all swept along in the great current of nationalism.

These two currents, science on the one hand, and nationalism on the other, are two systems of thought and institutions with major vitality to this day in all parts of the world, far beyond the domain where they were first seen in their modern form. Partly because they build upon older structures elsewhere, and partly because they have enough internal energy to live, grow and spread, they have been picked up everywhere.

Now, science has left a great deal in its wake. It has opened up a new universe to our insight. It has opened up new and powerful means of exploiting the natural resources of which we are all commonly endowed on the surface of the planet. I think it fair to say that it has been largely responsible for very large growth in the number of human beings. Nobody can deny its impact.

Nationalism, too, has now a pervasive role. We no longer speak in an ancient learned tongue; speech reflects the region where we live. The political, cultural, linguistic, everyday nature of existence for most people in the world today is formed by nation-states, with national territories, by national languages, by national schools. Of course, there are differences and disputes within borders, but overall I think this is a fair picture.

In the view of the Enlightenment, the achievement of the mind was much greater than the rule of any king, of any administration, or even of any leader. A harmony between science and nationalism was idealized by the writers of the 18th century, because they saw both under the sovereign domain of reason. Reasonable national institutions would respond properly by a show of "decent respect for the opinions" of all others and like

ringing phrases from all the national founding charters. But the fact of the
matter is that there lies between the two a certain structural conflict that
ought not be glossed over. Science is cosmopolitan, international, universal: The grass grows the same way everywhere. But nations are above all
parochial. Science speaks of unity, harmony and indivisibility. But nations, whether they're firmly in place or whether they are still in the
stages of formation or reformation, tend to be centered around divisive
and conflicting structures like tongue, religion, even race.

It is hard to remember now that during the Enlightenment even so respectable and conservative a republic as the United States of America
held remarkably subversive ideas about the independence of the new nation from the old forms of society. President Washington, writing to his
enemies in the Barbary Coast, beseeched them not to attack the ships of
the United States of America. Those marauders meant to conduct war
against the Christian infidels. But, said Washington, unlike all the states of
Europe, the United States is not a Christian country, nor indeed is it an Is-
lamic country, nor any other kind of a religious country. It is a state free of
religious affiliation. The nation had acquired a new and independent existence. That was good Enlightenment theory, right out of Locke. Today,
however, national institutions no longer carry with them the full enthusiasm of their rational founders; they have rather taken on the coloration of
that ancient surround of traditional parochiality.

The cosmopolitan and internationalist view of science, very common in
those days (and extended to public policy) is by no means the rule today,
even in quite new formations in human culture. For example, the great
states assembled around the ideas of Marx, Engels and Lenin have not
found it possible to maintain what was once a strong internationalist position. Indeed, we hear of what looks like national conflict between states
whose whole bent is to adhere to that system, like the People's Republic of
China and the Democratic Republic of Vietnam. The numbers of men who
have died for the idea of nationalism gives some measure of its power.

Science has been affected, too. In the time of the Enlightenment, both
the new US and royal France extended official friendship across the lines
of belligerence to Captain Cook as he set sail in a ship of the British Navy
on a peaceful mission to explore the Pacific and the islands of the South.
Documents offer the Captain safe conduct past all American and French
ships of war, because (as the Ministers put it), then "Captain Cook is the
common friend of mankind."

It's hard to see that attitude toward scientific discovery and development today. In fact, when I read of scientific development, I expect it to
refer to national advantage in war or, if not that, in commerce. The notion

that a disinterested general good will follow from discovery comes very far down the news column. That's the unhappy condition to which we have come. One of the most heartening stories of our time, a very small story in a large volume of bad news, told of a cease-fire for a few days that had been arranged in El Salvador to allow a national campaign for the vaccination of children against smallpox, a campaign to be carried out with impunity from military attack by either side of that civil war. Prevention of contagious disease in children is close to what people expected and wanted from learning in the old days. I assure you that today that is an exceptional case.

On the other hand, science has brought a gloomy material unity to the world through the weapons that people use. In almost every part of the world the strongest unifying idea is the automatic rifle! Different groups have different models, but they all have the same idea: a gas moving pistons to reload the barrel works in all climates. This is what is left of the universality of science. Between those two great systems, science and nationalism, there has slowly arisen an intolerable conflict. My thesis is that we will not survive very long under the simultaneous growth and flourishing of both systems. They create a fatal contradiction of great and growing strength. It will be up to us, but especially to the younger scientists, to do the work to resolve that contradiction.

Lessons from History

A few remarks about World War II and the origins of nuclear weapons will sharpen my point. In preface, I want to note a remarkable parallel in history. There were two discoveries in physics that were latent in its development. They were made too late in this sense: the moment that the discovery was made known, people all over the world were able to repeat the experiment and confirm it at once. They already had the equipment; they already had the background; they already understood what was going on, but somehow they hadn't quite put it together before. The first was the 1895 discovery of x-rays by Wilhelm Roentgen, which fascinated the world of the scientist and the nonscientist alike. Once he published his little paper, his work was reproduced within a week or so all over the world. It could have been done 10 maybe 20, years before, but no one had come up with the idea. It was a genuine discovery, unlike some modern experiments that require elaborate, unique preparations; it was like dropping a tiny seed to start a crystal.

The second one has a special place in history: it was the discovery of fission by Hahn and Strassman. They explained it to us around the end of 1938. Although the task was a little more difficult than making x-rays, many respectable departments of nuclear physics from Berkeley to Moscow and Tokyo had verified the existence and nature of fission before the end of January 1939. Again, the discovery was lying there waiting to be made. Even though I was then a naive graduate student, I was able with my friends to draw a funny chalk diagram of a bomb with heavy water and uranium and this and that. Even we knew that this was a very likely cause of fearful explosions. The implications of the discovery were drawn all over the world a few weeks after the first hint of such a thing.

This second discovery occurred under the shadow of approaching war, between Munich and the beginning of World War II. Clearly, it was no accident that everyone saw an explosion in it. People were thinking about war. The war meant that the exploitation of this in a technical way was going to lie within the powers of the nations in the ensuing years. As you know, that is just what happened. The United States, with the United Kingdom as a partner, was first, but in a matter of eight or 10 years, nations of some power all over the world took advantage of what was implied by the nature of fission and demonstrated over Hiroshima.

World War II engulfed everyone in this country, but to a degree only as economic and vicarious participants in that war. Only a small part of our population experienced what masses in Europe, European Russia, and the Far East went through. The cruelty of war and its terror were astonishing. Everyone worked hard here; the students were in uniform; there was rationing; but we saw the worst of war only through the newspapers.

I learned with convincing evidence between Thanksgiving and the middle of December of 1944 that the Germans could not field a nuclear weapon. Up until that very late time, I was uncertain. I knew before that—with very great misgivings—of the Peenemunde Laboratory, the developers of the V-1 cruise missile and the V-2 ballistic missile. These two systems made no sense to me unless the state that had expended so much money on them was going to have nuclear warheads to increase the damage of the missile. It may be that the V-1 had its own justification. It was much cheaper to make. It was very inaccurate. Yet if they dropped it anywhere in London, the chances were that they would get back its cost in damage. Not so with the V-2. The V-2 with a high-explosive warhead cost many times more than the block of buildings it destroyed. Therefore, we believed for a very long time that the Germans would have the nuclear warheads we were engaged in trying to build. As you know, they did not, and as 1945 began, Allied victory in Europe was inexorable.

Yet, this knowledge did not slow the advance of Los Alamos.

The sudden death of the President of the United States, Franklin Delano Roosevelt, is a key to understanding what happened. He was the most trusted and the most popular of our leaders. After the death of Roosevelt, there was no American leader whose conviction could immediately communicate itself to all of his advisers, and to the people as a whole. His successor, however firm, was not in a position to command the immediate acceptance of a strong new decision. I don't think he would have made one, partly for that reason. A decision that one cannot enforce is apt not to suggest itself. I have no reason to believe that Roosevelt would have acted any differently from Truman in his decision about the bomb. But the chance to change direction existed during Roosevelt's life and disappeared at his death. We felt this at the time: Things would go with the current and not in new response to a thinking, feeling person at the very top, to whom all gave immediate allegiance. We proceeded to work on the final details in order to complete the Trinity test, the existence theorem that was the triumph of Los Alamos. The combat bombs of Hiroshima and Nagasaki were rather anticlimaxes, tragedies, in which Los Alamos as a whole took no great part (although I and a few others did). Indeed, many on Capitol Hill were surprised by them. The subjective end of the project was the successful test on July 16, 1945.

The position of Niels Bohr stands in contrast to that of Truman. Bohr was a figure with whom we were well acquainted, as were all scientists of the Allied West and, indeed, those of the Axis states. Bohr desired a magnanimous treatment of this new weapon, in order to set up a peace that might have a chance to survive in the presence of a brand-new method of decisive warfare. That was his dream. He carried it out relatively poorly, but he influenced those of us at Los Alamos who saw him every day, and who were predisposed to favor the father of quantum mechanics. Bohr felt that it was imperative that the world should know of this radical change in warfare. If the existence of the bomb remained a secret (or even it remained potential, if there were no test) all the chancellories would still know, all the general staffs would know. Things had gone too far to expect otherwise. Any peace would have little chance if behind the scenes there was a struggle for superiority in an entirely new way of making war. No public could possibly comprehend such a struggle without having explanations that only publicity could give. Such a secret would be a very bad foundation for peace. I don't think Bohr wanted the bomb to be used in combat. But he felt the world must know if this thing could work. Once we knew that, it would be made visible to the world.

Of course, it was made visible in a cruel way. It was not revealed as an

THE AGE OF URANIUM

entry of a new type of warfare into the world, but instead as the climax to an equally cruel air attack upon the cities of Japan sustained from January to August of 1945. In those months, the 21st Bomber Command of General LeMay attacked and levelled about 170 square miles by fire and blast. The atomic bombs, terrible as they were, destroyed under one-tenth that area, and perhaps killed half as many people.

The difference, and this is what makes the conflict between nations and science so strong today, is not on the ground, but in the air. It's hard saying, but my overall judgment is that, while it is terrible to be at the target end of a fire storm caused by a 10 kiloton uranium 235 bomb, it is also terrible to be in a fire storm caused by General LeMay's million jelly-gasoline bombs. The district of Tokyo which was so affected on March 9, 1945, looked about the same as the district of Hiroshima that was so affected on August 6, 1945. There are important differences due to radiation on the ground, but not by an order of magnitude. In both cases on the ground there was terrible conflagration. But in the air in one case, there were 500 or 1,000 heavy bombers, all that a great country working for years could build, flying in all night long to lay that one city waste. In the other case, there was one essential airplane with one inexpensive device. Sure, the first bomb cost a lot, but it was fast amortized over many. That amortization has produced our world, in which destruction is between 100- and 1,000-fold cheaper. I hate to put life and death in such crass economic terms, but wars are, in fact, social. They are waged by large armed forces. The large forces are raised by states. The states must pay in labor and goods and life for the forces they raise. Science has cut the price of destruction easily by 100 times, yet nations have not reduced the power of their expenditures to take advantage of the new economics. They have instead spent more or less the same old sums, but now the damage that they can do is hundred fold greater. Yet the area of the people's houses and lands, and the volume of the air they breathe, have hardly gone up at all.

I think there is the physicist's sin that Robert Oppenheimer evoked. The physicists turned away from a search that was philosophical in nature for the ultimate nature of matter. With the very same techniques, the same people produced the desolation of Hiroshima and Nagasaki, and a new economics of destruction. Oppenheimer spoke very carefully, and he knew his terms. This is a sin we cannot forget, and it imposes a terrible moral obligation.

That obligation has been dismally met—although statesmen have learned something, since they have not tried nuclear war again. They have prepared for it. They have organized for it. They have planned it. From time to time, they gesture with it a little. But no statesman has been so imprudent as to try it.

Two other much less powerful entries into the new world of weapons merit mention. The first one was the US H-bomb decision, which many scientists opposed. A sensible alternative proposal was made then by Vannevar Bush. "We should announce," he said, "a policy of no first test." That is the most prudent position one could take. It allows research and preparations for test. Bush proposed our not proceeding with any test unless somebody else tested first. The opposition would come to the same point, and the two sides would achieve a reasonable symmetry about the point of understanding, but not realizing fearful arsenals of weapons. This was no public debate. This was a debate carried on mostly in secret between mid-September 1949, two weeks after the explosion of the first Soviet built atomic bomb, and mid-January 1950, when the President announced his decision. It was all an internal struggle. Had Bush's plan been adopted, as it might have been in a public forum, we would have saved a factor of 10 or 20 in potential world damage.

A less important one is the MIRVing of missiles, a process that ran from the late '60s until about 1981. The Multiple Independently Targeted Reentry Vehicle was first put forward to allow a missile to carry, in addition to a nuclear warhead, decoys and other devices that would counter ballistic-missile terminal defenses. That seems plausible; we couldn't afford to lose the deterrent, they said. Step-by-step, these "penetration aids" became warheads, until missiles contained three to a dozen warheads each. MIRVs put an important advantage into the hands of that side willing to be imprudent enough to strike first, because they might knock out up to five missiles by the expenditure of only one with its 10 warheads, which is a nice gain in some kind of strange calculus. We critics pointed out loudly at that time that the situation would be destabilized as soon as the other side had MIRVs. We argued that this would be very bad for everyone's nerves and our state of security, but no argument was heard. Finally, a decade later, after the Soviet Union had tested its MIRV and had made missiles larger than ours, with more warheads, we witnessed exactly what the critics had predicted. General Brendt Scowcroft and his commission were appointed to study the problem by President Reagan. The Scowcroft Commission concluded that our MIRVing has been a bad direction to follow, and that now we should try to return to one warhead per missile by agreement and example.

Our country is now at the point of yet another venture that promises to solve a problem of strategic balance, in which the fundamental difficulty is that the scale of nuclear destruction is too large compared to the scale of human life. I refer to the Strategic Defense Initiative. It looks to me as though the long technical road to ballistic-missile defense must lead first down many dangerous byways. For example, a simple early system good against satellites will threaten the present rather tolerant regime of reconnaissance and verification satellites in space. A new offensive arms race is the most likely result before the ballistic delivery system most favored today would lose its power. After 10 or 20 years of investigation, in the light of constantly changing technology, it seems quite likely that entirely new carriers for nuclear weapons would be found, just as they have been found in the past. It isn't the delivery system that bites; it's the terrible destructiveness of the cheap nuclear device; you can afford to lose 10 or 100 of them to get one through. The new delivery systems will be exactly those against which the powerful defensive systems are not effective, whatever they are. The SDI case seem hopeless to me; it guarantees a new arms race ending in no defense.

Individual Responsibility

All these examples come down to the same thing; if nations continue seeking an advantage out of the terrible destruction by nuclear weapons, through the constant improvement and modification of technology, we will not survive very long as a society. Nationalism and technology cannot both continue unimpeded. We cannot continue to seek every national advantage by the maximum new technology. We must either stop the search for national advantage by applied technology in war or stop technology application overall. I don't think we can do the latter, because it has so many other effects on the five billion soon to be alive. We will not reduce the population and go back to the war of an earlier time.

Earlier times were not freer from war; they were not freer from hate; they were not freer from injustice. On the contrary, they may have been less free from these things. But there was one great difference. Instead of "one head, one kiloton," it was then "one man, one sword." Never before has the world, that fragile globe on which we live, been in overall peril from conflict, and *now it is*. That is the real difference, the physical change that states must address, that people must address, and above all, that young scientists and engineers must address.

I am going to close with a few remarks of a personal kind that I include with diffidence. I don't like to go around giving people good advice; I was never able to take it myself! I will begin by saying that I shall not engage in the discussion of what things are right and wrong, what is the basis for moral behavior. I firmly believe these are important questions, but they are not my questions; they are not physics questions.

My morality, like that of most physicists, is largely a consequential morality. We know what the results of good intentions have been in the past, once we consider the problem of the scales of destruction that the physicists brought upon us all in 1945. The important thing is to avoid the consequences, and not to examine so much the hearts of those who bring good consequences or those who bring bad. The task is: not to find out how to act morally, but how to have good moral outcomes. And this is very difficult.

In the United States armed forces, around 100,000 people have been certified to handle nuclear weapons in the field. In 1945, we were some 50 young persons from Los Alamos in the same position. Those numbers represent the kind of institutional change which has occurred. When I reflect on that change, the sense of individual participation acquires a certain irrelevance. I believe firmly that institutions form and organize human life. We create those institutions. We do so slowly by a complex series of feedback loops which no person knows entirely. The very language that enables us to communicate with one another is a social structure. It is a social task to learn how to behave in the terrible crisis brought into our world by the combination of science, education, and the good intentions fostered by the splendid free institutions of the Enlightenment.

It's an old saw that it is old men who send young men out to war. I am afraid that it's almost as callous when elderly professors try to tell students how to behave. But, I am going to say a few things about it. First, judge your own behavior. Second, consider that abstinence from doing evil is not in itself absolution from responsibility. When Pontius Pilate called for the water to wash his hands, there was no doubt that the secular power rested in his troops. It will do no good if we have good people who take no part in the right kind of public debate, and the institutions roll on. It won't be enough. I am not saying that innocence is bad, but for me, it doesn't represent a commensurate answer of reasoning men and women to the problems we now have.

I recognize the appeal of our own nation for support as repayment for the symbols and for the realities that have nurtured and protected us from birth. Nationalism is the most powerful force in this stage of history. Nevertheless, I have argued as strongly as I can that the notion that the welfare of our own country can be served by stronger weapons better used has reached an end.

Ask yourself: Can you argue? Can you write? Can you stand as an example? Can you do anything to change the institutional structure, in small ways and in large ways, by defeating one single line in the budget, or by changing the attitude between two states? I don't know which is the more important, or the more realizable. But they both go in the right direction.

We need a new way of thinking, but I suspect that it will not come as a result of exhortation, however cleverly done. It will come by building up new experiences, new examples of institutions, through the rise and fall of new leaders. For the student of science and the young scientist, it is imperative to understand and explain, but not solely in a detached way. Explanations are not real unless those who listen can share in decisions. Those of us who have some advantage in learning these matters must find a way to engage others in the shared task of preservation. We should fulfill the promise of this remarkable development of the past centuries, not to lose it in the grasp of a nuclear winter. Approaching your work as a shared task, you will find those to share it. Eventually, not at once, but step-by-step, I have the hope that we will gain victory over what are our own worst experiences. Those currents that have flowed together so well together in the past don't buoy us up now. Change them.

You and I have a shared task. It is not yet complete. We are in a time when ahead lies the chance for widespread and important change that the world needs like water in the desert.

FRIENDS AND HEROES

That there is no such thing as one physicist is a favorite saying of mine. What could that person do without informed gossip, instructive news, novelties, and numbers, serious papers and books, and those enviable viewpoints occupied by others?

Even Fermi, who gave up on reading papers, wanted to know less what results had been offered then what had been chosen as questions to be answered. He relied on his own awesome powers to approach any answer that caught his interest. For the rest of us the case is all the stronger.

Thus some reminiscences about friends and heroes belong to this volume, even though long years of exchange with friends, both students and teachers, cannot be adequately reported. It is those encounters that provide the everyday furniture of any physicist's mind. Most of the narratives here of friends and colleagues are drawn at least in part from life and from living memory; only the last of them rests on the record of Charles Babbage in brass and paper alone.

Bruno Rossi

Against the harmonious backgrounds assembled of green gardens and carved stone that the princes and the people of Italy have made in their cities—in Venice, Padua, Bologna, Florence, and Rome—a vivid tale is told. Here is the rise of a new astronomy, the terror of tyrants and its harsh remedy in war, and the spread of the new ideas and tools of the devoted investigator to minds and lands far from the workshops of its origin.

You have read something like this somewhere before? A reader with any acquaintance with the history of science will catch the echo. But this is not another history of the seventeenth century. This is our time, our century, with its own ironies of wonder and of fear. Bruno Rossi is entirely a modern, as adept at electronics and quantum theory as he is in clear thought and sharp phrase, a man just as well known in New Mexico, Japan, Bolivia, or India as in the Tuscany of his first researches.

A young man eager to touch with newly-trained hand the mainspring of the world, he was among the few to found a new astronomy that displays an invisible part of the cosmos now as clear as the moons of Jupiter, and much more universal. It was in the years 1930 and 1931 that Rossi and his peers understood and first demonstrated that the mysterious radiation from on high, what we now call cosmic rays, was in fact particulate.

His telescopes were not fitted with glass lenses to augment the eye's grasp; they were metal tubes of gas whose signals were electrical pulses, invisible but rigorously countable if you had the right ingenious circuits. (The best-known of Rossi's circuit designs are still in use, though realized now in tiny silicon chips instead of with the glowing vacuum tubes of their beginnings.) Step by well-supported step, first one, two, three . . . then whole arrays of Geiger counters and heavy shields artfully deployed, they marked out the necessary path of the particles, not by images but merely by recording simultaneous pulses—and their absence—to lead logically to firm conclusions.

A new astronomy of charged particles and their secondaries was born. The particles flew in, not on straight lines but on long tortuous paths, from somewhere in distant space—where?—to the earth's surface. It was soon proved that the particles were so penetrating, so energetic, that their like had never been seen in the physics labs with radioactive sources at hand.

During the rise of the first astronomy, the telescopic one, its home Tuscany was at peace, though the stresses that brought the War of Thirty Years across Germany were already distorting life on both sides of the Alps. In the twentieth century, the German wars were more closely-spaced. A lifetime could not escape one or the other. Peaceable folk like the Rossis had perforce to leave behind the brand-new lab at Padua that Bruno had designed, to seek refuge overseas.

In a few years, the war had crossed to America. It is no surprise that Bruno Rossi soon found himself at Los Alamos, his mastery of submicrosecond timing directed to development and test of the atomic bomb. A faint trace of the record that his more than lightning-fast detector took at the first desert test shows the signal rising, rising as the chain reaction grows exponentially toward fateful maturity.

The postwar years found the cosmic ray physicists for a while happy monopolists of high-energy particle physics. It was then an opportunistic, even a serendipitous science, based on the chance infall of Nature's puzzling beam from afar. Cosmic rays were sought worldwide, up on mountain tops or in stratosphere balloons to look at the incoming beam, down in deep mines to filter out all but the most penetrating rays, under the open sky to strew a score of big counters over miles of prairie or forest land, the better to catch a great disc of speeding particles, the billion-fold progeny of one rare entrant particle of enormous energy. The hunt was open to players in rich countries and in poor. By the early fifties, the romantic days were mainly past. The time of the big particle accelerators had come. As a rarity, some single cosmic ray particle still carries much more energy than any you find in the latest magnetic tunnel, but for most purposes particle research has left mountain and mine for the great national and international labs.

The strong, well-controlled beams within the big labs are fine for the study of particles, but helpless to plumb the sky. Bruno became more and more of a astronomer, albeit an active intervenor in space, and less of a laboratory physicist. He pioneered the second new astronomy that he and his colleagues (and some friendly competitors, too) found up there: an astronomy not of particles, nor of light nor of radio, but of x-rays (and gamma rays, too). Out where those rays come from there are strange stars and disturbed galaxies spouting, spinning, and whirling, a turmoil little seen here in our quieter home, the solar system.

Perhaps it is the growth of a worldwide community of Rossi students, most of them walking on the stage, that best stands for those times. Yet growth continues into these times when satellite probes are important tools of the physicist. Rossi led this quest too, out into interplanetary space, there to sound the plasmas that blow from the sun to modulate the cosmic rays, to tickle the comet's tails, and to disturb the upper atmosphere and magnetism of the earth.

It has been a glorious time for cosmic rays and for those who puzzled over them, on paper or with intrepid searches. After all these years, an elderly physicist takes pleasure in recalling the years of the Rossi curve, when a new kind of digital logic, based on banks of counters, not only built but compelled a student's belief in extraordinary new physics, demonstrated *alla Rossi*.

Robert Noyce and the First Chip

Among physicists, this is the 87th year, and not the 83rd year, of the 20th century. The 20th century began for us in 1896 when X rays were discovered. It was the first month of 1896 that shook the physical world and started that extraordinary development which has led us farther and farther into the microcosm. I suppose nothing has been more important than the sense in which an understanding and control over the microcosm in the hands of physicists—and I would certainly include the chemists—have brought the microcosm and its properties consciously into the world of humanity. Of course we always were made of atoms—we always consumed molecules of atoms for the staff of life and the breath we inhaled—but we didn't know very much about them. The 20th century's contribution to the foundations laid for modern science since the 17th century has been the understanding of the small and the attempt to see what the small means for those phenomena like ourselves, very large compared to atoms.

If you want to try to periodize the world in which we live, to pick important events, to sum up in a few phrases a complicated piece of our history, I think many people would say, as I myself would have said ten years ago unquestioningly, that the most important technical outcome of our growth in understanding of the small world—the microcosm—was either the strange and terrifying cloud of a thermonuclear explosion or the terrifying flame with which an orbiting device enters into orbit. These are events which we have seen and thought about and observed spectacularly in films and ominously in our dreams for a good long time.

That's a commonplace. The newspaper editorials on New Year's Eve are likely to talk about the atomic age. But for the last two decades, and increasingly in the last few years, I find myself wavering in this simple account of what the 20th century is going to mean when our posterity looks back on it.

Ever since the 17th century it has been recognized that fine mechanisms—small, often elegant beautifully contrived, valuable, often precious devices—have been a delight to human beings. Such devices have attracted a good deal of attention, interest, and philosophical concern, as well as fortune and the world's acclaim. It would be hard to separate the creation of intricate mechanisms from the history of our times. Rome, Greece, the great Tang Dynasty in China—none of those had it. But ever since the 17th century we have had the idea of the watch. Nowadays a watch is something of a utilitarian object, but for many people that's not the case. Even to this day there are many collectors, appreciators of the watch, who hold that the watch is a symbol of human skill and insight that exemplifies itself in so many compact and elegant forms. Even those of us who are not so caught up can still appreciate it. Whenever I see a really fine watch displayed in a shop window, I find myself looking at it. I don't expect I want to buy it or to take it apart—neither of those. But it's satisfying just to see that accomplishment of hand and mind.

If I were to describe a watch in the terms of space and its partition, which is what a physicist likes to do, I would characterize a watch by saying, "Well, it's the kind of thing really very well made out of many metals, sometimes precious ones, often superb alloys and so on—such that a thimble could hold maybe two hundred carefully made parts of fine watchwork: the elegant cogs, the screws, the pinions, the jewels, all those things which make up the frame and the going movement of a watch. (Here I'm not talking for the moment about the case, the jewelry, the painter's work on the back . . . those are interesting things, but they don't belong so uniquely to the watch as the works. They could decorate eyeglass cases, or whatever you will.)

In the middle of the 17th century, people in the generations after Galileo and before Newton, seeing science opening up to them suddenly, came upon the microscope. There was a marvelous book on microscopy, written in England in the middle of that century by the brilliant investigator, Robert Hooke. In those days it was not yet the subject of the specialists, whose intimate knowledge of all microscopic techniques has revealed so much of the small world. It was still an exploratory affair, in which the commonplace objects in the world were put under the microscope and carefully examined and carefully drawn, because they had never been seen by the human eye in the way in which the lenses of the 17th century, much better our own, could reveal them.

Hooke figured very well the familiar poppy seed. To this day it's fascinating to look with a magnifier at the poppy seed that so frequently decorates the dinner roll. You'll see the same reticular network growing on it that Hooke first described in 1665, when he was among the first people to see such a thing. He was quite impressed by the intricacy of the carving, so to speak, the natural sculpturing on that rather small object, the poppy seed.

Somewhat smaller than a poppy seed is a grain of salt. If you look at a few grains under a magnifier, you'll see that they try hard to be cubes, though they don't quite succeed. They are more or less similar cubes, all of a kind, perhaps not as alike as peas in a pod, but not very distinct. Their enormous numbers and the fact that no human effort has shaped them make them natural products that we don't in fact pay much attention to. They're not artifacts in the usual sense.

About twenty years ago a well-known physicist in Southern California, an old friend of ours, Richard Feynman, made a proposal. He announced a prize; that's always a stimulating way to get attention at a dinner. He was thinking in those days of a small mechanism. (I tell you this story as a foil to what I really am talking about. I think it's important, because until we grasp the difference between our ordinary experience and the experience implicit in the work of Robert Noyce and his colleagues, I think we cannot appreciate it.) Feynman offered a prize—his own money—of one thousand dollars to the first person who would show him a spinning electric motor, like that which drives an electric fan, but one made truly small. Feynman reckoned in those days a motor could be built inside a cube 1/64th of an inch on an edge.

For those who are not familiar with small fractions of inches, as few of us are really, one of the grains of salt among the few that you might sprinkle out of the salt shaker on the table might be about 1/64th of an inch cubed. Now you must imagine that somebody has the task of making a smoothly spinning little shaft with a magnetic ring on it, somehow feeling the magnetic force of little coils of very tiny wire, all arranged neatly into that 1/64th-of-an-inch cube! Feynman was careful to say the lead-in wires which bring the power from the outside to that motor need not remain within that small cube.

Of course, this got into the papers, the thousand-dollar prize by a professor for this crazy thing! All the Los Angeles papers carried it. For weeks Feynman was not left alone. Nobody came in on the spur of the moment to offer this object, because who would have made such a thing? But many people who read the paper in their vague way were enthusiasts of one kind or another. They said, "Well, here's a crazy professor who will pay one thousand dollars for a tiny motor, whatever that is. . . . Let me show him what I have that's equally wonderful. Maybe he'll buy that, too!" A whole succession of people came in with the Eiffel Tower made in redwood splints or any such curiosity; one after another they trooped to his office in Pasadena. He got a little tired of it, because nobody had paid the slightest attention to his meticulously-described task, one he felt was just at the limit of technical accomplishment and might lead people to thinking about small-scale mechanism.

After a couple of months went by, and nobody had fairly claimed the prize, he got rather bored with the whole thing. Indeed, it all quieted down. Six months later—of course, that is significant—a telephone call came from a Mr. William McLellan in Pasadena. He said, "I'd like to show you your motor, Dr. Feynman." "Well," Feynman said, "Don't worry about it." He has a very breezy, Far Rockaway manner of speech, and he said, "Don't worry about it. I don't know . . . maybe that prize doesn't exist anymore. I got tired of it." "That's all right," the man said. "I don't care. I really didn't make it for the prize. I just wanted to show you that I did make it, and I can show it to you." Feynman said, "OK, come around." The man duly came to his door sometime later. As he walked into Feynman's office, Feynman's heart sank, because this man who was going to bring him the tiny precious object was toting a rather large box with him, wrapped up in brown paper. Feynman felt: "It's another mad-man who's going to show me the *Monitor* carved in Swiss cheese or something like that. . . . It will have nothing to do with my task."

But McLellan put the box on Feynman's desk and unwrapped the paper. Once Feynman glimpsed even the very outside of the box, he knew at once he was before a man who had truly solved this difficult problem. What he saw once the brown paper was torn off the package, was the familiar polished wooden case of a microscope. The point is, if anybody honestly offers to show you a motor in a 1/64th-of-an-inch cube, he *has* to bring a microscope, because you can't *see* its details with the naked eye.

But the way of success in the practical exploitation of the microcosm to make technologically intricate, valuable, and beautiful devices was not to make electric motors. That was a naive view. Dick Feynman was a marvelous physicist, I admired him intensely, but he would have been the first to agree with me that his was an early and naive view of how the world would work. You had to be naive to take a big motor and make it small. To this day there has not been anything better done along that line, I think, than the motor made in a few months' spare time by Mr. McLellan and his technicians in his shop in Pasadena just to satisfy the challenge. No, that wasn't the right idea. But we do know now what the right idea was. The idea has been exemplified, not only by a few examples, not only by a prototype or two, but as everyone knows, by the hundred million.

Between 1959 and 1961 a small group of people led by Robert Noyce made the first paper-thin transistor. When I say paper-thin I exaggerate—it was five or ten times thinner than any sheet of ordinary paper. And then this group of people found how to carve their paper-thin device. At first their single device was on a chip the size of half of your fingernail, a small flat piece of single crystal. But they became skilled enough to put these paper-thin objects side by side and fit 1000 of them in the area of a postage stamp. After twenty years, a chip the size of that same portion of your little fingernail carries about 300,000 subtly interrelated elements of a circuit! We all realize that 300,000 is a big number for a tiny piece of material. Remember that the thimble held only two hundred watch parts.

Let me try to put it another way, because I want to speak to people who are not electronics people. We all know boxes that you would carry about called transistor radios. Typically you could buy one for a few tens of dollars, each with a small loudspeaker. A transistor radio could get the AM radio stations; if you got a good one, it might get FM stations as well, or maybe short-wave. It is a fairly complicated device, quite powerful, quite successful. Even if the tone was not very good, you got many stations; it was convenient and splendid. I don't suppose there is anybody in the room who hasn't at one time or another owned and used such a portable device. If you go out into the Third World, say to the markets in the interior of central Africa, you'll find the batteries and the radios themselves for sale, so remarkably are they distributed around the world.

To accomplish its intricate functions—picking out the minute electrical energy of a radio station that pervades every room, amplifying it, making it audible—a transistor radio contains maybe fifty circuit elements. On a chip one finds (and of course it's not the end) perhaps 300,000 circuit elements. This is the kind of extraordinary intricacy in fine mechanism that has been achieved in our time in the chip: 5000 radios on a fingernail!

Perhaps you are cool. "OK, it's a mere quantitative change in compressed complexity." I agree, it can be described in that way. But like every large change of quantity, it is an extraordinary one, probably the most important juncture in our century. There's only one comparison with those artifacts. You can't compare them to print; you can't compare it to the tiny inscription of the Lord's Prayer on a pinhead, a typical example of compressed workmanship of no great meaning in the past. Even if I were to place letters with this same enormous density, 1000 times denser than writing the Lord's Prayer on a pinhead—even if I were to do that, each letter stands alone. They don't do anything with each other. They are simply isolated neighbors. But the objects on a chip are precisely and elaborately interrelated. Each one of them can send some kind of signal or receive some kind of signal from many of the other ones. What they do is manage together a kind of cooperative behavior in small dimensions.

There's only one part of the world that we know, only one part of the universe that presents such small-scale interrelated complexity. We all know very well what it is: it is the tissues of living forms. Yes, these evolved collections of interrelated cells still have an advantage over the best work of the silicon-chip-makers. In size, the typical cell, instead of having a fairly complicated component in fifty square microns (to use a certain unit of size), might have it in only ten or twenty, rather smaller. The chip has another factor of ten to go. Of course, the cell's internal complexity is great. But on the whole, we can compare these artifacts we now make with nothing under the sun except complex tissues—the retina, the inner ear, yes, the very cortex of the brain itself. It must also be said that cells operate much more slowly. They are watery objects working mainly by chemical diffusion, not by the swifter progression of electrical signals. It is true that the complexities of cells are greater than chips can now handle. But we are only twenty years into a technology undreamed of before 1958 or so. The chip has already flowered into enormous success, even in everyday life. It is surely a parallel to the subtle complexity that parts living structures from non-living structures in every corner of the universe.

It's curious that the way in which these chips have worked is not to carry out the common functions of all living forms—metabolism, the secretion of special substances, motion, and so on. That's not what we find ourselves first able to do with these chips. What we find ourselves able to do is something more subtle, harder to have guessed. It is cognition and sensing, the manipulating of symbols of the external world. That is what the chip does preeminently well. It begins to rival not the motion of cells, not the reproduction of cells, not the metabolism of cells, not the chemical fermentations they induce, not the elaboration of enzymes, none of those things, but rather the strange and unusual phenomena we associate with our own consciousness and the sensory inputs to our consciousness. That's what chips so far do best.

I'm not prepared to make a prophesy—that isn't our task. I am willing to say that the future will bring as much more. The only moving substance now within these objects is electrons. Probably photons of light will flow around there as well, and then still more elaborate structures, growing into a full three dimensions, will arise one day out of this technology.

In a sense we see dimly the future rise of a second kind of life—a new kind of life that we will have given birth to—not through the long evolutionary chain which connects us physically, so to speak, by blood, to the primeval shores, but through a second level—our understanding and our conscious control of the microcosm.

The name that is inscribed most strongly in this chronicle will be that of Robert Noyce. If we have any sort of luck at all, and if we invoke the sense of survival which every cunning species certainly ought to have, that's what our 20th century will be remembered for, not the mushroom cloud and the great flame, which shadow all our fears.

Niels Bohr:
A Glimpse of the Other Side

I t is fitting to ask what limits can be placed on the utility of Niels Bohr's legacy to physicists, and to a wider circle still.

Let me begin with some anecdotal history, to establish my stance. I am not one of the generation of the Golden Age, that group of physicists who found themselves between the wars in Copenhagen, at Blegdamsvej, enchanted by the subtle quantum music of the north. (I did not visit Bohr's institute until 1959, and at that time Bohr himself was not in the city.) But I am one of the immediate generational successors to those heroic physicists; my own teachers, Robert Oppenheimer and Hans Bethe chief among them, shared the Copenhagen spirit, even if perhaps they were not steadily under its spell. During 1937, my first eye-opening year as a naive graduate student at Berkeley, Bohr himself came to that campus to give a series of lectures for the learned public. Of course, he attended Oppenheimer's series of theoretical seminars at which I was still an awed and generally uncomprehending kid in the back row.

I took careful notes at Bohr's public lectures, a bit intimidated by their content—but I wonder if I ever studied them with care? My ideas of quantum mechanics came mainly from the problems and style set by Oppenheimer, and from long hours of study of the famous Pauli article. However, I have one strong conviction that does date back to Bohr's Berkeley lectures: the experimental arrangement, the classical limit that makes the inevitable cut away from description through the quantum state, is best represented by the pseudorealism of Bohr, those wonderful drawings of slits milled out of heavy brass, held to an optical bench by serious nuts and bolts. Measurement theory for me will always have that decisive background.

It was at Los Alamos that I came into more direct contact with Bohr. Again I was no intimate; but since I was close to Oppenheimer and to Berkeley classmates (such as Robert Wilson and Bob Christy) I found myself quite often in a circle around the Bohrs, Niels and Aage, both at work and at evening gatherings. My political stance about the bomb and its international meaning owed much to Bohr's ideas, filtered for most of us through Oppenheimer. Bohr's public views persisted a little time into the postwar world, and the idea of an international agency charged not only with the policing task but also with the positive development of nuclear science and technology caught me up for a while. It is surely a conception that goes back to the interaction of Bohr and Oppenheimer. The chill of the Cold War caused me, an old campus radical, to shiver rather early. The idealism of the war's end withered in the frosty political context of the late forties, and soon the urgent claims of national security, East and West, put a stop to Bohr's "open world," the hope for universal nuclear community.

One Sunday afternoon in the late summer of 1944, many newcomers to the mesa country made the modest drive over the Jemez Mountains from Los Alamos to Jemez Pueblo, to witness the marvelous ancient dances held there. Bohr came too, and I watched him within that scene of unpredictable contrasts with the sense of wonder so often induced by the omens and incongruities of wartime Los Alamos. It was not long afterward that my first few hours of personal contact with Bohr took place. Somehow it fell to me to offer to accompany Bohr to a Sunday evening film at the post theater. It was a large-scale Western, perhaps with John Wayne or some similar star of gun and saddle. I undertook to interpret the peculiar genre of the Hollywood Western to Bohr the European, and we sat together as I told him how to watch for white hats and black, how to part the good guys from the bad. He seemed to enjoy the film, and to appreciate my whispered interpretations. It may have been a decade before I read recollections by others of their years in Copenhagen, and realized that Bohr possessed substantial critical expertise on Westerns, arising out of long study. My help had been as superfluous as it was ingenuous, but his warmth and insight led him to keep his secret: subtle as always.

A Generation of Natural Science Since Bohr

Niels Bohr died in 1962. It is not hard, in retrospect at least, to list the chief discoveries in the natural sciences that have grown clearly visible in the years since Bohr's work ended.

The oldest of these—the key findings go back to the early fifties—was the rise of modern biology, especially the genetic biology of information storage and transfer, based upon the double helix and its template method of reproduction. A generation after the double helix, the science has acquired great powers, both intellectual and technological. It has opened the way to the understanding of the development of multicellular organisms, a goal still not close at hand.

Next, as the sixties opened, came knowledge of sea-floor spreading. Eventually it was convincingly shown that the motion of distinct plates of continental size, which together form the earth's outer crust, must be a steady feature of earth history. For the first time, geology became open to a broad theoretical account, more or less as a century ago paleontology became clarified as the consequence of Darwin's theory of evolution. Whereas previously a special history had had to be invoked for every region and for every period, even though the physical processes were roughly similar over time, now the whole globe had a unitary albeit complex history. Geologists who hardly knew that the world beyond the domain of their special interest existed now saw the earth as one interactive system.

Also in the sixties, the ideas of relativistic cosmology were extrapolated through aeons of history, back to a time when the universe appeared qualitatively different from what it is today. The discovery of quasars—discrete galaxy-like bodies bright enough to be seen at distances corresponding to red-shifts several times greater than any known for ordinary galaxies—made plain that the universe had a long history. By the end of the decade there remained little doubt that the distant microwave background was a residue of the ubiquitous early presence of space-filling plasma—matter directly observable that showed complete uniformity and isotropy, and gave no sign of discrete condensed objects. This provided evidence for the long-predicted early hot universe, seat of the synthesis of helium in cosmic space, whereas all subsequent nucleosynthesis would be within stars and galaxies not yet gravitationally assembled out of the uniform hydrogen-helium plasma.

Bohr himself would probably have rejoiced most over the great advances in particle physics. From the end of World War II until about 1960, experiments disclosed a plethora of candidates for designation as fundamental particles. But for some time there was little understanding of these results. Then, beginning with the successes of the algebraic grouping called the Eight-fold Way, theorists rapidly developed schemes to explain the entire array of particle states we now know. That understanding rests squarely upon advances in quantum field theory—advances that have brought about the demonstrated unification of the electromagnetic and the weak nuclear interactions. Then came the recognition, perhaps less well established but compelling all the same, that the strong nuclear interactions too could be understood on the basis of quarks, which are believed to be the inner constituents of neutrons, protons, and other strongly interacting particles. Quarks now seem so real (despite the fact that, according to the theorists, they can never be found free) that their involvement at energies far beyond experimental reach, as described by the so-called grand unified theories, lays widely recognized claim to good sense. Today the edifice of particle physics is quite impressive, with only a few hints from experiment that closure of the physics of accessible energy might not be final.

Assessing Complementarity

How have Bohr's general views—his principle of complementarity, his concern with correct language and the recognition of its limits—stood up throughout the decades of such scientific novelty?

The answer seems, fittingly enough, to be itself dual. The new biology, wonderfully apt, and growing by leaps and bounds in scope and depth, is positively anti-Copenhagen. We can see this with some confidence, for in his lecture "Light and Life" Bohr gave us insight into how he expected a biology of power to look. He knew, of course, that it would require molecular analysis if it were to join the rest of science; life depends on fine structure. But certainly, at the time Bohr gave the lecture, he believed that we would require a new viewpoint on chemistry, on reactions and structures. That viewpoint, he suggested, would come out of the exploitation of the complementarity between the microdescription of an organism and its living function.

Before Bohr died, he came to realize that his initial view was wrong. Something must be added to chemistry. That novelty lies in two directions. Theoretically, we need to follow not only free energies and structures in the usual way, but we must keep in mind the category information. Replication of pattern and an intricate set of variations on that theme are key matters. Moreover, there is a structural level beyond the simple collections of sequenced monomers. Polymer chemistry is evident, but probably the work of membranes and other components of cells goes beyond that, toward surface structure, closer to solid-state physics than to test-tube reactions. Yet nothing appears that is more subtle: no new interactions, no dynamics achieved only at the expense of form. It looks as though simple close fit of enzyme molecules and substrates, sometimes with ancillary molecular partners, will account for the reaction chains and cycles. It looks, too, as though the origin of secondary and tertiary structure arises from the long sequences that make up the key polymers. Normal collisional processes allow eventual use of secondary and tertiary bonding to build the architecture beyond the long chains. Cunning evolutionary design is enough to construct such molecular mechanisms as ribosomes out of molecules specified only chemically, although at tedious length and detail. Schrüodinger's metaphor of the aperiodic crystal seems a better hint than any of Bohr's more dynamic suggestions—or so we think today. Life ought to exhibit complementarity, if anything does, but that subtlety enters perhaps beyond the molecular level, in evolutionary processes and in learning systems associated with neural networks.

It is almost self-evident that plate tectonics, even though we do not know much about its driving forces, is a mechanism of a fully Newtonian kind. At that scale, there is not much room for subtler views. Yet how slow we were to recognize these worldwide processes that can be described in Cartesian language.

But in the physics of particles, quantum theory—quantum field theory—conceptually complete by the 1930s, seems to have found flawless application up to energies of a hundred thousand electron masses at least. The theorists of today have been original and ingenious; their daring in extending the old ways is a marvel. Plainly, dynamical quantities go well beyond ordinary space-time, but they fit the ideas of quantum field theory like a hand in a snug glove. When Bohr and Rosenfeld made their painstaking study of the measurement physics of the vacuum fluctuations implied by quantum electrodynamics, they showed how quantum complementarity ideas held in strength, and how these ideas were required by, and in turn illuminated, the meaning of the formal theory. That all still holds true. It is even the case that the dynamics of the vacuum—a true quantum ether—is physically more and more important. Some day we will extend the wonderful complementary relationship that Bohr and Rosenfeld showed to exist between the assignment of sharp particle numbers and the description of the fields.

Today the cosmology of the early universe is thoroughly integrated with particle physics. Not only do the conjectured high temperatures of that early state make the relationship inevitable (if the extraordinary extrapolation of the expansion holds as expected), but even more the union leads to testable explanations for the symmetries of the universe we see. Here the ground is so newly entered that we must be tentative. But I can recall from the dim past of 1937 Bohr's firm opinion, when he was asked for comment upon the then entirely geometrical relativistic cosmologies of the day, that we would understand the universe as a whole only when we understood in depth the nature of the fundamental particles; the trouble was that there was as yet no recognized context which the two subjects shared. Now we have found that context (if we are right) in the early expansion and—a straightforward interpretation of evidence now at hand— in the unexpected uniformity and isotropy of early matter as a whole.

The history of particle physics is an account of splendid successes, old and new, and of shortcomings and limitations too, to be expected in any single work of our finite species. They built a city shining upon a hill, they who half a century ago founded the quantum mechanics of particles and fields, and the design of this city is understood best through Niels Bohr.

Richard Feynman: An Old Friend

Millions of his fellow citizens saw Richard Feynman just once, but in an absolutely characteristic moment. There he was, a theoretical physicist unknown outside of the world of science, visible on all the versions of the nightly news. He owed this attention to his appointment to the commission set up to study the tragic end of the space shuttle *Challenger*. Before your eyes Feynman chilled a small neoprene O-ring in ice water to turn it hard and inflexible; thus one of the proximate causes of the disaster became a homely experience sifted out of the engineering complexities. Such was his lifelong delight: grasping the world by particular example, even if drawn from the most abstract of generalizations. He had no peer in this style, whether with O-rings, the contemporary response to Isaac Newton's prism experiments, Mayan almanacs or the state vectors of fermions and bosons.

I think it was in March of the hard wartime year of 1943 that Feynman stopped at the University of Chicago, on his way out to the laboratory just being established on the mesas of Los Alamos. Perhaps his ingenuity and expertise would be of use to the group of young theorists lakeside, also engaged in the secret enterprises of the Manhattan Project. He was not yet twenty-five, his highly original Ph.D. thesis still unpublished, but he brought with him an open reputation for unmatched quickness at the hard integrals that arise in mathematical physics. We all came to meet this brash champion of analysis, who for the first time was entering a circle of physicists wider than his long-impressed fellow students and teachers. Feynman was patently not struck in the prewar mold of most young academics. He had the flowing, expressive postures of the dancer, the quick speech we thought of as Broadway, the pat phrases of the hustler and the conversational energy of a finger-snapper. But he did not disappoint us; he explained on the spot how to gain a quick result that had evaded one of our clever calculators for a month. Such a display was a shallow way to judge a superb mind, but it made a lasting impression. To this day physicists of the postwar generation recommend their best students with the mock-modest concession that he or she "is no Feynman, but . . ."

One of Feynman's last publications, the text of a lecture given at Cambridge University in 1986 as a memorial to the great English physicist P.A.M. Dirac, is another typical performance: it begins with a creditable drawing of Dirac that Feynman himself sketched in 1965, and it includes a photo of the ebullient Feynman engaging the attentive but warily reserved Dirac. The lecture is a tour-de-force of exposition, seeking to derive a couple of the most important results of quantum field theory not "in the spirit of Dirac with lots of symbols and operators" but by explicit arguments that flow from those summary zigzag sketches known everywhere as Feynman diagrams. True to his own way, Feynman explains here too that he will set out some "very simple examples . . . because if you do you will understand the generalities at once—that's the way to understand things anyhow."

Adept at the formalism, inventor of more than one mathematical device of sweep and power, Richard Feynman nevertheless fought for an understanding that was explicit, concrete, without the constraints of jargon, usage, decorum and precedent. In that way he was like a poet, at home in metaphor and image, cheerfully exhibiting the relation between a dance figure and the rotation of a two-valued amplitude. The Feynman diagrams began as a genuine aid to intricate formalism, for they clearly index the bewildering alternatives that beset any calculation of what happens to interacting particles. Once Feynman tried to convey the difficulties of such a calculation to someone utterly strange to such tasks: "You know how it is with daylight saving time? Well, physics has a dozen kinds of daylight saving."

A few examples are painted like pictographs on the family van that stands parked outside his suburban house. The diagrams are a form of shorthand analysis, but they are far from a geometrization of the events. Their fidelity relies on careful and highly original mathematical rules that accompany each straight or wiggly line, usually generating at every vertex an exercise in matrix algebra; between intersections they imply a clever nesting of integrals that serve to sum over the space-time excursions that intervene.

Certainly Feynman diagrams have become as useful to the quantum field theorist as circuit diagrams ever were to the electronics designer. Like circuit diagrams they are rather less than maps, rather more than logical outlines. The diagrams allow a transparent ordering that makes clear the intricate calculations that would have seemed inhumanly beyond masters of the same theory.

The diagrams helped Feynman gain the understanding that in his and other hands allowed quantum electrodynamics to be made into the most accurate of physical theories, even though it is plagued by deep inconsistencies, logical icebergs that can be skirted but not removed. Like quantum field theory itself, Feynman diagrams transcend any single set of forces and particles. They guide the computation of the intercourse of quarks and gluons, just as they have the computation of the electromagnetic processes for which Feynman devised them in the early postwar years.

When the feeling was abroad that the teaching of physics to beginners was growing stilted, Richard created a series of lectures that grew into a multivolume text. It is tough, but both nourishing and full of flavor, and after 25 years it is an everyday guide for teachers, and for the best of beginning students. Everyone has heard of his varied talents: raconteur, artist, musician, dancer, gambler, puzzler, cipherer, locksmith, sometime chemist, rolling with the good times at the carnival in Rio or in the casinos of Las Vegas. Most of the anecdotes that cluster around him are true!

What he accomplished in physics is well-known. Besides the insights concerning quantum electrodynamics for which he shared the Nobel Prize, he made a strong mark at the right time on the nature of the weak interactions, on the conceptual wave functions that describe liquid helium, on the foundations of quantum mechanics. It was his encouragement and his fresh interpretations of the experiments with electron scattering that signalled to all that free-moving subunits, his "partons," were deep within proton and neutron. That opened one main experimental path toward quarks.

For him learning was never passive. He was no omnivorous reader. Yet one evening long ago he ran across for the first time Prescott's famous accounts of the conquests of Mexico and Peru. Fascinated, he borrowed the heavy volume, to read it steadily all night and day. He bubbled over with questions. The interest never left him; 30 years later he worked for a while, with the early success of a gifted amateur, at the decipherment of Mayan documents. Frustrated with anatomical terms on one of his excursions into the brain, he asked the librarian for "a map of the cat." At the time of his death he had left half-a-dozen tracks behind in topics from molecular biology to computer algorithms to the psychology of hallucination.

The arts of the theater are inherently two-fold. The actor on the stage pretends to be who he is not, by artful empathy and the words of another. That was not Richard's way. His theater—and it is impossible to evoke him without the word "theatrical"—was on the other side, that stage where dancers and wire walkers and magicians daringly perform. What they do is striking, but it is not dissembled or illusory. It is real, whole, the expression of mastery of some real challenge, trivial or urgent, posed by nature and by human perceptions. On that stage he performed most of the time in four real dimensions.

In Far Rockaway days, his remarkable father once explained to the questioning boy that the name of a bird was certainly interesting, part of human beings and their rich languages. Yet almost nothing of the wood thrush itself lies within its many names. What you must do to know more of the thrush is not to name it but to listen to it, watch it, think of what you notice. No powerful theorist in our time has so conspicuously kept a keen eye on the great pied bird of Nature; the example Richard Feynman set us each day is no longer renewed. We miss him deeply, and we all learned from him while he was here.

The Exploratorium:
Frank & Jackie Oppenheimer

The dim and lofty arc leads the visitor in, to wander from one exhibit to the next, hundreds of them set within the acres spanned by this celestial attic. When it first began to grow, it was already a delight; today it has matured to what must be the most original and for me is the most brilliant museum of science in the world.

Of course the judgment is subjective, the opinion of one physicist. The museum itself is subjective in that very sense: The Exploratorium is styled "a museum of science, art, and perception."

Curiously enough, a great many scientists who have visited the place share my pleasure in it. The fact argues for a community of opinion; the subjectivity of one scientist is widely shared. What is even more exciting is that a million visitors, as various in their interests as visitors to a beach, largely agree. The researcher, the teen-ager, and the little family group alike enjoy what they perceive here. The crux is perception; here learning is direct, experiential. Indeed, no one can convince you by words alone that the random dot pattern displays a three-dimensional form. You must see it—or perhaps miss it—for yourself. It is that feeling of shared experience that dominates the subjectivity of the research worker and the serious student of science, and here it is offered over and over again to everyone who will try.

There is not much of a collection of famous artifacts here, not many models of imposing works, nor mementos of the great. Most of what is here was made right in the building, drawing of course upon the richness of a twentieth-century world city. The collection rests on a set of ideas, all right, ideas that have been made concrete by artist and artisan, with eye, hand and tool, of course animated by and interacting upon purpose and concept. The machine shop is wide open, right by the door. Every visitor can see there the working of wood, metal, and glass to produce the artful systems on display, the spectrum as real as the vortex, and even some dance of symbols on the computer screen as simple rules of manipulation play out their complex consequences.

The place of the artist here is secure, too. For art and science deal alike in the minute particular, with the thing itself, the key phenomenon behind the function unfailingly in place. These artists-in-residence have worked in the same shops using the same tools as the experiment designers, but they have employed the phenomena to present something beautiful of their own. The unity of art and science is at once achieved through perception, an experience necessarily mediated by the senses and sensibilities of the onlooker, who becomes a participant in the work.

There are challenges still unmet. If art and science are manifest in the concrete, through direct participation by the visitor, how can abstractions, the concepts of science that go beyond mere perception, be conveyed? The issue is given sharp expression in the fact that the electronics shop, a place of major importance in this milieu, is not open to the visitor. For indeed there is not much to be seen there. Consider how much more the uninitiated but interested observer can learn from opening the case of a ticking clock than by staring at the chips of a digital timepiece. That is parallel to the distinction between watching machinist and circuit-builder at work.

The electrical exhibits make the point well, the author explains. Here perception is itself gained only through instruments. Sensing has become a process once removed. The Exploratorium, like most scientists themselves, needs to work by inference and analogy. For are not our own eyes and ears themselves but inborn instruments? Their proper use and interpretation, like that of a microammeter, had once to be learned by test and re-test, sometimes reasoned, sometimes only implicit. If that view is at all correct, there will one day open a broad public way into energy and charge and gene frequency and exponential change through elaborated experience, analogous to but beyond common sense. The Exploratorium is well on the road, but the journey is not a short one. The hope behind this display of friendly marvels is the unity of human beings, a unity that lies below all their differences, that rests in their common ability to share in reason and delight. That egalitarian hope is not new in science. Perhaps it rises in all times of transition, as it appeared in the days of the tireless traveler Alexander von Humboldt, who took his thermometer and his search for unity to the cold current off Peru. He was a democrat in the service of kings, a man between two eras; he journeyed romantically afar, but he always took along with his instruments the critical spirit of the Enlightenment.

There is at work here a fundamental but playful healing. Its subtle prescriptions are not covert but wide open. Values lie deep within these actions, from what I know of the late founders of the Exploratorium, Frank and Jackie Oppenheimer. We were close friends for fifty years. Their vision was ". . . that human understanding will cease to be an instrument of power . . . for the benefit of a few, and will instead become a source of empowerment and pleasure to all."

Ours is a time of transition. Science and technology are pressed to the service of the powerful as never before. The rise of abstract concepts filled with practical power bids to partition society into the few who know and the many who watch. But the unities between play and purpose, between science and art, between hand and mind, bonds that transcend human diversity, shine there on the Exploratorium floor like the grand rainbow of the Sun Painting itself.

Heaven and Earth
One Substance: Bernard Peters
and the Heavy Primaries

The Celestial and the Earthy

Long ago the cosmologist Anaxagoras came from the east to teach in Athens, before that city had attained its glory. One of his theories proposed that "the sun was a flaming stone larger than the Peleponnesus." For such blasphemy he was expelled from Athens almost 2500 years ago, escaping a capital sentence only through the intercession of the statesman Pericles, whose teacher he had been.

Only around 1450 in his final years, would Copernicus risk open publication of his arguments that the earth was an orbiting satellite of the sun, like the other planets in the sky. Then physicist and astronomer Galileo Galilei about 1610 recognized with his telescope, the first of all telescopes to succeed in astronomy, that the silvery moon was an earthlike mountainous ball. One of its big central craters was ringed with high rocky peaks, looking very much, he wrote, "like a plain in Bohemia." He too was punished during those years of religious conflict; the blame directed at him was in part for his explicit unification of the celestial with the merely earthy.

Isaac Newton too would in the 1690s demonstrate an astonishing unity in the world. The sun and the planets danced out their complex steps under the simple rules of gravitation. Within a few decades it was shown that a mountain in Scotland could pull aside the surveyor's plumb-bob just as the moon pulled the tides; gravitation was as earthy as it was celestial. Then laboratory instruments were made sensitive enough to detect the tiny gravitational forces from heavy weights right in the lab. By the early decades of the nineteenth century the astronomers had shown that double stars themselves circulated in orbits fully obedient to universal gravitation. Newton's prescriptions extended to regions so distant that they had never been under the sun; Newton's force was universal indeed.

Beginning in the mid-nineteenth century, another material unity appeared at the slit of the astronomer's spectroscope. The surfaces of the sun and the stars were chemically analyzed by their light, the spectral lines, at first only quantitatively, by now quantitatively as well. The results are profound: all the matter whose form we can see is solid like the moon's craters or gaseous like the solar surface (liquids are rare), and all of it gravitates just as here below. All of it shares a common constitution: with the human hand—your hand—or any common pebble on the beach, celestial matter too is made up entirely of a somewhat varying mix of a hundred more or less familiar earthy atoms. Here below as there above, the recipes of all matter call for no more than that; atoms comprise the entire material world.

The Cosmic-Ray Primaries

By the 1930s, the incoming cosmic rays—their very name hints at their story—were demonstrated to be mainly positively charged nuclear particles, the protons. They formed a unique gas out in space, particles moving uniformly in all directions, with speeds relative to earth that were far greater than the speeds of planets, comets, or even stars, close to the ultimate speed, that of light. The cosmic-ray primary particles seemed to be a simple gas, the simplest of stable particles, in uniform isotropic flow at relativistic energies still encountered nowhere else in nature.

Those incoming primaries might well be truly cosmic, for the first time a kind of matter that was distinct from the rest of the astronomical world. Perhaps they were so strangely simple because they were the sign of an earlier, distinct, phase of the universe, relics of a physics of creation long past, of some novel, never-repeated processes we could never grasp. The Oxford cosmologist E.A. Milne said so explicitly. (Once, much later, when I was fortunate enough to talk about the matter with elderly Albert Einstein, he too—he easily conceded that he was far from well-informed in that field—still felt that such an exceptional point of view was possible for so remarkable a phenomenon as cosmic rays).

Most of us hoped instead for universality. Cosmic rays too might some day be explained as strange samples from our own world of matter, hydrogen stirred into random motion and promoted in energy well beyond the energy of any nuclear decay. They would in effect be familiar matter in a new condition, some kind of "flaming stone," just as the amazing sun is itself in reality no novel material but is a fiery ball of impure hydrogen, even though it is much larger than the Peleponnesus.

The label *cosmic* seems apt enough for that remarkable unceasing radiant flow from space, even though they are not literally cosmic, but astronomical; we are pretty sure that nearly all the rays are as galactic as the starry sky itself, yet the term sticks. It was in the year 1948 that we were shown for the first time the hard evidence that primary cosmic rays were not that idealized, uniform, primitive "cosmic" gas, matter made up solely of the most primitive of atoms in headlong motion. They were in fact a complicated sampling of the ordinary matter of sun and stars, right here in the disk of the Milky Way.

Starstuff is itself rather uniform. It is mainly hydrogen like the cosmic rays. But it is not at all chemically pure, mysteriously simple, reagent hydrogen; rather it is hydrogen of technical grade, clearly mixed with five or ten percent by weight of a dominant helium impurity, plus a little more of many, many heavier nuclear species: starstuff. The cosmic-ray chemical samples first put into evidence were brought back from stratospheric balloons floating high above 98 or 99 percent of our shielding atmosphere. The big balloons carried a sensitive detecting payload, thin films of photographic emulsion. Once recovered, developed, and painstakingly scanned under the microscope, long coherent wakes of developed silver grains in the emulsion layer unmistakably recorded the passage of fast, multiply-charged ions of many familiar elements. Recognition of unity was quick and it was firm. By now those results are very much elaborated, in as much quantitative detail as the label on a bottle of first-rate laboratory reagent.

The 1948 papers[1], brilliantly conceived and technically expert, include the first postwar work of physicist Bernard Peters. For me it opened a new world, into which I moved slowly, step by step from theoretical nuclear physics to the astronomy of high energies that enfolds me to this day. The two teams had given every investigator who came after them the full courage to study cosmic rays as another part of the vast astronomical world, curious wanderers to be sure, but only an unusual constituent of the inclusive, familiar whole.

Heaven and earth are of one stuff indeed.

---

[1] P. Freier, E.J. Lofgren, E.P. Ney and F. Oppenheimer; *Phys. Rev.* 74,1818 (1948); H.L. Bradt and B. Peters; *Phys. Rev.* 74,1828 (1948), 80, 943 (1950).

By 1950 the first few experimenters sampling the nearest edge of space could use the sample to set limits on the history of the wandering particles. They could estimate from the composition of the incomers the time of storage and the distance made good since those particles were first accelerated to cosmic-ray energy. For the rays must traverse interstellar space, which is not utterly empty, and there from time to time they collide with the dilute thermal gas in space, itself mainly hydrogen.

The secondary products of any such collisions far out in space become a part of the incoming "primaries" here on earth. The number and nature of the products—other nuclear isotopes and electrons and even positrons—shows that the wandering protons must have been stored in space for a time orders of magnitude longer than the mean time for the arrival of light from the stars, which is something like the averaged transit time for light across the galaxy, some tens of thousands of light years. That makes sense, for it reconciles the fact that the energy density of starlight and of cosmic-rays are about the same, although cosmic rays surely have unusual sources, while starlight pours from every star. So began the dating of cosmic rays and the long and difficult search for their regions of origin.

Cosmic rays are stored up in space; they do not march out of the galaxy into all of space at full speed, as starlight does. The reason is evident. Charged particles cannot move in space along straight lines, but diffuse waywardly from the sources through the intricate magnetic fields of our magnetic galaxy, as photons diffuse through fog. Dust-free space is transparent to light, but a dense fog to particles that feel magnetic forces. We cannot expect to see any cosmic-ray stars show up as point-peaks of intensity; cosmic-ray astronomers have to try to build up their source maps without appeal to direct lines of sight. (Only the sun is so close to us that we recognize it as an occasional transient source of a small portion of the cosmic rays.)

Small variations of intensity and direction can and do show diffusion gradients in the galactic cosmic-ray gas, but for the most part we cannot trace back a proton to any particular distant source as we can follow a ray of light or of any other uncharged component to its real origin.

These ideas, all still under quantitative study, flowed from the postwar results that first showed the primary rays to be worldstuff like ourselves. I shall leave to others—perhaps to Bernard himself—to talk about the exciting part taken by the Indian physicists in the worldwide unravelling of the rays and in the search for their origins, after fifty years by no means a settled matter, one probably more complicated than we like to admit.

Most clues point to magnetic interactions to accelerate the charged rays, energy added steadily by moving macroscopic magnetic fields, not impulsively by some unknown events. In the frame of the particles a magnetic field carried by any moving plasma is seen as an electric field, and is thus an energy-giver. The individual particle is in a way trying to come into energy equipartition with an enormously larger moving mass, whose mass and kinetic energy are incomparably larger than that of any single cosmic ray, a mass that carries magnetic lines of force on macroscopic spatial scale. The "high energy" of a single cosmic-ray proton is tiny on such a scale; the process of equipartition is limited so that the proton takes up plenty of energy for one lone particle, as much as a few joules for the highest-energy cosmic rays, so that we call the rays high energy. It is the energy per particle we describe as high; a joule is not much energy for any astronomical structure. The total power of the cosmic rays that enter earth is no more than the power in ordinary starlight. It is the energy concentration that astonishes.

There are many possibilities for energy exchange between the macroscopic world and some lone proton. Perhaps the magnetic field that pushes the proton, whether only once, or time and time again, is generated by the electric currents that fill a vast slow-moving cloud of magnetized plasma, some large feature of the orbiting galactic gas. Perhaps it is found in a smaller, but still enormous, moving shock front light years across, generated by some stellar explosion, likely a supernova. Perhaps it is a smaller, star-sized shock from a magnetic flare on the surface of some star, like those we see close by on the sun. Perhaps it is an entire spinning, magnetized neutron star the size of a mountain that can share rotational kinetic energy as its rotating pattern of magnetic field sweeps past the proton. We have evidence for distant spinning quasar discs as big as the solar system. Those might be the energy-rich magnetized sources of the relativistic electron pools that we know to power the giant radio sources far away from our galaxy; possibly the cosmic-ray protons are out there as well. (We cannot detect the protons by their radio emission, for particles so massive accelerate slowly and radiate too little. Big enough pools of cosmic-ray protons can be mapped at a distance for example in the disk of our own galaxy, by detection of the gamma-rays they make as the products of occasional collision against the dilute gases of space. The gamma-rays bring us signs of far-off cosmic rays along the straight-line path that is denied to the protons themselves.

All of these places, and more too, are among the sources of the cosmic rays we study; we are not yet able to assess fully their several roles in the whole story. But for all that variety of astronomical source environment the primary rays themselves are more or less ordinary, familiar ions in the natural world, mostly, but by no means all of them, protons. They gain their astonishing energy, not in some unique mystery of creation, but by pushes they underwent wherever they happened to be in the evolving context. Some are the products of secondary collision. The cosmic rays are part and parcel of the material whole, as much as the stars themselves.

The foundation for that unifying insight was well-laid by young Peters and his young friends back in 1948.

The Perils of the Skeptic

We were pretty surely right to seek the unity of the material world, and to deny that cosmic rays must be broken away from the rest of astrophysics. But that insistence can also be wrong, however satisfying to our hopes; the physicist should not be beholden to any special metaphysics, even to simplicity. Nature decides. Newton himself violated the fine physical intuition and high sense of unity of his great colleagues by his blithe willingness to use the daring notion of action at a distance, his simple unexplained law of the inverse square for gravitation. He would make no hypotheses, he said, about the state of the medium between the attracting masses. The great Huygens was openly a little scandalized, and even said aloud that he hoped for a real mechanical theory from Newton, not merely talk of occult forces. But Newton enabled so much progress with a simple formula, and his universal force was so plainly demonstrated here on earth, that his law still leads us. That incomplete but amazingly precise description far outlasts the vague, impractical vortices his less incisive contemporaries sought out, however metaphysically correct they may have been as viewed in the light of the geometrical theory that came much later.

As much as we enjoy the unity of matter that has since 1948 placed the once-ineffable cosmic rays squarely within the real, impure astronomical world, we have ourselves come to a new time of metaphysical strangeness that seems all the same to have ample support from theory and experiment. It remains short of final demonstration, yet the amazing simplicity of the cosmic microwave background has placed us once again into a stance where the simplest material unity we hoped for the world may have to be denied.

The relativistic early cosmological stage that is dubbed inflation gives a clear dynamical account of the best-measured results in all of cosmology, the thermal background. The featureless hot radiating plasma that we record at far distances, well behind all the irregular, clumpy matter we see, is fully interpreted by inflation (itself outside the scope of these lines). We have long recognized that some dark unseen matter of some sort, perhaps only dark, cold little failed stars, was copious in the galaxies and their halos. Now the inflation theory all but demands a cosmos that has one or two orders of magnitude more dark matter than all that we can see radiating through any of the electromagnetic channels we now master, much more of it even than the best estimates of how much ordinary baryonic matter can be present, much heavier than all the stars, the gas, and the dust.

The good predictions of the pre-stellar, primordial production of helium and deuterium is what limits baryonic matter strongly. That prediction too stands on a strong base. Very old "metal-poor" stars and gaseous clouds are directly seen to contain 25% of helium by weight, close to the amount we impute to the sun. But those same samples may have only a fiftieth part of the sun's small content of the heavier elements! That minimal, ubiquitous helium was formed before any of the galaxies or stars; the oldest stars are rich in it. The cosmologists can say just when and how the helium was made in a smooth, bland cosmic volume filled with a very hot gas of neutrons and protons, electrons and photons. (A few light isotopes are made there as well in trace amounts that fit the best measurements.) Hydrogen and helium are of cosmological origins; the stars made all the rest—which is not much.

We have come to an odd metaphysical stance. Old Copernicus could see the planets easily; he showed that our green earth, so close to us that it looked much different, was just one more shining planet among the planets, nothing so special. So for gravitating matter, and so also for the atoms. They are universal. So also for the cosmic rays. They are just matter like ourselves, only much energized, just as shining, artificial satellites are highly-energized machine-shop artifacts, made by human hands of samples of a single familiar mix of atoms.

Now we are told that most of the mass of the cosmos is a new kind of matter whose very name we do not know; it gravitates, but does not interact detectably either with electromagnetic fields or with the forces that bind the nucleons. For the first time we find that it is not the matter of everyday experience that fills the cosmos; the situation is anti-Copernican for the first time since Anaxagoras. Space does not appear to be filled with dark cold stones or with anything like them, but with axions or photinos or some other still conjectural particles, predicted by field theory from new and tempting symmetries that so far have found no experimental support.

The most palatable proposal for such novelties, made very early by Ramanath Cowsik,[2] fills our world with unseen massive neutrinos of one

---

[2] Reviewed nicely in *Cosmic Pathways,* editor, Ramanath Cowsik, pp. 289–310, 1986 (Tata Mc-Graw Hill Publishing Co. Ltd, New Delhi).

or another flavor, though so far that proposal does not seem to work in detail. Neutrinos, while not exactly everyday matter, are at least no strangers in the laboratory or to the astronomers. If they were the dark matter, we could see a unity in matter as we see the chemistry of our earth: silicon and oxygen and iron are plentiful here in the rocks, though in sun and stars those atoms are found only in parts per thousand. Still, they are present, to imply a deep unity within. That unity is simply locally disturbed, but not destroyed, by those complex physico-chemical processes that formed our unusual, extra-dense home planet, collecting up oxygen, even iron, but losing light fugitive hydrogen.

Experience must one day decide. Something of the spirit of old Copernicus will hold: our earthly home seemed special, but was not. So too our kind of matter, made of nucleons and electrons, seemed prevalent, but perhaps it is not. The real unity we seek is not of the familiar alone; that seems too narrow for an unbiased view. It is rather a unity of the demonstrable, a unity that joins all the processes we can grasp. The next decade or so may begin to bring an answer.

Felix: An Extended Acknowledgment

We were a dozen young graduate students of theoretical physics in the group around Robert Oppenheimer at Berkeley during the late thirties. I can easily recall my first encounter with a new student, Bernard Peters, now more than fifty years ago. He seemed a little different from us, more mature, marked with a special seriousness and intensity. He was a little older than most of us, but his experience went far beyond ours. He was a European, who had seen and felt the barbarous darkness that mantled Nazi Germany; then he had worked for a living among the longshoremen in San Francisco Bay. In contrast, we were only North American students, largely innocent, save by hearsay and our eager reading, of most of life beyond the campus and overseas.

Bernard's wit and playful spirit, personal affection, and deep generosity were in no way muffled by his patent maturity. The time came when I had to leave Berkeley to take up a first teaching post across the Bay in San Francisco. I was a poor enough graduate student; we had no car, though plainly it would become very welcome, once my wife and I lived so far from the Berkeley campus that visits would be long journeys.

Sure enough, Bernard had a answer, generous, delightful, and effective. He would give us an automobile! It had long served Bernard and Hannah, and they could manage a change. They had named their loyal old vehicle, Felix. Felix, a hardworking veteran Ford, became ours. Up and down the roads he carried us in pleasure, happy as the Latin meaning of his name.

Felix had one failing, a bad drinking habit. Let me expand on the analogy: all cars live on a liquid diet. But it is clear that their nutritive food is their fuel, gasoline (petrol). The lubricating oil they must also sparingly use is something extra, not unlike the alcoholic beverages many humans take moderately, not for nutrition but for other internal reasons. Poor Felix was a lubriholic. Where most automobiles take a few quarts of engine oil each month or two of work, Felix consumed the thick delicious stuff at easily ten times that rate, a quart of oil to enrich every few gallons of gas. (1 US gallon = 4 US quarts = 3.785 liters.)

Whenever my wife Emily, who was then the driver in our family, guided hungry Felix into a petrol station, my work was routine. Once the attendant had fed Felix, and gotten his pay, I would lift off the back-seat floor a big square multigallon can we kept always at hand. It was the economy-sized package of the cheapest lubricating oil we could buy. I would open the hood, lift Felix's filler cap, and satisfy poor Felix's habit from the big can, until the next time we had to stop for gas. Only so could we support thirsty Felix as he puffed his way along the roads of Northern California, a thin blue stream of tailpipe smoke always behind.

Our student days are long gone. The Berkeley friends were dispersed by the usual academic pilgrimages, then scattered by the exigent years of World War II, and parted indeed by decades of Cold War that led Bernard to a physicist's life in Bombay. I am happy to evoke our fifty years of friendship inside the diffuse little republic of physics, and to wish for him the high hopes we share: a better physics to come than we old-timers have yet seen, and a century of change far beyond the laboratories towards a world wherein reason will be more at home, and justice widespread.

Charles Babbage: Far Ahead of His Time

WITH EMILY MORRISON

Charles Babbage is a name known fairly widely today; in his own time the value of his work was recognized by few of his contemporaries, and he was held a crackpot by his London neighbors. His name has emerged from obscurity in the past generation because it has become increasingly clear that he was a man far ahead of his time. That knowing and critical figure, J. M. Keynes, his students still recall, admired Babbage highly. Today there are applied mathematicians everywhere who share his passion for developing calculating machines; their technical resources have vastly improved, but the fundamental principles of design remain very similar. The British magazine *Nature* entitled a discussion of one of the first large American calculators "Babbage's Dream Comes True," and described that Harvard relay computer, Mark I, as a realization of Babbage's project in principle, but with the benefit of twentieth-century mechanical engineering and mass production methods for its physical form. A few years ago B. V. Bowden in his excellent book on calculating machines, *Faster than Thought,* said that Babbage enunciated the principles on which all modern computing machines are based. Unquestioned pioneer in the field of large-scale mathematical machines, Babbage was, in a sense, the unheralded prophet of the even newer field now known as operational research, foreshadowed in his book *Economy of Manufactures and Machinery.* This study of scientific manufacturing processes of all kinds, written as a by-product of his interest in mathematical machines, was in fact the only major undertaking he actually completed. He was ahead of his contemporaries in still a third way: he made a determined campaign for Government subsidy of scientific research and education at a time when research was still, to a large extent, a gentleman's hobby.

Life

Born in Devonshire in 1792, Charles Babbage was the son of a banker who later left him a considerable fortune. Because of poor health, he was privately educated until he entered Trinity College at Cambridge in 1810. He was already passionately fond of mathematics before coming to college and was discouraged to find then that he knew more than his tutor. He soon had a great circle of friends, or rather, a number of circles—chess- and whist-playing groups, fellow members of the Ghost Club, and boating companions, all in addition to mathematical colleagues. Of the latter, his most intimate friends were the younger Herschel (later Sir John) and George Peacock (later Dean of Ely). The three undergraduates entered into a compact that they would "do their best to leave the world wiser than they found it." In 1812, as their first step toward the achievement of this goal, the three, together with several others, founded the Analytical Society, hired rooms for it, read mathematical papers, even published transactions. Babbage, Herschel, and Peacock translated Lacroix's *Differential and Integral Calculus* and published two volumes of examples. In spite of considerable opposition, the Society fought valiantly to put "English mathematicians on an equal basis with their Continental rivals," and actually had a profound effect on the future development of English mathematics. Babbage believed that he was certain to be beaten in the tripos examinations by both Herschel and Peacock, and preferring to be first at Peterhouse rather than third at Trinity, he transferred in his third year. In fact, he stood first in Peterhouse in 1814, and received his M.A. in 1817. For about ten years after his graduation he published a variety of mathematical and physical papers, mostly on the calculus of functions, but also including one on Euler's study of the knight's move in chess and one on barometric altitude measurements.

Babbage, Herschel, and Peacock continued to be friends after they left school. Each in his own way lived up to their joint compact, though their careers were very different. Peacock devoted himself to mathematics and astronomy until finally he decided to join the ministry. He took his D.D. in 1839, and shortly thereafter became Dean of Ely, a post he filled with great vigor and success. Herschel, after a brief apprenticeship at law, decided to follow his great father into astronomy. As a crowning achievement, to supplement his father's work on stars of the Northern Hemisphere, Herschel went to the Cape of Good Hope in 1833, and in four years completed his observations of the southern stars. After his triumphant return to England, his main work was compiling his great catalogues of nebulae and stars. He was knighted by the Crown, served as Master of the Mint, and avoided all scientific feuds; his biographers all report that his was a life full of serenity and innocence. In contrast, Babbage published at thirty his "Observations on the Application of Machinery to the Computation of Mathematical Tables," and was received with general acclaim and presented the first award of the Gold Medal ever given by the Astronomical Society—but spent the rest of his life fruitlessly trying to bring his machines to completion. He was led by his broad interests into many byways, from a vigorous campaign against the policies of the Royal Society to the study of ciphers, and from speculative geology to the design of tools for lathes and shapers, but always his work centered on his beloved engines. His career was a long series of disappointments, and to friends who visited him in 1861 he said that he had never had a happy day in his life and spoke "as though he hated mankind in general, Englishmen in particular, and the English government and organ-grinders most of all . . . in truth Mr. Babbage was a mathematical Timon."

Actually Charles Babbage was a most social and gregarious fellow, with a considerable sense of humor, as one can see in his autobiography. Charles Darwin wrote, "I used to call pretty often on Babbage and regularly attended his famous evening parties." Babbage was an enthusiastic conference man, instrumental in founding the Astronomical Society (1820), the British Association for the Advancement of Science (1831), and the Statistical Society of London (1834). Darwin also recalled a brilliant dinner at his brother's house at which even Babbage who "liked to talk" was outdone by Thomas Carlyle; and an Edinburgh professor who was asked to dinner by Babbage reported that "it was with the greatest difficulty that I escaped from him at two in the morning after a most delightful evening." Encouraged to travel for his health, he made many trips to the continent of Europe, and was equally interested in meeting members of the aristocracy, fellow mathematicians, and skilled mechanics. He was a friend of Laplace and of Alexander Humboldt and knew Poisson, Fourier, and Biot. "It is always advantageous," he advised, "for a traveller to carry with him anything of use in science or in art if it is of a portable nature, and still more so if it has also the advantage of novelty." Among his most useful objects of this sort were some gold buttons stamped by steel dies with ruled parallel lines 4/10,000 of an inch apart, produced by a designer of machine tools named Sir John Barton, who was Comptroller of the Mint. The rainbow patterns playing on these small diffraction gratings, which indeed they were, provided a splendid opening gambit in conversations with strangers.

Machines

Babbage's transformation from a cheerful young man into a bitter old one was to a large extent the result of his devotion to his mathematical machines. He has provided us with two versions of the origin of his ideas about machines, but the one written in 1822 seems more plausible than the other, which appeared in his autobiography some forty years later. According to the first story, Herschel brought in some calculations done by computers for the Astronomical Society. In the course of their tedious checking, Herschel and Babbage found a number of errors, and at one point Babbage said "I wish to God these calculations had been executed by steam." "It is quite possible," remarked Herschel. From this chance conversation came the obsession that was to rule Babbage for the rest of his life. The more he thought about it, the more convinced he became that it was possible to make machinery to compute by successive differences and set type for mathematical tables. He set down a rough outline of his first idea, and made a small model consisting of 96 wheels and 24 axes, which he later reduced to 18 wheels and 3 axes. In 1822, in addition to his above-mentioned original note published by the Astronomical Society, he wrote an article "On the Theoretical Principles of the Machinery for Calculating Tables" for *Brewster's Journal of Science,* and a letter on the general subject to the President of the Royal Society, Sir Humphry Davy. In this letter, Babbage pointed out the advantages such a machine would have for the Government in producing the lengthy tables for navigation and astronomy, and proposed to construct a machine on an enlarged scale for the Government's use. There had been machines since the time of Pascal for carrying out single arithmetical operations, but they afforded little saving of time or security against mistakes. The Astronomical Society received Babbage's proposal with the highest enthusiasm, and the Royal Society reported favorably on his project for building what he called a Difference Engine. In an interview held in 1823 between Babbage and the Chancellor of the Exchequer, a rather vague verbal agreement was made whereby the Government would grant funds for the enterprise which was expected to take three years. Work proceeded actively for four years although Babbage was constantly having new ideas about the machine and scrapping all that had been done before.

In 1827 Babbage went abroad for a year on the advice of his physician, and during this period made a study of foreign workshops and factories to supplement his considerable familiarity with British manufacturing processes. He later used this information in his book *Economy of Manufactures and Machinery,* published in 1832, about which we shall have more to say later. While still abroad, he learned that he was to be appointed Lucasian Professor at Cambridge, the chair once held by Isaac Newton. Although he hesitated because of his work on the Difference Engine, he decided to accept and hold the position for a few years. He remarks in his autobiography that this was the only honor he received in his own country. He resigned in 1839 to devote himself completely to his machines, even though during his entire term of office he neither resided at the college nor taught there. His income as Lucasian Professor was between eighty and ninety pounds a year!

Upon returning to London in 1828, Babbage made a new application for funds to the Treasury. The Royal Society again reported favorably, the Duke of Wellington inspected the model, and once more the Government granted liberal funds for the work, and decided to build a fireproof building and workshop on land leased next to his home. In 1833, when arrangements were made for moving the engine and the work to the new shops, Babbage and his excellent engineer Clement reached a crisis. There had for years been differences about the various delays in salary payments and Clement refused to continue work in the new buildings without new and expensive arrangements. At this point Clement abruptly stopped work and dismissed those of his men who were working on the Babbage job. After months of dispute he allowed the drawings and the parts of the engine to be moved to the new building, but he was legal owner of his tools and retained all those tools which had been so laboriously built in his shop at Babbage's and the Government's expense over six or eight years of effort.

Twelve months after work on the Difference Engine had stopped, Babbage thought of an entirely new principle for a machine which would wholly supersede and transcend the Difference Engine. The Analytical Engine, as he called it, would have far more extensive powers, more rapid operation, and yet a simpler mode of construction than his original design. In 1834 Babbage requested an interview with the First Lord of the Treasury to explain his new idea and get an official decision on whether to continue and complete the original Difference Engine, or to suspend work on it until the new idea could be further developed. For eight years he pressed for an answer either from the First Lord of the Treasury or from the Chancellor of the Exchequer, with whom he corresponded at some length. At last he was advised that the Chancellor and the Prime Minister, Sir Robert Peel, concluded that the Government must abandon the project because of the expense involved. The Government had already spent £17,000 and Babbage had contributed a comparable amount from his private fortune. The parts of the machine already completed and the drawings for the whole machine were deposited in the Museum of King's College, London, were shown at the International Exhibition of 1862, and were eventually delivered to the South Kensington Museum, where they are now. The part on exhibit is in working order and has recently been taken apart, thoroughly cleaned, and reassembled so that an exact copy could be made for the International Business Machine Corporation's museum. The copy was built by the firm of R. W. Munro, who built the "mill" for Babbage's son.

Much embittered by the Government's withdrawal, Babbage turned his attention to the development of the Analytical Engine, and maintained a staff of draftsmen and workmen to work on drawings and experimental machinery for future construction. As always, Babbage would start work on a model and then abandon it in an unfinished state to start work on a new one. In 1848, after working for several years on the Analytical Engine, he decided to make a complete set of drawings for a second Difference Engine, which would include all the improvements and simplifications suggested by his work on the Analytical Engine. He again offered to give the completed drawings and notations to the Government provided they would build it, and again his offer was turned down by the Chancellor of the Exchequer, whom Babbage termed "the Herostratus of Science, [who] if he escape oblivion, will be linked with the destroyer of the Ephesian Temple."

Difference Engine

In 1834 the *Edinburgh Review* had published an account of the principles of Babbage's Difference Engine. Inspired by this article, a well-to-do Stockholm printer, George Scheutz, undertook to make a machine of his own. With a little belated financial assistance from his own Government and from members of the Swedish Academy, Scheutz and his son Edward completed, after many years of work, a difference engine of their own, which Scheutz brought to England for exhibition in 1854. Somewhat to Scheutz' surprise, Babbage did everything in his power to help, and in a speech before the Royal Society recommended Scheutz and his son for one of the Society's medals. Babbage's own son Henry used the Scheutz machine to demonstrate his father's pet system of "mechanical notation." The Swedish machine won a Gold Medal in Paris in 1855. Babbage and his son had prepared a series of drawings to accompany the machine and explain its operation. The Scheutz machine was bought for $5,000 in 1856 by an American businessman for the Dudley Observatory in Albany, New York, whose first building was to be dedicated that same year. G. W. Hough, first Director of the Dudley Observatory, was certainly a man to appreciate the machine, for he himself developed, with much effort, both a printing barometer and a very early form of recording chronograph for daily use at the Observatory. The Scheutz machine computed four orders of differences and displayed its results by setting type to eight decimal places. An engraving of the machine standing on its four fluted hardwood legs appears as plate 4 of Volume I of the *Annals of the Dudley Observatory*. The machine was used in Albany for many years in computing Ephemerides and various correction tables, and exists today in a private collection in Chicago. In 1863 an exact copy was made for the British Government and used by W. Farr for the computations for the *English Tables of Lifetimes, Annuities, and Premiums* published by the Registrar-General.

Calculating machines were by no means new; they had been devised by such luminaries as Napier, Pascal, and Leibnitz. Their devices were meant to serve as "desk calculators," like the wonderful and ubiquitous machines whirring on so many desks today, or like the humble slide rule. But detailed mechanical realization of the designer's ideas had not yet been skillful enough to make the machines more than mere curiosities in 1830. Babbage had higher ambitions; he planned to make a machine suited for the computation and direct setting-up in type of lengthy mathematical tables. He was greatly concerned about the errors introduced in the processes of printing and publishing tables, and listed and analyzed many repeated errors. He remarked in the first enthusiastic year of his public campaign to see his machine initiated with government aid: "Machinery which will perform . . . common arithmetic . . . will never be of that utility which must arise from an engine which calculated tables."

The Difference Engine was an embodiment in wheels and cranks of the principle of constant differences. The non-mathematical reader may find Babbage's own account in the *Life of a Philosopher* entirely clear. We summarize it here more compactly.

Let us actually construct a table of the squares of the successive integers: $1^2, 2^2, 3^2, 4^2$, etc. . . . , using exactly the method of the machine. We set up three columns: the first two columns are work columns, and the an-

A	B	C
		1
	1	
2→		
	$\overline{3}$→	
2→		$\overline{4}$Ä
	$\overline{5}$→	
2→		9
	$\overline{7}$→	
		$\overline{16}$

swers will appear in column C. To begin we have only to specify their three initial entries and a fixed pattern of procedure. From this we can generate the table as far up as we have the patience to go, using the single mathematical operation of addition. The three staggered columns are shown in the above with the pattern of procedure. The three given numbers are a 2 for column A, 1 for B, and 1 for C. The function of the digit 2 in A is to show that we are constructing a table of the second powers, while the first entries for columns B and C simply tell where the table begins. Complete column A by filling in 2's as far down as you like. Now complete column B by adding in the entries from column A again and again as shown by the arrows. Once column B is constructed its values are fed in turn into column C by addition in exactly the same way. The result in column C is automatic and almost painless construction of a table of the squares of integers, made wholly by this very repetitive pattern of simple addition.

More mathematical readers will recognize the procedure in an obvious notation of the difference calculus: $\Delta_{2n}y = 2$; $\Delta_1 y = 1$; $y_0 = 1$. This gives the general rule and the specific initial conditions.

All the work here performed consisted of additions (including the carrying of units to the next higher place), storage (or memory) of previous results, and the repeated addition of the column entries in a certain simple and invariable order. In 1822 Babbage made himself a small machine which could do exactly the operation given above, up to five-place numbers. This was the harbinger of Difference Engine No. 1. The frontispiece to the *Life of a Philosopher* is a facsimile of a woodcut of a very similar fragment of Difference Engine No. 1 itself, which has been preserved in the Science Museum in London. By the use of toothed wheels on shafts, not much different in principle from the familiar figure wheels of the mileage indicator on an auto speedometer, the Difference Engine can carry out the operations exemplified above. In the frontispiece, the first table entry is set at the bottom of the column of figure wheels at the right; column B is the central vertical set of wheels, column A the left-hand vertical shaft. Turning the crank once presents a new figure in the table column, the other columns taking their proper values. In the machine as designed, but never constructed, the number indicated in the table column would each time be transmitted through a set of levers and cams to a collection of steel punches, which would then be in position for stamping the number on a copper engraver's plate. The plate was moved with each turn of the mechanism so that the punched number would appear in the proper place on the printed page. Mechanically all this was far from simple. Recall that standardized machine parts, without hand-fitting, were yet a rarity. Clocks, which most closely resembled this type of mechanism, were still principally hand-fitted. Babbage's plans were on a grand scale—one of his most conspicuous failings—and called for no less than twenty-place capacity, up to differences of the sixth order. The variety and number of bolts and nuts, claws, ratchets, cams, links, shafts, and wheels may be imagined! All of these parts were designed with skill and care, with supplementary mechanism intended to minimize wear, prevent improper registration, and so on. Some of the modern practices of instrument design were foreshadowed, and there is no doubt that the technical devices used were superior for their time. The presence of gauges, of a shaper, of a kind of embryonic turret lathe, of die-cast pewter gear wheels and the pressure molds in which they were made, is evidence enough of that. Babbage even studied the action of cutting tools, and rationalized tool-grinding. But perhaps the very care and thoroughness of the design was its greatest weakness, for it was far from completion when the controversy which ended its financial support became the cause of its postponement after years of work. And the rise of Babbage's own interest in a far grander (and still more unrealizable) project at last killed the Difference Engine. Its state at death was most incomplete, but all the drawings, and a considerable number of the tools, gauges, jigs, and a quite respectable amount of development of methods and machinery had been completed. Precise information cannot be found, but a reasonable estimate would seem to show that Babbage's engine would have cost about fifty times what the similar, though more modest, Swedish version sold for in the fifties. It would have been some two tons of novel brass, steel, and pewter clockwork, made, as nothing before it, to gauged standards.

Analytical Engine

What Babbage saw after his work with the Difference Engine was a really grand vision. He had early conceived the notion he picturesquely called

"the Engine eating its own tail" by which the results of the calculation appearing in the table column might be made to affect the other columns,
and thus change the instructions set into the machine. On this insight, and
after a striking mathematical digression into difference functions new to
mathematics, and suggested only by the operation of the engine, he built a
great program. It was nothing less than a machine capable of carrying out
any mathematical operation instead of only the simple routine of differences we have inspected. Such a machine would need instructions both by
setting in initial numbers, as in the Difference Engine, and also far more
generally by literally telling it what operations to carry out, and in what
order. Capable of repeated additions, of multiplication which is hardly
more than that, and of reversing the procedure for subtraction and division, the arithmetical unit would do these operations upon command. It
would work on previously obtained intermediate results, stored in the
memory section of the Engine, or upon freshly found numbers. It could
use auxiliary functions, logarithms, or similar tabular numbers, of which
it would possess its own library. It could make judgments by comparing
numbers and then act upon the result of its comparisons—thus proceeding
upon lines *not* uniquely specified in advance by the machine's instructions. All this, which forms the backbone of modern computing development, was to be carried out wholly mechanically with not even a simple
electrical contact anywhere in the machine, nor, of course, a tube or a relay. The scale, as usual, was grand. The memory was to have a capacity of
a thousand numbers of fifty digits—respectable even by today's standards. Of course the speed of today was wanting. The multiplication which
takes not a millisecond in the fast electronic giants of today, and some
seconds in a punch-card business machine installation, would have taken
the Analytical Engine two or three minutes.

This operation depended upon punched cards. They were not the fast-
shuffled Hollerith cards moving over handy electrical-switch feelers, but
cards modeled on the already well-worked-out scheme of the Jacquard
loom. Punched holes in these cards would supply the machinery with numerical constants and directions for operation. The cards would be interposed into long lines of linkages within the engine. Whether or not holes
came in the right places would determine the passage of feeler wires capable of linking together the notion of "chains" of columns and whole
sub-assemblies. Thus, numbers or even arithmetical processes, transfer
from column to column, storage, inspection of given columns already in
the store, intercomparison of results, and so on—could be told to the machine. All this was done purely mechanically, and the process was elaborately safeguarded against the perils of friction, wear, jamming, and even

errors by human attendants who, at the signal of the machine, were to set in cards at programmed points in the process. This is the barest sketch of the machine. Only looking at the visible complications of a modern machine, and translating them into the still self-conscious machinery of more than a century ago, can do justice to the plan. Charles Babbage would be proud to see how completely the logical structure of his Analytical Engine remains visible in today's big electronic computers.

Babbage was too much concerned with the development of his engines to publish any description of them, but in 1840 he was invited to Turin to discuss his Analytical Engine. In the audience was L. F. Menabrea (later a general in Garibaldi's army) who summarized Babbage's ideas in a paper published in 1842 in the *Bibliothéque Universelle de Genéve*. This paper was translated into English and extensively annotated by the Countess Lovelace (daughter of Lord Byron) and published in *Taylor's Scientific Memoirs*. The Countess thoroughly understood and appreciated Babbage's machine, and has provided us with the best contemporary account—an account which even Babbage recognized to be clearer than his own. Miss Byron studied mathematics and, with the encouragement of her mother's various intellectual friends, her interest continued after her marriage. The Countess often visited Babbage's workshop, and listened to his explanations of the structure and use of his Engines. She shared with her husband an interest in horse racing, and with Babbage she tried to develop a system for backing horses; Babbage and the Earl apparently stopped in time, but the Countess lost so heavily that she had to pawn her family jewels. Apparently Babbage was willing to try anything once in an effort to raise funds for his Engine. He once designed a tit-tat-toe machine, which he intended to send round the countryside as a travelling exhibit to raise money for his serious machines. The tit-tat-toe machine was designed to recall a splendid eighteenth-century automaton, with the figures of two children, a lamb, and a cock, alternately clapping, crying, bleating, and crowing. Underneath the bric-á-brac was to be the mechanism for a genuinely automatic machine, slow-moving but unbeatable at tit-tat-toe. Babbage was persuaded to abandon the exhibition of this machine as an unprofitable venture by someone wise in the ways of the theater, who advised him that it was impossible to compete with General Tom Thumb, the reigning favorite of the day.

Mathematical Interests

Although Babbage never strayed very long from his calculating Engines, his tremendous scientific curiosity led him into many byways—some stemming directly from the main line of his machines, and some that were far afield. The machines themselves were, of course, a direct result of Babbage's great interest in mathematical tables, and he was much impressed with the importance of having them easy to read as well as accurate. In 1826, after a vast amount of labor, he published a table of logarithms from 1 to 108,000 in which he paid great attention to the convenience of calculators who would be using the tables. His work was much appreciated by computers both in England and abroad, and several foreign editions were published from his stereotype plates, with translated preface. In the same year he published a short book called *A Comparative View of the Different Institutions for the Assurance of Life,* which was one of the first clear, popular accounts of the theory of life insurance. He was led into this field as a result of his interest in calculating tables of mortality, and his tables were adopted by several German companies from the German edition of his book. In England his life tables were used by life insurance companies until a new set of tables was compiled by the Government in about 1870 on a Difference Engine built especially for the purpose. In 1831, in an effort to determine which was easiest to read, Babbage printed a single copy of his tables of logarithms in 21 volumes on 151 variously colored papers with ten different colors of ink, and also in gold, silver, and copper on vellum and on various thicknesses of paper.

He was constantly calling to the attention of scientific societies and government offices the number and importance of errors in astronomical tables and other calculations. At one of the first meetings of the British Association for the Advancement of Science, Babbage recommended a calculation of tables of all those facts which could be expressed by numbers in the various sciences and arts, which he called "the Constants of Nature and Art." At another BAAS meeting, in remarking on the vital statistics of an Irish parish, he said "to discover those principles which will enable the greatest number of people by their combined exertions to exist in a state of physical comfort and of moral and intellectual happiness is the legitimate object of statistical science."

He even extended his demand for statistical accuracy to poetry; it is said that he sent the following letter to Alfred, Lord Tennyson about a couplet in "The Vision of Sin":

"Every minute dies a man, / Every minute one is born": I need hardly point out to you that this calculation would tend to keep the sum total of the world's population in a state of perpetual equipoise, whereas it is a well-known fact that the said sum total is constantly on the increase. I would therefore take the liberty of suggesting that in the next edition of your excellent poem the erroneous calculation to which I refer should be corrected as follows: "Every moment dies a man / And one and a sixteenth is born." I may add that the exact figures are 1.167, but something must, of course, be conceded to the laws of metre.

It is a fact that the couplet in all editions up to and including that of 1850 read "Every minute dies a man, / Every minute one is born," while all later editions read "Every moment dies a man, / Every moment one is born."

Like many mathematicians he was fascinated by the art of deciphering, and believed firmly that every cipher could be deciphered with sufficient time, ingenuity, and patience. He began composing a series of dictionaries in which words were arranged according to the number of letters they contained, then alphabetically by the initial letter, then alphabetically by the second letter, etc. This work was never finished, nor were the grammar and dictionary he began to write when, as a young man, he first heard of the idea of a universal language. He also wrote a paper, never published, "On the Art of Opening all Locks," and then made a plan to defeat his own method. During all of his travels he never missed an opportunity to measure the pulse and breathing rate of any animals he happened to encounter, and prepared in skeleton form a "Table of Constants of the Class Mammalia."

Babbage made one excursion into the field of apologetics with an incomplete work entitled, *The Ninth Bridgewater Treatise, A Fragment,* published in 1837. The regular Bridgewater series had been supported by a bequest which called for the preparation of eight treatises which would give evidence in favor of natural religion. Babbage decided to add a ninth, at his own expense, on the same general subject, but with particular arguments against the prejudice, which he felt was implied in the first volume of the series, that the pursuits of science, and of mathematics in particular, are unfavorable to religion. He used his experiences with his calculating engine to bolster his arguments on the nature of miracles and in favor of design. As B. V. Bowden says, "he thought of God as a Programmer." He repeatedly alludes to the possibility of defining a series by such a complicated rule that the first hundred million terms might proceed according to an obvious scheme, and the next number violate it, while the rest of the sequence continued according to the first plan. He described the programming of a calculator for generating such a series. In this argument he felt he had shown a possible origin of miracles in a world otherwise controlled under God by orderly natural law.

In his autobiography, Babbage jokingly traces his ancestry to the pre-historic flint workers because of his "inveterate habit of contriving tools." He reports that as a child he had a great desire to inquire into the causes of all events, and that his invariable question on receiving any new toy was "Mamma, what is inside of it?" If the answer did not satisfy him, the toy was broken open. He remembers as a small boy being fascinated by an exhibition of clockwork automata, especially one of a small figure of a silver lady dancing. He ran across the silver lady many years later, and ac-quired her for his drawing room, where she was dressed in elaborate robes and displayed on a pedestal in a glass case. She held a place of honor next to the portion of his Difference Engine which he also had on exhibit at home. He would set either or both of them into operation for the entertainment of his guests: the lady to dance, and the Engine to print a small table, and noted ruefully that on one occasion his English friends were gathered about the silver lady while an American and a Hollander studied the Difference Engine. Babbage recounts a conversation he once had with the Countess of Wilton and the Duke of Wellington, who had called at his home to see the Difference Engine. The Countess asked what he considered was his greatest difficulty in designing the machine. He re-plied that his greatest difficulty was not that of "contriving mechanism to execute each individual movement . . . but it really arose from the almost innumerable *combinations* amongst all these contrivances," and compared his problems to those of a general commanding a vast army in battle. He was pleased to have Wellington confirm his analysis.

Machinery

Just as the mathematical machine-designing team of today soon becomes involved in a welter of problems about the properties of vacuum tubes and electronic circuits, so Babbage became deeply involved in the prob-lems of the machine shop and the drafting room. During the course of the work many ingenious mechanical devices were perfected and even some of those that were rejected for use on the calculator were not entirely wasted, but were introduced with success into other machinery as, for ex-ample, into a spinning factory at Manchester. To create the great variety of new and complex forms with the required precision for the Difference Engine, a number of new tools to use with a lathe were invented. To test the "steadiness and truth" of the tool-holders used in making some key gun-metal plates of the Difference Engine, Babbage reports that he "had some dozen of the plates turned with a diamond point," and he was delighted to observe the resulting grating spectra or "Frauenhofer images," as he called them. With his own sketches and with drawings prepared by a full-time draftsman, Babbage placed the construction of the machine in the hands of the engineer Joseph Clement. Clement was one of the great machine-tool builders of the century, who earlier had been a draftsman for Henry Maudslay, the introducer of the slide rest and the screw cutting lathe. Babbage tells of an order Clement once received from America to construct a large screw in "the best possible manner." This he proceeded to do, according to his standards, with a precision, and consequently a bill, far greater than his customer expected. The customer was required to pay some hundreds of pounds, although he had anticipated a bill of twenty pounds at most! The mainstay of Clement's custom shop was the first large planing machine, although from the sums expended by Babbage it appears that the Difference Engine was one of the largest single jobs in Clement's shop, somewhere between one fifth and one third of its whole effort. In Babbage's own writings it tends to appear as though Clement were in fact a full-time employee of Babbage, but this seems to be inaccurate. Babbage broke with Clement in 1833 and seems never to have carried any further projects beyond the stage of drawings and experimental parts.

Among the workmen in Clement's shop was one J. Whitworth, who became Sir Joseph Whitworth, Bart., leader in the machine-tool industry in the nineteenth century. It was Whitworth who first brought about the standardization of screw threads, and the Whitworth thread remained the British standard until 1948. He insisted upon the use of gauge blocks, recognizing end measurement as better than measurement between scratch lines. Even in the 1830's he could work to standards of a micro-inch. He is frequently credited with the familiar machinist's scheme of preparing plane standard surfaces three at a time by hand scraping, but this ascription seems to be doubtful. Whitworth's independent career began when he left Clement, probably because of the curtailment of the Babbage contract, and set up his own shop at Manchester. His was probably the first shop to build machine tools mainly for sale to other tool manufacturers.

Babbage wrote a paper "On the Principles of Tools for Turning and Planing Metals" for a three-volume reference work on the lathe, published in 1846 by Holtzapffel & Co. The publisher acknowledges that "The cultivation of Mechanics by Gentlemen . . . has given rise to many ideas and suggestions on their part, which have led to valuable practical improvements," and offers instruction to amateurs in "Turning or Mechanical Manipulation generally," either at Holtzapffel's shop or at the gentlemen's private residences. The parallel is complete: like today's mathematical-machine builder, today's scientific hobbyist is apt to use vacuum tubes; he is more likely to build amplifiers than do ornamental turning on a lathe. The drawings which Babbage had for his machines covered over 400 square feet of surface. They were described by experts at that time as perhaps the best specimens of mechanical drawings ever executed, done with extraordinary ability and precision. In the course of preparing them, Babbage invented a scheme of mechanical notation to make clear in a drawing the action of all the moving parts of a piece of machinery. Since Babbage's machinery was particularly complex to describe in motion, he was extremely proud of his notation. He prepared and had printed a short paper describing his principles of mechanical notation, which he gave away in considerable numbers during the Exhibition of 1851, requesting readers to send him any criticisms or suggestions.

Babbage's most successful book, and in fact the only work of any consequence which he ever *completed,* was the *Economy of Manufactures and Machinery,* published in 1832. Although originally intended as a series of lectures at Cambridge, Babbage published the work in book form, with a condensed version prepared as a prefix to the appropriate volume of the *Encyclopedia Metropolitana.* It ran through several editions, was reprinted in the United States, and was translated into German, French, Italian, and Spanish, in spite of the trouble Babbage reports having with booksellers because of his chapter analyzing the book trade. As a result of supervising the construction of his own Engine, he became interested in the general problems of manufacturing and visited factories in England and on the Continent. He learned from a workman how to punch a hole in a sheet of glass without breaking it, and found a demonstration of this skill a useful method of winning the confidence of the various craftsmen with whom he spoke. The book includes a detailed description and classification of the tools and machinery used in various manufacturing operations which he observed, together with a discussion of the "economical principles of manufacturing." In the mood of an operational research man of today, Babbage takes to pieces the manufacture of pins—the operations involved, the kinds of skill required, the expense of each process, and the direction for improvements in the then current practices. He makes a number of suggestions about methods for analyzing factories and processes and finding the proper size and location of factories, and stresses the need for studying the work of contemporary inventors in other countries. He points out that the division of labor, so important for manufacturing, can be applied also to mental operations, and cites as an example the work of G. F. Prony, director of the École des Ponts et Chaussés, who successfully organized three groups of workers—skilled, semi-skilled, and un-skilled—to prepare a great set of mathematical tables. Prony began work in 1784 under the same impetus which led to the establishment of the metric system in revolutionary France. He undertook to construct elaborate trigonometric tables based on the division of the quadrant into a hundred parts. A necessary auxiliary work was an unprecedented table of logarithms. This tabulated the logarithms of the natural numbers up to 200,000, carried out to fourteen decimal places. Prony realized that life was too short for such an effort (one sixth of that work had cost Briggs six or eight years). The story goes that, happening to read the new book of Adam Smith on the division of labor, he proceeded to organize the computations on this basis. His most skilled handful included mathematicians of the stature of Legendre. Their task was, of course, no mere computation but the choice of the best analytical expressions for numerical evaluation. Their formulae were transmitted to a group of about eight well-trained computers who put them into the appropriate numerical form. The computers of unskilled kind varied in number from 60 to 80. Their task was nothing more than addition and subtraction, according to the rules that were specified. It seems that nine tenths of them literally knew no more than addition and subtraction, and these turned out to be the best computers. Two teams worked independently and in duplicate and finished in about two years. The final "Tables du Cadastre" were never published, but remained in two copies, each of seventeen manuscript folio volumes. These were frequently consulted as checks by other computers, including Babbage himself who visited the Observatory in Paris on this errand.

Also included in the *Economy of Manufactures* is a panegyric for the "Science of Calculation . . . which must ultimately govern the whole of the application of Science to the Arts of Life." Babbage reports as one of the best compliments he ever received on the book a remark by an English workman he met, who said "that book made me think." One profoundly practical result of his operational research method was the introduction of the penny post in England; Sir Rowland Hill was encouraged to do this by Babbage's analysis of postal operations, which showed that the cost of handling the mail in the post office was greater than the cost of transportation. This pioneer work, *Economy of Manufactures,* is good reading even today.

Other Interests

Babbage made a number of suggestions for practical inventions of various kinds. Much interested in railroads, he attended the opening of the Manchester and Liverpool Railway and made several suggestions for ways of preventing accidents, including a method for separating a derailed engine from a train. In 1838 he was consulted by Isambard Brunel and the directors of the Great Western Railroad and spent five months doing experiments, which consisted largely of tracing on paper the curves of motion made by the special car in which he worked. On the basis of these experiments, he recommended use of a broad gauge and proposed an automatic speed-recording device for every engine. He suggested a numerical system of occulting lighthouses, and sent a description of his scheme to the authorities of twelve maritime countries. The United States Congress appropriated $5,000 to try his scheme experimentally, and the results of these experiments were published in 1861 in an extremely favorable report recommending adoption by the U.S. Lighthouse Board. An experience in a diving bell in 1818 led Babbage to consider the question of submarine navigation, and he prepared drawings and a description of an open submarine vessel with air for four persons for two days. Such a vessel, he thought, could be screw-propelled, and might enter a harbor and destroy even iron ships! He also suggested the use of a rocket apparatus to boost projectiles and the use of mirrors for indirect fire for artillery. Once at an opera, much bored by the performance, Babbage had the notion of using colored lights in the theater. He did some experiments, using cells formed by pieces of parallel glass filled with solutions of various colored salts, and even devised a rainbow dance to demonstrate this new technique.

Babbage did some physics in Cambridge, and published a paper with Herschel in 1825 on magnetization arising during rotation, based on experiments of Arago. Babbage was also much interested in geology and astronomy. After the eclipse of 1851 he suggested the germ of the idea of the coronagraph for seeing the sun's prominences without an eclipse, but he had not at all analyzed the problems of scattered light. It seemed possible to him that one could get a record of the succession of hot and cold years in the past by examining and comparing tree rings in ancient forests; this method was rediscovered early in this century and used to great advantage in southwestern United States. He wandered once again into the field of archaeology in his last scientific paper, entitled "On Remains of Human Art, mixed with the Bones of Extinct Races of Animals," published in 1859. Babbage once proposed to write a novel in order to help finance the completion of his Analytical Engine. He planned to devote a year to preparing a three-volume novel with illustrations, which was to earn 5,000 pounds. He was discouraged from pursuing this project by a poet friend wise in the pitfalls of literary fortune.

Unlike Darwin, who wrote his autobiography for his own children, with no thought of publication, Babbage says that he wrote his *Passages from the Life of a Philosopher* to "render . . . less unpalatable" the history of his calculating machines by an account of his own "experience amongst various classes of society." He was over seventy when he prepared this collection of anecdotes and ideas, and more obsessed than ever with his beloved engines. The book's characteristic combination of peevishness and humor is apparent as early as the title page. Babbage was a bitter man, and his autobiography is as much a record of his disappointments as of his achievements. The largest part of the book is, of course, devoted to his engines, with an account of their theory and principles of construction, and the sad tale of their neglect by the British Government. In 1832, and again in 1834, he ran unsuccessfully for Parliament on a Liberal (Whig) platform, and he includes in his autobiography parts of an amusing electioneering play used in connection with his campaign. In the play, entitled *Politics and Poetry* or *The Decline of Science,* Babbage is characterized as Turnstile, "a fellow of some spirit; and devilish proud," and again as "a sort of a philosopher—that wants to be a man of the world." In a chapter entitled "Street Nuisances," Babbage describes his one-man battle against street musicians, which brought him as much fame, in London at least, as all his scientific accomplishments combined. Babbage maintained that his ideas vanished when the organ-grinder began to play, and calculated that such interruptions destroyed one fourth of his working power. He waged a vigorous campaign of letters to newspapers and to members of Parliament, and personally hauled many individual offenders before a magistrate. One of the latter once asked Babbage whether a man's brain would be injured by listening to a hand organ, and Babbage replied "certainly not, for the obvious reason that no man having a brain ever listened to street musicians." That this particular battle of Babbage's was not taken seriously is made clear by his obituary notice in the sober London *Times* in 1871, which remarks somewhat cruelly in the first paragraph that Babbage lived to be almost 80 "in spite of organ-grinding persecutions."

Defense of Science

Even before he had a personal grievance against the British Government for its failure to support his own machine, Babbage had sharply criticized the Government for its neglect of science and scientists. Never a man to avoid a fight for what he considered a good cause, Babbage had for years led an assault on the decline of science in England. He published two stinging tracts, *Reflections on the Decline of Science in England, and on some of its Causes* (1830), and *The Exposition of 1851; or Views of the Industry, the Science and the Government of England* (1851). Bitter because few Englishmen pursued science for its own sake, he attacked the neglect of science in the educational system and urged Government subsidies for scientists. He felt that scientists should hold many Government posts and that pure science should be encouraged. "It is of the very nature of knowledge that the recondite and apparently useless acquisition of today becomes part of the popular food of a succeeding generation," he wrote. Chief target of his diatribes on the neglect of science in England was the Royal Society, to which he had been elected while still at Cambridge. He submitted a plan for sweeping reforms to the Society, which rejected it without discussion. His plan included such items as requirements for publication of scientific articles as a test for membership, instituting democratic election procedures, and free discussion of policies at meetings. Infuriated by the Society's refusal to consider his plan, he continued to condemn what he termed the intrigues of the Society, pronounced its secretaries third-rate, and its president elected on the basis of rank rather than scientific interests. Babbage would perhaps be disappointed to find that within the last decade the BAAS which he helped to found was headed, nominally at least, by the Duke of Edinburgh! Babbage described the Council of the Royal Society as "a collection of men who elect each other to office and then dine together at the expense of the society to praise each other over wine and to give each other medals." Although some of Babbage's accusations against the Royal Society and the Government for their neglect of science were exaggerated, his position was fundamentally sound, and he found a good deal of support. Babbage has been aptly characterized as "a scientific gadfly" who "successfully needled his contemporaries into general agreement."

Babbage blamed the Royal Society for conditions at the Royal Observatory at Greenwich; on one occasion he had been refused a copy of some of the Greenwich observations, and later located five tons of the Greenwich tables in a shop which had bought them by the pound for making pasteboard. Babbage remarked that the Astronomer Royal was certainly the man best fitted to decide what should be done with his own publications, but did not think it possible to invent a more extravagant way of compensating a public servant than to establish an observatory and computing center for the production and printing of astronomical tables simply as a source of wastepaper! He had no great love for the then Astronomer Royal, Sir George Airy, in any case, for that official had recommended no further Government support for the Difference Engine, and had refused to consider the possibility of mechanizing his own computations.

Conclusion

Babbage once said that he would gladly give up the remainder of his life if he could be allowed to live three days five hundred years hence and be provided with a scientific guide to explain the discoveries made since his death. He judged that the progress to be recorded would be immense, since science tends to go on with constantly increasing rapidity, and Babbage always took a confident view about human progress. The wide range of his practical and scientific interests and his clear commitment to the notion that careful analysis, mathematical procedures, and statistical calculations could be reliable guides in almost all facets of practical and productive life give him still a wonderful modernity.

More than one spiritual contemporary of Babbage is flying today from site to site on the missions of the Atomic Energy Commission and the Rand Corporation. His whole story bears witness to the strong interaction between purely scientific innovation, on the one hand, and the social fabric of current technology, public understanding, and support on the other. His great engines never cranked out answers, for ingenuity can transcend but it cannot ignore its context. Yet Charles Babbage's monument is not the dusty controversy of the books, nor priority in a mushrooming branch of science, nor the few wheels in the museum. His monument, not wholly beautiful, but very grand, is the kind of coupled research and development that is epitomized today, as it was foreshadowed in his time, by the big digital computers.

Sources and Acknowledgments

A guide to when, where, and why these pieces were first published in full, along with my gratitude to all those who granted permission to reprint them here in edited form:

The PREFACE and the paragraphs that announce each section were written in 1993 for first publication in this volume.

SOMETHING PERSONAL

RADIO DAYS appeared in 1984 in the MIT student yearbook *Technique.*

ENGINEERS IN KINDERGARTEN was written as an op-ed piece for the Portland *Oregonian* in November 1989 (perhaps unpublished?)

THEORIES GREAT AND SMALL

SEARCHING FOR OUR ANCESTORS is the prologue I wrote for the small book *Search for the Universal Ancestors,* editors: P. Morrison, H. Hartman, and J.F. Lawless, Special Publication 477, Scientific and Technical Information Branch, NASA, 1986.

THE WONDER OF TIME, an Afterword to *Nature and Life,* an essay of the 1930s by philosopher Alfred N. Whitehead, University of Chicago Press, 1980.

THE FABRIC OF THE ATOM, a brochure to accompany my TV lecture series on introductory quantum mechanics, BBC, London, 1964.

WHY MAN EXPLORES: part of a panel discussion held at Cal Tech in July 1976, sponsored by NASA.

TWO DIALS: article based on a speech to the annual meeting of the NY Academy of Sciences, 1980, published in their journal *The Sciences,* July–August 1981.

SCIENCE AND THE NATION, the statement I made as Chairman of the Federation of American Scientists before the Committee on Science and Astronautics, US House of Representatives, July 11, 1974.

ON THE CAUSES OF WONDERFUL THINGS, chapter 20 from the book *Science and the Paranormal,* edited by George O. Abell and Barry Singer. Scribners', 1981.

THE SIMULATION OF INTELLIGENCE, talk given in a conference at the Institute of Advanced Study, Princeton, in 1973, published in *Journal of Altered States of Consciousness,* vol 1, no. 2, Spring 1974.

THE ACTUARY OF OUR SPECIES, a chapter in the book *Visions of Apocalypse—End or Rebirth,* Holmes & Meier, Publishers, 1985. The profound effects of cosmic impacts are by now a commonplace; ten years ago, those events were dubious.

CAUSE, CHANCE AND CREATION, an article in the weekly Saturday Evening Post, republished in their book *Adventures of the Mind,* Second Series, Saturday Evening Post, 1961.

ON BROKEN SYMMETRIES, an MIT lecture published in the book *On Aesthetics in Science,* editor Judith Wechsler, MIT Press, 1979.

LOOKING AT THE WORLD, an essay from the book *Powers of Ten,* by Philip and Phylis Morrison, and the Office of Charles and Ray Eames. Scientific American Books, 1982.

ASTRONOMY SHINES OUT

WHAT IS ASTRONOMY? from a book in celebration of Cornell astrophysicist Ed Salpeter, *Highlights of Modern Astrophysics: Concepts and Controversies,* edited by Stuart L. Shapiro and Saul A. Teukolsky. John Wiley & Sons, Inc., 1986.

THE EXPLOSIVE CORE, active galactic nuclei as they then seemed (on the right track?) from the annual supplement to the Encyclopedia Britannica, *Science Year* 1977.

IS M82 REALLY EXPLODING? No, not at the core, but it is indeed now the closest of all galaxies that host a starburst. From the monthly *Sky and Telescope,* vol. 57, January 1979.

A WHISPER FROM SPACE is the script for an hour-long TV show that I wrote and presented about the thermal cosmic background radiation. The show was co-produced in 1978 by BBC with PBS, under BBC producer Peter Jones. Available from Nova, WGBH Educational Foundation, Boston MA.

SEARCHING FOR INTERSTELLAR COMMUNICATIONS

LIFE BEYOND EARTH AND THE MIND OF MAN, edited by Richard Berendzen, Scientific and Technical Information Office, NASA, 1973. Reports my part of the symposium held under that title at Boston University, 20 Nov 1972.

TWENTY YEARS AFTER . . . a look-back at our 1959 paper (jointly with Cocconi) on seeking microwave transmitters among the stars. The memoir appeared in the first issue of a magazine meant for friends of the search, called *Cosmic Search,* vol. 1, no. 1, January 1979, published for some years at Columbus, Ohio.

LIFE IN THE UNIVERSE, my reflections as a summary of a December 1980 meeting on that topic at NASA-Ames Research Center, Mountain View, CA. Published in *Life in the Universe,* editor, John Billingham, MIT Press, 1981.

TALK WITH PHILIP MORRISON, by Charlene Anderson, reports an interview with two staff members of The Planetary Society about the history and the future of the search. Published in their newsletter, *The Planetary Report,* vol. III, no. 2, March/April 1983, Pasadena CA.

THE SEARCH FOR EXTRATERRESTRIAL COMMUNICATIONS, an overview presented in Montreal at the first official symposium of the International Astronomical Union on the topic, published as *The Search for Extraterrestrial Life: Recent Developments,* editor, Michael Papagiannis. IAU, with D. Reidel, 1985.

ON LEARNING AND TEACHING

LESS MAY BE MORE, edited version of a talk at a conference on college courses in science for non-science majors, given in October 1963. It was published in *American Journal of Physics,* vol. 22, no. 6, 441–452, June 1964.

ICE THAT SINKS, a brief account drawn from a real lecture-demonstration, given at an international meeting of historians of science at Cornell to commemorate the great discoveries in nuclear and particle physics of the year 1932. Published in *The Physics Teacher,* November 1963, 209–211.

THE NEW GENERAL PHYSICS, a reminiscent account of developments of the sixties in physics teaching, given at the fiftieth anniversary of an earlier famous college physics textbook. From *AAPT Pathways,* published by the American Association of Physics Teachers, November 1981.

THE FULL AND OPEN CLASSROOM, a talk at an education colloquium at MIT, extended and published as Occasional Paper No. 10, MIT Education Research Center, 1971. The Center has long since been closed; it may well have a successor some time soon.

PRIMARY SCIENCE: SYMBOL OR SUBSTANCE?, a joint discussion by Philip and Phylis Morrison. Delivered as the 1984 Catherine Molony Memorial Lecture, City College of New York Workshop Center, 1984.

KNOWING WHERE YOU ARE, a commentary on a study by David Hawkins on a key educational phenomenon: it is not at all a blank slate, but familiar "common-sense" theories, older than Aristotle, that children often bring to early study of the world. Published in *Critical Barriers Phenomenon in Elementary Science,* North Dakota Study Group on Evaluation: Center for Teaching and Learning, University of North Dakota, Grand Forks ND, 1985.

WAR AND PEACE IN THE AGE OF URANIUM

IF THE BOMB GETS OUT OF HAND, an early account of the effect one atomic bomb might have on mid-Manhattan. From *One World or None,* editors, Dexter Masters and Katherine Way, McGraw-Hill Book Co., Inc., 1946.

PHYSICS OF THE BOMB, an early primer for general readers, published in *Science News 2,* Atomic Energy Number, edited by R.E. Peierls and John Enogat. Penguin Books, 1947, Harmondsworth and New York.

ACCIDENTS WITH ATOMIC WEAPONS, an appraisal of unlikely calamities. From *Scientific World,* vol. II, No. 4, 1958, The World Federation of Scientific Workers.

CAUGHT BETWEEN ASYMPTOTES, a brief but wide perspective on the nuclear arms race for physicist readers. From *Nuclear Weapons and Nuclear War,* American Association of Physics Teachers, SUNY, Stony Brook, LI, NY, 1982.

THE SPIRAL OF PERIL, a similar piece with more historical focus, *Bulletin of the Atomic Scientists,* January 1983.

INSECURITY THROUGH TECHNICAL PROWESS, a set of lessons from arms race history, published in the anniversary brochure *Toward a New Security: Lessons of the Forty Years Since Trinity,* Union of Concerned Scientists, Cambridge MA, July 16, 1985.

NATIONALISM, SCIENCE, AND INDIVIDUAL RESPONSIBILITY, the re-worked form of a talk before the American Student Pugwash annual meeting. Published in *Technology in Society,* vol. 8, 319–327, Pergamon Journals Ltd, London, 1986

FRIENDS AND HEROES

BRUNO ROSSI, a foreword to his fascinating autobiography *Moments in the Life of a Scientist,* by Bruno Rossi, Cambridge University Press, 1990. (Professor Rossi died in January 1994.)

ROBERT NOYCE, a talk about the first silicon chip, its background and its inventor, from *The Exploratorium,* vol. 8, Issue 1, San Francisco 1984.

NIELS BOHR . . . THE OTHER SIDE, a friendly estimate of the few limita-tions of this paragon among physicists and men, one balancing item in the celebratory collection *Niels Bohr: A Centenary Volume,* edited by A.D. French and D.J. Kennedy, Harvard University Press, 1985.

IN MEMORY OF RICHARD FEYNMAN, written by me but signed only as "by an old friend," published in *Scientific American,* June 1988.

FRANK AND JACKIE OPPENHEIMER, a commentary on my two late friends and the unique "museum of art, science, and perception" they built, a Foreword to *The Exploratorium: The Museum as Laboratory,* by Hilde Hein, Smithsonian Institution Press, 1990.

BERNARD PETERS AND THE HEAVY PRIMARIES, memoir on a fellow graduate student who helped to knit together both heaven and earth and East and West, from *Current Science,* Bangalore, December 1991. (Pro-fessor Peters died in late 1992.)

CHARLES BABBAGE, an introduction to an anthology, *Charles Babbage on the Principles and Development of the Calculator and Other Seminal Writings by Babbage and Others,* with co-author Emily Morrison. Dover Publications, New York, 1960.

Index

About the Author

Philip Morrison is one of the most admired physicists of our day. Scholar, author, educator, TV personality, Morrison has not only played a starring role on the American science stage, but has also emerged as one of the most articulate and knowledgeable leaders in the struggle for arms control. Along the way, he and his wife Phylis Morrison have championed the need for more sensitive and skilled science teaching in our schools.

In physics, Professor Morrison's work spans widely across the panorama of advanced science—quantum electrodynamics, nuclear reactions, high-energy astrophysics and cosmology. With Nobel laureate Hans Bethe, he is the author of a fundamental textbook in nuclear physics and with Ray Eames and Phylis Morrison, he wrote the startlingly original *Powers of Ten*. Morrison is also the author of works on Charles Babbage, arms control and defense, and the search for extraterrestrial communication, among others.

Morrison's career has been deeply engaged in the power of the media to give shape and meaning to his ideas. Since the mid-sixties, his polymath book reviews on nearly every facet of science and technology have been published monthly in *Scientific American*. Beginning in 1961, with his first BBC show, *Fabric of the Atom*, Morrison has appeared frequently on the BBC, *Nova* and commercial television, culminating in the acclaimed six-part PBS series, *The Ring of Truth*, which first aired in 1979 and which was accompanied by a best-selling companion volume.

In the early forties, Professor Morrison taught physics at San Francisco State College and the University of Illinois at Urbana. His teaching career was interrupted when he joined the Manhattan Project at the University of Chicago. A physicist and group leader at Los Alamos, he participated in the first desert test of the atomic bomb.

After the War, Morrison returned to academic life at Cornell. Since 1965, he has been at MIT, presently as emeritus professor of physics. A member of the National Academy of Sciences, Morrison has been awarded numerous prizes and honorary degrees in recognition of his illustrious and distinguished career.